RENEWALS 458-4574
DATE DUE

**WITHDRAWN
UTSA Libraries**

Desert Regions
Population, Migration and Environment

Springer

*Berlin
Heidelberg
New York
Barcelona
Hong Kong
London
Milan
Paris
Singapore
Tokyo*

Boris A. Portnov   A. Paul Hare (Eds.)

Desert Regions

Population, Migration and Environment

With 110 Figures and 33 Tables

 Springer

Editors

DR. BORIS A. PORTNOV
Ben-Gurion University of the Negev
Center for Desert Architecture and Urban Planning
J. Blaustein Institute for Desert Research
84990 Sede-Boker Campus
Israel
E-mail: portnov@bgumail.bgu.ac.il

PROFESSOR DR. A. PAUL HARE
Ben-Gurion University of the Negev
Social Studies Unit
J. Blaustein Institute for Desert Research
84990 Sede-Boker Campus
Israel
E-mail: paulhare@bgumail.bgu.ac.il

ISBN 3-540-65780-0 Springer-Verlag Berlin Heidelberg New York

Library of Congress Cataloging-in-Publication Data
Desert regions: population, migration, and environment / Boris A. Portnov, A. Paul Hare, eds.
    p.   cm.
  Includes bibliographical references and index.
  ISBN 3-540-65780-0 (hc.)
  1. City planning -- Arid regions. 2. Desert resources development. 3. Migration, Internal -- Planning.   I. Portnov, B. A. (Boris Adolfovich) II. Hare, A. Paul (Alexander Paul), 1923-   .
HT166.D386 1999
307.1'2'09154--dc21   99-30636
                                                                                    CIP

This work is subject to copyright. All rights are reserved, whether the whole or part of the material is concerned, specifically the rights of translation, reprinting, reuse of illustrations, recitations, broadcasting, reproduction on microfilm or in any other way, and storage in data banks. Duplication of this publication or parts thereof is permitted only under the provisions of the German Copyright Law of September 9, 1965, in its current version, and permission for use must always be obtained from Springer-Verlag. Violations are liable for prosecution under the German Copyright Law.

© Springer-Verlag Berlin Heidelberg 1999
Printed in Germany

The use of general descriptive names, registered names, trademarks, etc. in this publication does not imply, even in the absence of a specific statement, that such names are exempt from the relevant protective laws and regulations and therefore free general use.

Photographs: Wolfgang R. Motzafi-Haller
Cover Design: Erich Kirchner, Heidelberg
Typesetting: Camera-ready by the editors

SPIN: 10701187   30/3136 - 5 4 3 2 1 0 - Printed on acid free paper

Preface

Despite the common understanding of the importance of desert development in the contemporary world, there are relatively few books published to date on this subject. The books and collective volumes published in this field deal primarily with environmental and physical aspects of desert development such as soil, agriculture, vegetation, water resources, etc.

In contrast, this book addresses the issues of regional and urban development in desert areas, which have not been given sufficient attention. The present book is socially oriented. It considers physical development of desert regions not as an end in itself, but rather as an essential precondition for creating socially attractive and desirable environments for human settlement. The book addresses the issues of desert development at three distinctive conceptual levels – region, urban environment, and building – and deals with both cold and hot deserts.

Approximately half of the chapters in this book are original contributions that have not been published elsewhere. The remaining chapters fall into two groups: 1) chapters which have been reprinted from various refereed journals, and 2) chapters initially printed elsewhere and revised by their respective authors specifically for this collective volume. In the former case, permission to reproduce the material has been obtained from the respective copyright holders, and the details of original publication and names of copyright holders are indicated in footnotes. The chapters which are revised versions of previously published material include citation of the original publication details in footnotes. The authors of these chapters are solely responsible for the revised versions of their articles in matters pertaining to copyrights.

We would like to acknowledge the generous assistance of the Center for Desert Architecture and Urban Planning at J. Blaustein Institute for Desert Research, Ben-Gurion University of the Negev that has made possible to publish this book. We would also like to thank the authors of individual contributions for helping us to prepare this collective volume and Wolfgang R. Mozafi-Haller for his help in preparing the camera-ready copy for this book.

Dr. Boris A. Portnov	Prof. A. Paul Hare
The Center for Desert Architecture and Urban Planning	The Social Studies Center

Jacob Blaustein Institute for Desert Research,
Ben-Gurion University of the Negev, Israel

Contents

Preface	V
List of Contributors	XV
1 Introduction	1
B. A. Portnov	
1.1 Climatic Causes of Aridity	2
1.2 Criteria for Aridity	3
1.3 Geographic Extent of Deserts	4
1.4 The Process of Desertification	5
1.5 Deserts and Urban Growth	6
1.6 Scope of the Book	10
References	13

Part One

REGIONAL DEVELOPMENT AND POPULATION CHANGE

2 Long-term Development Patterns of Peripheral Desert Settlements	17
B. A. Portnov, E. Erell	
2.1 Introduction	17
2.2 Desert Urbanization in Israel: Prerequisites and Historical Background	18
2.3 Desert Settlements: Exogenous Factors	20
2.4 Development Paradigms	21
2.4.1 Economic Development of Peripheral Desert Areas	21
2.4.2 Population Growth	23
2.5 Research Method	24
2.6 Development Peculiarities of Desert Settlements	26
2.7 Influence of the Desert	30
2.8 Conclusions and Policy Implications	31
References	34

3 Sustainable Population Growth of Urban Settlements: Preconditions and Criteria 37

B. A. Portnov, D. Pearlmutter

3.1 Introduction 37
3.2 Defining Sustainability 37
3.3 Measuring Population Growth 38
3.4 Urban Settlements in Israel: Inequalities of Population Growth 39
3.5 Patterns of Urbanization in Israel and
    General Development Policies 40
3.6 A Generalized Model of In-country Migrations 44
3.7 Research Method 46
3.8 Components of Population Growth 48
3.9 The MB/NG Ratio as an Integrated Indicator for
    Measuring the Sustainability of Population Growth 50
3.10 Factors Influencing Migration Attractiveness of Urban Areas 53
3.11 Conclusions and Policy Implications 54
References 58

4 Private Construction as a General Indicator of Urban Development 61

B. A. Portnov, D. Pearlmutter

4.1 Introduction 61
4.2 Private Construction as a Development Indicator 62
4.3 Private Construction in Israel:
    Historical Background and Spatial Trends 64
    4.3.1 History of Private Construction in Israel 65
    4.3.2 Geographic Distribution of Private Construction in Israel:
          General Trends 65
4.4 Private Construction in the Hierarchy of Development Data 69
4.5 Location Paradigm 71
4.6 Case Study 74
4.7 Research Results 79
4.8 Applications in Planning 80
References 82

5 The Effect of Remoteness and Isolation on Development of Peripheral Settlements 87

B. A. Portnov, E. Erell

5.1 Introduction 87
5.2 Sustainable Population Growth of Urban Settlements
    Components and Research Paradigms 88

		5.2.1 Population Growth	88
		5.2.2 Measuring Economic Development	89
	5.3	Spatial Characteristics of Urban Development in Peripheral Areas	91
		5.3.1 Distribution of Population and Settlement Location	91
	5.4	Research Method	92
	5.5	Controls	95
	5.6	Analysis Procedure	97
	5.7	Research Results	98
		5.7.1 Population Growth	98
		5.7.2 Index of Clustering	101
		5.7.3 Climatic Harshness	103
	5.8	Economic Development	104
	5.9	Conclusions and Policy Implications	108
	References		109

6 Modeling the Migration Attractiveness of a Region ... 111

B. A. Portnov

6.1	Introduction	111
6.2	Employment and Housing Factors of Interregional Migration	112
6.3	Modeling the Migration Behavior	113
6.4	Housing-employment Paradigm of Interregional Migration	114
6.5	Israel and Japan: General Patterns of Regional Development	116
	6.5.1 Patterns of Urbanization in Israel and Recent Development Policies	117
	6.5.2 Current Issues of Regional and Urban Development in Japan	117
6.6	General Patterns of Interregional Migration	118
	6.6.1 Israel	118
	6.6.2 Japan	120
6.7	Research Method	122
6.8	Influencing Factors	125
6.9	Employment-housing Balance	127
6.10	Conclusion and Policy Implications	129
References		130

7 Investigating the Effect of Public Policy on Population Growth in Peripheral Areas ... 133

B. A. Portnov

7.1	Introduction	133
7.2	Regional Policy Evaluation: Contemporary Trends	134
7.3	Research Methodology	136
7.4	Research Approach	138

7.5	Preliminary Results and Discussion	140
7.6	Modeling Procedure	143
7.7	Influencing Factors	144
7.8	Alternative Scenarios	145
7.9	Conclusion	147
	References	148

8 Ecological-oriented Options for the Sustainable Development of Drylands ... 153

U. N. Safriel

8.1	Desert and Development	153
8.2	Development of Hyperarid Drylands	154
8.3	Development of Arid Drylands	155
8.4	Development of Semiarid Drylands	155
8.5	Development of the Dry-Subhumid Drylands	157
	References	158

Part Two

CITIES OF COLD AND HOT DESERTS

9 Physical Environment and Social Attractiveness of Frontier Settlements: Cities of Siberia, Russia ... 161

B. A. Portnov

9.1	Introduction	161
9.2	Previous Research	163
9.3	The Region	163
9.4	The Cities	165
9.5	Economics of Transition	166
9.6	Research Method	167
9.7	Spatial Patterns of District Attractiveness	170
9.8	Components of Attractiveness	173
9.9	Relative Importance of Influencing Factors	174
9.10	"Experts" and "Residents": Different Visions	176
9.11	District Attractiveness to Business Activity	179
9.12	IP and the Market Value of Residential Land	180
9.13	Social Factors	182
9.14	Applications in Planning	183
9.15	Conclusion	184
	References	185

10 Planning Theories versus Reality: A Desert Study Case 187

I. A. Meir

10.1 Introduction	187
10.2 A Short Note on Settlement in the Past	188
10.3 The Modern Era	188
10.3.1 Ottoman Period and the European Influences (1900-1917)	188
10.3.2 Colonialism and British Mandate (1917-1948)	189
10.3.3 Suburban Agriculture of the First Israeli Period (1948-1950)	190
10.3.4 Garden City and Neighborhood Unit (1950s)	191
10.3.5 The New Master Plan and Design Experimentation (1960s)	192
10.3.6 Prefabrication, High Rise Buildings and Traffic Separation (1970s)	193
10.3.7 Satellite Rururban Development (1980s)	194
10.3.8 Emergency Planning for Immigrants (the Early 1990s)	195
10.4 Microclimatic Variability	196
10.5 Conclusions	200
10.5.1 Positive Intervention in the Existing Fabric	201
10.5.2 Creating Desert Responsive Urban Forms	202
10.6 Theory and Implementation	202
References	203

11 An Experimental Evaluation of Strategies for Reducing Airborne Dust in Desert Cities 205

E. Erell, H. Tsoar

11.1 Abstract	205
11.2 Introduction	205
11.3 Background	206
11.3.1 The Transport and Deposition of Dust in the Urban Environment	206
11.4 The Urban Climate	207
11.4.1 Temperature	208
11.4.2 Rainfall	208
11.4.3 Wind Regime	208
11.5 Experimental Sampling, Dust Deposits	209
11.5.1 Description of Sampling Locations	211
11.5.2 Field Methods	213
11.5.3 Laboratory Methods	213
11.6 Results	213
11.6.1 The Dust Deposition Rate	214
11.6.2 Grain Size Characteristics	215
11.6.3 Chemical and Mineralogical Composition	216

11.6.4 The Effects of a Major Dust Storm	217
11.7 Discussion	218
11.7.1 The Sources of Urban Dust in Desert Cities	218
11.7.2 The Effect of Common Design Strategies for Reducing Exposure to Airborne Dust	220
11.7.3 Reducing Dust in Desert Cities - A Comprehensive Approach	223
11.8 Conclusion	224
References	225

12 Planning in Desert Environments: Three Cases of Responsive Planning ... 227

Y. Gradus

12.1 Introduction	227
12.2 Israel: Ideology and Planning	228
12.3 Settlement System	228
12.4 Town-planning	231
12.5 Bedouin Towns	234
12.6 Summary and Applications in Planning	238
References	239

13 The Past as a Key for the Future in Resettling the Desert ... 241

A.S. Issar

13.1 Introduction	241
13.2 A Lost Paradigm	241
13.3 The Rise and Decline of the Deterministic Paradigm	243
13.4 The Breaking Down of the Consensus	244
13.5 Conclusions with Regard to the Future	247
Bibliography	248

Part Three

BUILDING AND DESIGN

14 A Desert Solar Neighborhood in Sede Boker, Israel ... 251

Y. Etzion

14.1 Introduction	251
14.2 The Neighborhood	254
14.3 Orientation	254
14.4 Circulation	255

14.5 Building Clusters	258
14.6 Setback Lines	260
14.7 Water Heating	261
14.8 Conclusion	262
References	262

15 A Bio-Climatic Approach to Desert Architecture ... 263

Y. Etzion

15.1 The Climate of the Negev	263
15.2 Building Design: Sealing the Envelope	264
15.3 Windows: Opening the Envelope by Design	269
15.4 The Value of the Courtyard	273
15.5 Performance Monitoring	274

16 Urban Microclimate in the Desert: Planning for Outdoor Comfort under Arid Conditions ... 279

D. Pearlmutter, P. Berliner

16.1 Urban Attractiveness and the Desert Climate	279
16.2.1 Preconceived Planning in the Negev	280
16.2.2 Microclimatic Considerations	281
16.2 Case Study: Analyzing the Urban Microclimate	282
16.2.1 Summary of Case Study Results	284
16.2.2 Discussion: Creating the Urban "Cool Island"	287
16.3 Conclusions	288
References	289

17 Adaptive Architecture: Low-Energy Technologies for Climate Control in the Desert ... 291

Y. Etzion, D. Pearlmutter, E. Erell, I. Meir

17.1 Introduction	291
17.2 The Problem - Local Climatic Conditions	291
17.3 The Response: Project Overview	293
17.4 Experimental Evaluation of the Building's Thermal Performance	296
17.4.1 The Sunken Atrium	296
17.4.2 The Evaporative Down Draft Cool Tower	300
17.4.3 Indirect Space Heating from Solar Heated Air	302
17.5 Conclusions	303
References	304

Part Four

CASE STUDIES

18 Desert Settlements in Israel: 307
Socio-Economic and Physical Data

B. A. Portnov, W. Motzafi-Haller

18.1 Be'er-Sheva	307
18.2 Eilat	310
18.3 Dimona	311
18.4 Arad	313
18.5 Yeroham	315
18.6 Mitzpe-Ramon	317
References	319

Subject Index 323

Author Index 329

List of Contributors

Berliner, Pedro
Senior Researcher, Els Wyler Department of Dryland Agriculture, J.Blaustein Institute for Desert Research, Ben-Gurion University of the Negev, Sede-Boker Campus, 84990, Israel
berliner@bgumail.bgu.ac.il

Erell, Evyatar
Senior Researcher, Center for Desert Architecture and Urban Planning, J.Blaustein Institute for Desert Research, Ben-Gurion University of the Negev, Sede-Boker Campus, 84990, Israel
erell@bgumail.bgu.ac.il

Etzion, Yair
Research Professor, Head of the Center for Desert Architecture and Urban Planning, J.Blaustein Institute for Desert Research, Ben-Gurion University of the Negev, Sede-Boker Campus, 84990, Israel
etzion@bgumail.bgu.ac.il

Gradus, Yehuda
Professor, Department of Geography and Environmental Development, Ben-Gurion University of the Negev, Be'er Sheva, 84105, Israel
ncrd@river.bgu.ac.il

Issar, Arie
Professor Emeritus, J.Blaustein Institute for Desert Research, Ben-Gurion University of the Negev, Sede-Boker Campus, 84990, Israel
issar@bgumail.bgu.ac.il

Meir, Isaac A.
Senior Researcher, Center for Desert Architecture and Urban Planning, J.Blaustein Institute for Desert Research, Ben-Gurion University of the Negev, Sede-Boker Campus, 84990, Israel
sakis@bgumail.bgu.ac.il

Motzafi-Haller, Walfgang R.
Research Assistant, Center for Desert Architecture and Urban Planning, J.Blaustein Institute for Desert Research, Ben-Gurion University of the Negev, Sede-Boker Campus, 84990, Israel
womoha@bgumail.bgu.ac.il

Pearlmutter, David
Senior Researcher, Center for Desert Architecture and Urban Planning, J.Blaustein Institute for Desert Research, Ben-Gurion University of the Negev, Sede-Boker Campus, 84990, Israel
davidp@bgumail.bgu.ac.il

Portnov, Boris A.
Senior Researcher, Center for Desert Architecture and Urban Planning, J.Blaustein Institute for Desert Research, Ben-Gurion University of the Negev, Sede-Boker Campus, 84990, Israel
portnov@bgumail.bgu.ac.il

Safriel, Uriel
Professor, Director of J.Blaustein Institute for Desert Research, Ben-Gurion University of the Negev, Sede-Boker Campus, 84990, Israel
urielsf@bgumail.bgu.ac.il

Tsoar, Haim
Professor, Department of Geography and Environmental Development, Ben-Gurion University of the Negev, Be'er Sheva, 84105, Israel
tsoar@bgumail.bgu.ac.il

1   Introduction

Boris A. Portnov
J.Blaustein Institute for Desert Research, Ben-Gurion University of the Negev, Sede-Boker Campus, 84990, Israel

People have lived in arid lands since pre-historic times. There have always been deserts, and there have always been droughts. Only in comparatively recent years, as Cloudsley-Thompson (1986) justly notes, has the man-desert interaction led to such dramatic social and environmental consequences as those that were witnessed during the Sahel droughts of 1968-73 and 1982-85. These droughts caused the collapse of the entire agricultural base of five countries (Mauritania, Upper Volta, Mali, Niger and Chad), already among the poorest nations in the world, and severe damage to the agricultural base of two others: Senegal and Gambia (UNCOD 1977). By some estimations, between 100,000 and 250,000 died as a result of the drought. In addition, some refugees never returned to their homelands (Kates et al. 1977)

Mass migrations and refugee camps during the Sahelian draughts, brought the attention of the world community to the problem of desertification and became an immediate stimulus for international actions.

In 1974, the United Nations General Assembly passed a resolution calling for an international conference on desertification, to be held in 1977. It was specified that areas vulnerable to desertification should be identified, all available information on desertification and its consequences for development should be gathered and assessed, and a plan of action to combat desertification should be prepared with emphasis on the development of indigenous science and technology (UNEP 1996).

The report prepared for the 1977 United Nation Conference on Desertification (UNECOD) showed that in the mid-1970s, approximately 14 per cent of the world's population, 628 million people, lived in the areas exposed to the danger of desertification. Of these, the majority, 72 per cent, lived within semi-arid zones, 27 per cent inhabited arid zones, and 1 per cent lived in extremely arid regions (Kates et al. 1977).

More recent assessments show that in the early 1990s, more than 61.0 million km^2, nearly 40 per cent of the Earth's land area, is dryland. Out of this, about 9.0 million km^2 are hyper-arid deserts. The remaining 52.0 million km^2 are arid, semi-arid and dry sub-humid lands, part of which have become desert as a result of human activity. These lands are the habitat and the source of livelihood for about one fifth of the world's population (UNEP 1996).

The extent desertification varies by continent. In the Mediterranean Basin, Asia, and the Pacific, almost 500 million people are affected (Table 1.1).

Table 1.1. Population estimates of desert regions (millions of residents)

Region	Overall population	Urban-based population
Sub-Saharan Africa	75.5	11.7(15%)
North and South America	68.1	33.7(50%)
Mediterranean basin	106.8	42.0(39%)
Asia and the Pacific	378.0	106.8 (28%)
Total:	628.4	194.2 (31%)

Source: UNCOD 1977

1.1
Climatic Causes of Aridity

Hare (1977:74-75) singles out four distinctive climatic causes of aridity. These include:
- widespread, persistent atmospheric subsidence, which results from the general circulation of the atmosphere;
- localized subsidence induced by mountain barriers or other special physio-geographic features;
- absence of rain-inducing disturbances causes dry weather even in areas of moist air; and
- absence of humid airstreams.

It is suggested these four controls are interdependent, but their relative effect depends on the locality. Hare (*ibid.*) distinguishes between:

- almost continuously dry climates, leading to desert surface conditions, in which there is no season of appreciable rainfall;
- semi-arid or sub-humid climates with a short wet season of varying intensity; and
- sub-humid areas in which rainfall is infrequent, but not confined to special season.

In general, deserts characterize the earth's subtropical zones: Global patterns of air circulation dictate that the subtropics can be regions of subsiding air. When air subsides it warms up and its capacity to hold moisture increases, so inhibiting the formation of rain (UNCOD 1977). This accounts for prevalence of dry climates between latitudes 15 and 30 degrees north and south of the Equator (Fig. 1.1). However, dry climates extend into other latitudes and their patterns are caused by additional factors, such as distance from the rain-supplying oceans, seasonal high-pressure zones of large continental areas linked with monsoon systems, or the presence of mountain barriers down which air spills on their lee sides, creating rain shadows (*ibid.*).

1.2
Criteria for Aridity

What is a desert? The answer to this question is not straightforward since there is little agreement concerning the criteria for aridity (Middleton and Thomas 1997, Hare 1977).

The Russian-born climatologist, Köppen, devised one of the most commonly used classification methods for deserts in 1918 (NSF 1977). He defined a desert as a region where the moisture that could evaporate, if it were available, is at least double the amount of actual precipitation. In many deserts, Koppen's criterion is greatly exceeded. Vast areas of California, Arizona, and Northern Mexico are so dry that the ratio of potential evaporation to actual precipitation is more than 100: 1 (*ibid.*).

According to another index, developed in 1948 by the American climatologist Thornthwaite, deserts can be classified on a scale that describes the extent of their moisture deficits. Thornthwaite's moisture index (I_m) is defined as follows:

$$I_m = 100(P/PET - 1),$$

where P = precipitation, and PET = potential evapotranspiration.

If the annual precipitation is exactly equal to the maximum potential evapotranspiration (evaporation plus moisture loss from the leaves of plants), the moisture index equals zero. On Thornthwaite's relative scale, climates with an index between zero and minus 20 are considered subhumid. Between minus 20 and minus 40 the definition is semiarid. Below minus 40 the area is classified as arid (NSF 1977).

While precipitation data for Thornthwaite's formula are widely available, the values of PET require some calculation. A practical approach was suggested by Thornthwaite himself and uses an empirical relationship between measured PET and values of more readily gained environmental variables. This approach allows PET to be calculated from just two parameters -- mean monthly temperature and the average number of daylight hours per month (Middleton and Thomas 1997).

Using this method, Meigs mapped for UNESCO the arid lands of the world. He estimated the area of the world's semiarid lands at 20,5 million km^2; the area of arid lands at 21 million km^2; and the area of extremely arid lands at 5,5 million km^2. The total of 47 million km^2 of moisture-deficient lands is nearly one-third of the Earth's land surface (NSF 1977). These estimations do not, however, include cold deserts.

Another method of PET calculation is based on Penman's formula (Middleton and Thomas 1997). Compared to the Thornthwaite approach, the Penman method requires a larger body of directly measured meteorological data, including solar radiation, wind velocity, relative humidity, and temperature. In 1977, this method was used by the National Center of Scientific Research, France (NCSR) in preparing an updated map of arid regions (LCT 1977).

The classification of arid areas suggested by NCSR is given in Table 1.2. Although Petman's method for the calculation of the aridity index and that of Thornthwaite are based on somewhat different research techniques and require different sets of data, the geographic limits of arid zones produced by these two techniques exhibit, in general, considerable similarities (Middleton and Thomas 1997).

Table 1.2. The categories of aridity according to the P/PET ratio, and bio-climatic characteristics of arid areas

Category	P/PET ratio	Bio-climatic characteristic
Hyperarid	Less than 0.03	Corresponds to a true desert, where there can be one or several years without rainfall. These areas either do not have any vegetation or have only ephemerals, and some shrubs in the river beds
Arid	From 0.03 to 0.20	Comprises areas with thorny or succulent species and sparse annual vegetation.
Semiarid	From 0.20 to 0.50	Includes steppes or tropical shrublands, with a more or less discontinuous herbaceous layer and increased frequency of perennial grasses.
Subhumid	From 0.50 to 0.75	Includes certain types of tropical savannas, Mediterranean maquis and chaparral, and chernozem steppes.

Source: LCT 1977

The desert regions can also be classified by the vegetation they harbor. In 1956, Shantz (NSF 1977) suggested that all desert plants could either be characterized as xerophytes – those adapted to dry conditions – or were short-lived annuals. According to Shantz, the presence of these plants, to the exclusion of other plants, is indicative of aridity. According to Shantz, deserts can also be partially defined based on soil types, specifically by the presence of pedocals -- soils with dry subsoils. Not all areas of pedocals are arid; in many such places the surface soils receive enough rainfall to produce rich grasslands The areas are fragile, however; a short drought can turn them into wastelands (*ibid.*).

1.3
Geographic Extent of Deserts

The boundaries of the world's drylands are not fixed. There is, for instance, some evidence that the African Sahel extended southwards some 20,000 years ago (UNCOD 1977: 18). It is also believed that some 8,000 years ago, the Rajasthan region in India was 1500 km east of the arid zone (*ibid.*). Despite these changes, five major desert belts can be can be singled out. These are:

- the Sonoran desert of north-western Mexico and its continuation in the desert basins of the south-western United States;
- the Atacama desert, a thin coastal strip running west of the Andes from southern Ecuador to central Chile, from where dry climates extend eastwards into Patagonia;
- a vast belt running from the Atlantic Ocean to China and including the Sahara, the Arabian desert, the deserts of Iran, Turkmenistan and Tajikistan, the Rajasthan desert of Pakistan and India, and the Takia-Makan and Gobi deserts in China and Mongolia,
- the Kalahari and its surrounding arid lands in southern Africa,
- most of the continent of Australia (UNCOD 1977: 17).

There are also isolated areas of arid lands in other parts of the world, such as the Guajira peninsula in Colombia, southwestern Madagascar, northeastern Brazil, and 'pockets' of cold deserts in Russian Siberia (see Fig. 1.1).

The extent of desert regions is shown in Table 1.3.

Table 1.3. Aridity zones by region (million km^2)

Aridity zone	Africa	Asia	Australia	Europe	North America	South America	Total
Cold	0.0	10.8	0.0	0.3	6.2	0.4	17.7
Humid	10.1	12.2	2.2	6.2	8.4	11.9	51.0
Dry subhumid	2.7	3.5	0.5	1.8	2.3	2.1	12.9
Semiarid	5.1	6.9	3.1	1.1	4.2	2.7	23.1
Arid	5.0	6.3	3.0	0.1	0.8	0.4	15.6
Hyperarid	6.7	2.8	0.0	0.0	0.0	0.3	9.8
Total	29.6	42.5	8.8	9.4	21.9	17.8	130.1

Source: Middleton and Thomas 1997

1.4
The Process of Desertification

There is agreement that the rate of land degradation has increased significantly during the last decades: According to some estimations, desertification affects at least 50,000 km^2 per year, primarily in the developing world (UNECOD 1977).

The immediate causes of desertification in the developing countries are well known – overgrazing, the felling of trees for fuel, waterlogging, salinization, and inferior agricultural practices (Cloudsley-Thompson 1986). Desertification is thus not a mysterious or detached technological problem: As Jain (1986) suggests, desertification, in most cases, indicates that the existing path of development fails to provide people with a non-destructive way to make a living.

As revealed by satellite data, desertification affects nearly one quarter of the total land area of the world, and about one sixth of the world's population. These figures exclude natural hyper-arid deserts (UNEP 1996).

The urgency of addressing the global problem of desertification is due to the following considerations:

- Socio-economically, the process of desertification constitutes the main cause and mechanism of global loss of productive land resources, causes economic instability and political unrest in areas affected, brings pressures on the economy and the stability of societies outside the affected areas, prevents achievements of sustainable development in affected areas and countries;
- Environmentally, desertification contributes to loss of global biodiversity, loss of the biomass and bioproductivity of the planet, and global climate change (*ibid.*).

The strategy developed in the framework of the United Nation Environmental Program (UNEP) suggests the following immediate actions to combat desertification:

- Social, economic, cultural and political development with emphasis on solving problems of food, poverty, housing, employment, health, education, population pressures and demographic imbalance;
- Conservation of natural resources with emphasis on water, energy, soil, minerals, plant and animal resources in arid, semi-arid and dry sub-humid areas;
- Environmental control with special emphasis on protection against decline of soil fertility, soil loss, water, soil and air pollution as well as deforestation (UNEP 1996).

1.5
Deserts and Urban Growth

The agricultural potential of deserts is extremely limited (see Chaps. 8 and 13 of this book). Without enough rain to support sustainable agriculture (which is possible in arid regions only with special adaptations, as, for example, by the development of irrigation), deserts remain, in most cases, sparsely populated (UNCOD 1977).

Population densities in desert areas vary according to the productivity of the local environment, but they generally remain below 15 person per km^2 in the extremely arid zones, below 20 persons per km^2 in the arid zones and below 70 persons per km^2 in semi-arid zones (Middleton and Thomas 1997). Except for portions of Mexico, dry sectors of northern West Africa, Spain, the Fertile Crescent and much of western Iran, there are few arid lands with population densities of 100-250, the densities that are so common in adjoining humid and sub-humid regions (White 1966).

Dryland aridity stimulates the formation of spatially isolated (nucleated) settlements (Fig. 1.2) since their necessary supports, such as water and agricultural land, tend to be localized in deserts and dry lands (Kates et al. 1977).

Fig. 1.1. Arid regions and major population centers of the world

Historically, dryland settlements have served as commercial and administrative centers, grown up around mines, and other local amenities (Saini 1980; Golany 1978; Issar, Chap.13 of this volume), or were established as strategic outposts in response to various geo-political and security considerations (Portnov and Erell, Chap.2 of this book). Today, desert settlements function as irrigation centers (including oasis settlements), garrisons and communications centers, political, administrative and regional centers; they also may be focused on tourism, recreation, mining or other industries (Kates et al. 1977). As Table 1.1 shows, about one-third of the population of dry lands live in urban areas.

Portnov and Erell (Chap. 2 of this book) single out six distinctive reasons contributing to the considerable increase in the pace of desert urbanization in recent years. These include:

- Relocation of territory-consuming industries, military and research installations from overpopulated core regions to underdeveloped peripheral desert areas;
- Mining and power engineering, whose resources are often depleted in traditional mining centers and less remote locations;
- Development of transport infrastructure which extends the commuting frontier of existing population centers into a more remote periphery;
- Development of means of pumping fresh water considerable distances from its natural sources;
- Socio-economic, political and ecological considerations stemming from both overpopulation of the core and underdevelopment of the periphery, and
- The acceleration of desertification processes.

Kates et al (1977) distinguish between two types of urban-environment interactions in desert areas. The first type of interaction is intrinsic and relates to problems that might arise from any concentration of population in the arid lands: the need for fuel, shelter, water, and food that may not be available in sufficient quantities in the immediate hinterland of the urban center. In the developed world, these needs are satisfied within a larger, modern system of exchange. In the developing world, the needs of urban dwellers often cause the overexploitation of the local hinterland that, in turn, often lead to deforestation around urban centers. The second type of interaction is indirect. It arises from the fact that during the years of drought, desert settlements may become the focus of a continued flow of rural migrants. Although towns provide a temporary employment and housing solution to refugees, burdens on urban housing and services tend to intensify the adverse environmental impacts that desert cities and towns already exert.

Understanding the population-environment interactions in desert areas is an important challenge for contemporary urban and regional planning since this knowledge may lead to the development of more sustainable urban planning policies for existing desert regions and the areas affected by desertification.

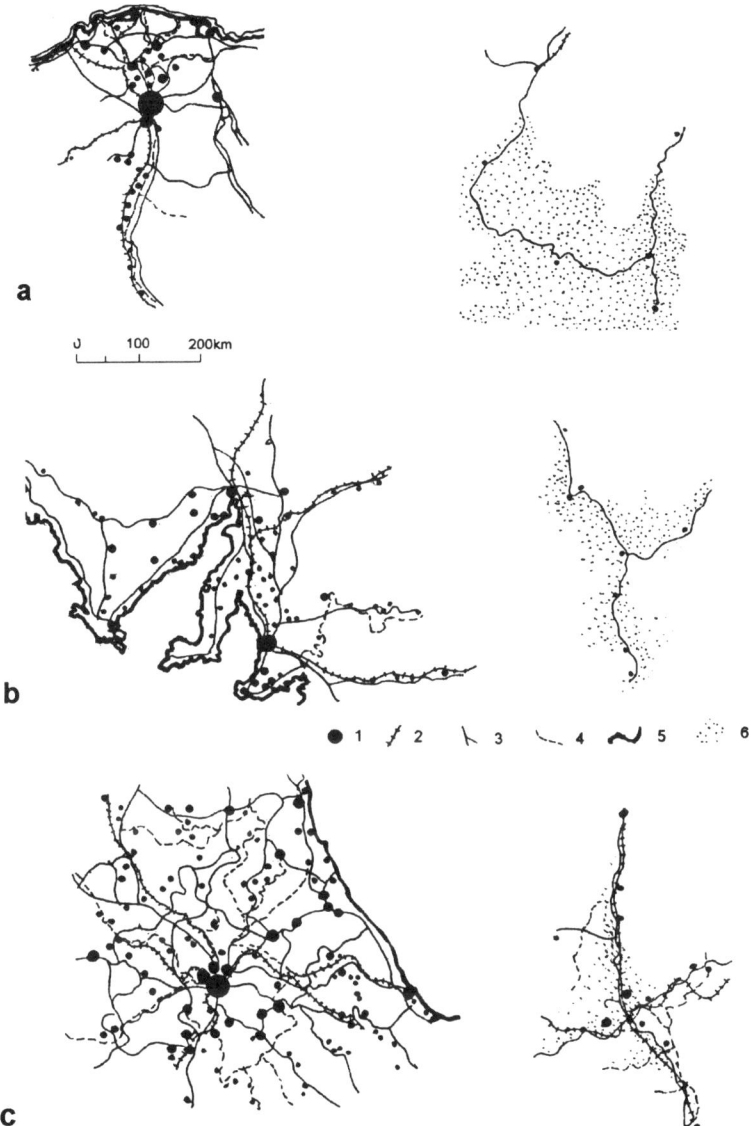

Fig. 1.2. Patterns of urban settlements in core areas and peripheral desert regions: examples from some countries (a, Egypt; b, Australia; c, Mexico. 1 – urban settlements; 2 – railroad; 3 – main road; 4 –river; 5 – sea/ocean shore; 6 – desert/sand area)

The diagram shows that settlement patterns in desert areas are usually more scattered than elsewhere. This is primarily due to the fact that necessary supports, such as water and agricultural land, tend to be localized in deserts and drylands

The interactions between man and the desert environment form the main focus of the present collective volume. In contrast to other books on desert development, the book considers *physical development of desert regions not as an end in itself, but rather as an essential precondition for creating socially attractive and desirable environments for human settlement.*

1.6
Scope of the Book

Following this introductory chapter *(Chap. 1)*, the book consists of four parts, each of which considers different conceptual levels of desert development: I. Regional Development and Population Change; II. Cities of Cold and Hot Deserts, III. Building and Design, and IV Case Studies. In addition to the Israeli experience, the book includes research and design from other countries which face acute problems of regional development in climatically extreme areas, i.e. either cold or hot deserts.

In *Chap. 2*, Portnov and Erell compare long-term patterns of both economic development and population growth in desert and non-desert urban settlements of Israel using three major criteria: the overall rate of population growth, the structure of population growth, and the rate of private construction in the locality. It is argued that in comparison with urban settlements located in central, "non-desert" districts of the country, peripheral desert communities exhibit wider fluctuations of economic activity, unstable population growth, and an attenuation of general urbanization trends that manifest themselves elsewhere in the country's urban areas.

Portnov and Pearlmutter (*Chap. 3*) analyze the potential of the 'migration balance – natural growth' ratio (the MB/NG ratio) as a general indicator of sustainability exhibited by an urban settlement in its population growth. It is argued that as long as a locality maintains its attractiveness to in-country migrants and foreign immigrants, its population growth can be considered as sustainable. Using extensive statistical data for urban settlements in Israel, the authors argue that after reaching a particular size (on the average, 20-30,000 residents), urban localities in the country tend to experience substantial changes in the components of their annual population growth. Starting with this inflection point, the growth of settlements gradually becomes less dependent on natural causes (fertility and mortality rates) than on the ability to attract newcomers and retain current residents. On the basis of this conclusion, a strategy of 'redirecting priorities' to developing the peripheral regions of the country is suggested. This strategy proposes the concentration of state and local financial resources on selected development settlements until they reach the above population threshold and become more attractive for newcomers, followed by the sequential transfer of this support to other small urban localities in frontier areas.

Another criterion for gauging the degree of sustainability exhibited by urban settlements – the rate of private construction – is discussed by Portnov and

Pearlmutter in *Chap. 4*. The authors suggest that the extent to which a particular urban locality is able to attract private developers can be considered as a prominent indicator of its socio-economic prosperity. In the case of Israel, the highest per capita rates of private construction are found in settlements of a particular size (70,000-80,000 residents).

The set of criteria developed in Chaps. 3 and 4 (the MB/NG ratio, and the rate of private construction) is used by Portnov and Erell in *Chap.5* for studying the effect of remoteness and isolation on the development of peripheral settlements. In this study, three sets of small urban communities in Israel – core settlements, desert localities, and small towns in the Galilee – are mutually compared. Sustained growth was found to be related to the location of the settlement, and in particular to the spatial characteristics of a cluster of urban localities of which the town may be a part. An index of clustering was defined, which allows an analysis of the combined effect on population growth of spatial isolation and distance from major metropolitan centers of the country, an a minimal population size of settlement cluster is determined, which is conducive to the sustainable growth of peripheral urban localities.

Portnov (*Chap. 6*) analyses the factors and forces affecting the degree of attractiveness of various geographic areas to migrants. He argues that interregional migration is a function of the interrelationship between employment growth and housing availability in the area: Migration occurs when, because of a scarcity of land, a large influx of immigrants, or a government policy, housing and employment in a region are not in balance.

Portnov (*Chap. 7*) discuss a methodological approach for evaluating the effect of regional policy on patterns of inter-regional population growth. It is suggested that the effect of a policy can be investigated by comparing the actual disparity in population between core and periphery regions to the disparity that would have been achieved in the absence of policy intervention. To test this hypothesis, the policy of population dispersal in Israel was considered. The analysis indicated that although the national policy of population dispersal, aimed at achieving a more even distribution of the country's population, generally failed to reduce the population imbalance between central core and periphery, this policy appears to have prevented the population gap from becoming even wider.

Safriel (*Chap. 8*) examines the role of ecology in the development of desert regions. Profitable and sustainable development options are, he argues, the export of desert assets: solar energy, cash crops that grow better in desert than elsewhere, and wilderness for tourism. The role of ecology is to prescribe optimal land allocation for the various uses, depending on ecological conditions along the aridity gradient of hyper-arid, arid, semiarid and dry-subhumid drylands.

Portnov (*Chap. 9*) investigates intra-city divergences in the level of social attractiveness of urban areas in the major population centers of Russian Siberia. Two different approaches – sociological poll and expert survey – are used to investigate urban inequalities in this unique geographic region of cold deserts.

The evolution of the urban environment in another climatically extreme region – the Negev desert of Israel – is analyzed by Meir in *Chap. 10*. Using the case of Be'er Sheva, the capital of the Negev desert, the author argues that abstract planning and design theories, disconnected from the environment and its constraints are often costly and problematic in such extreme environments as the desert. A number of alternatives, such as locally-oriented planning and design solutions are suggested to improve the quality of the urban environment in the area.

The effect of building shapes on the dry deposition of dust in urban areas is investigated by Erell and Tsoar in *Chap. 11*. The study suggests that strategies commonly employed in the design of buildings and urban spaces to reduce exposure to dust, such as the construction of walled courtyards, are not effective. A significant reduction in the concentration of dust near buildings in desert cities may require a comprehensive approach that deals with the entire urban area and its immediate surroundings. This approach should lead to reducing the availability of erodible particles by means of planting or paving all exposed land surfaces.

The idea of environment-responsive urban design and planning for extreme desert environments is discussed by Gradus in *Chap. 12*. The author argues that preconceived urban models cannot simply be 'applied' to arid zones, since cultural and environmental considerations are essential for the implementation of such projects. To achieve a better quality of urban environment in the desert, moving from the preconceived to responsive planning is needed. The concept of responsive planning is illustrated by three development projects in the Negev desert of Israel.

The effect of climatic changes on long-term regional development is studied by Issar (*Chap. 13*). The author traces the historical patterns of settlement and migration in parallel with evidence for temperature change and humidity, using the Middle East as an example. In the region, the warmest and driest periods of the past 5000 years are coincident with the largest invasions of desert tribes into agricultural lands and the desertion of cities along the margins of the desert.

Etzion (*Chap. 14*) discusses the design approaches and principles laid in the foundation of a desert solar neighborhood in Sede-Boker, Israel. In this design, a special effort was made to respond to questions of orientation of building, solar rights, air circulation, building clustering, and the relationship between open and closed spaces.

A bio-climatic approach to desert architecture is investigated by Etzion in *Chap. 15*. Desert architecture is perceived as "Architecture of the Extremes," being similar to architecture in other regions but differentiated from it by its obligation to address needs and problems of an extreme character. The author argues that such architecture should be able to improve the thermal performance of buildings without the use of artificial means and expendable energy. This approach is illustrated by the Etzion House in the Neve-Zin neighborhood in Sede-Boker, Israel. The house was designed and built by the author of this chapter.

Pearlmutter and Berliner (*Chap. 16*) analyses the influence which built form may have on urban microclimate, and on the resulting comfort conditions for pedestrians within a desert city. Rather than relying on accepted "myths" concerning the benefit or liability of compact urban planning under hot-dry conditions, the author presents a study which combines physical microclimatic observation with an integrated energy-exchange model to determine the relative impact which densely-built "street canyon" spaces have on a person's thermal relationship to his urban environment. Carried out in one of several "example neighborhoods" which were built in the arid Negev region of Israel in response to the perceived inadequacies of previous European-style planning, the study concludes that a compact urban fabric, with proper attention to a range of physical details, may provide certain amenities which are unique to the cities of desert regions.

A "climatically adaptive" approach to intelligent building in which a variety of technologies are integrated is discussed by Etzion et al. (Chap. 17). This concept is illustrated by the design of a multi-use building complex in the Negev region of Israel. In this design, a number of strategies were developed to exploit natural energy for heating and cooling: earth berming of major parts of the building, "selective glazing" for seasonal shading and energy collection, and a downdraft "cool-tower" for evaporative cooling.

Six case studies on desert towns in the Negev region of Israel are presented by Portnov and Motzafi-Haller in *Chap. 18*. Statistical data are listed for comparing the state of development of these towns to the region in which they are located, and the rest of the country. Appearances typical to the towns are shown in photographs.

References

Cloudsley-Thompson JL (1986) Foreward. In: Jain JK (ed) Combating desertification in developing countries. Scientific Publishers, Jodhpur, India, pp v-viii

Golany G (1978) Planning urban sites in arid zones: the basic considerations. In: Golany G (ed) Urban planning for arid zones. John Wiley & Sons, New York, pp 3-21

Hare KF (1977) Climate and desertification. In: UNCOD Desertification: its causes and consequences. Pergamon Press, Oxford, pp 63-168

Jain JK (1986) Preface. In Jain JK (ed) Combating desertification in developing countries. Scientific Publishers, Jodhpur, India, pp ix-xiv

Kates RW, Johnson DL, Johnson-Haring K (1977) Population, society and desertification. In: UNCOD Desertification: its causes and consequences. Pergamon Press, Oxford, pp 261-318

LCT (1977) Map of arid zones. Laboratoire de Cartographie Thematique, Paris

Middleton N, Thomas D, Eds. (1997) Word atlas of desertification. Arnold, London.

NSF (1977) Mosaic 8(1), National Science Foundation, Washington

Saini BS (1980) Building in hot dry climates. John Wiley & Sons, Chichester

Thornthwaite CW (1948) An approach towards a rational classification of climate. Geographic Review 38:55-94

UNCOD (1977) Desertification: its causes and consequences. Pergamon Press, Oxford

UNEP (1996) Status of desertification an implementation of the United Nations plan of action to combat desertification. United Nation Environment Program. Report of the Executive Director, Internet Edition

White GF (1966) The world's arid zones. In: Hills ES (ed) Arid lands: a geographical appraisal. Methuen & Co Ltd, London, pp 15-30

Part One

REGIONAL DEVELOPMENT AND POPULATION CHANGE

2 Long-term Development Patterns of Peripheral Desert Settlements

Boris A. Portnov and Evyatar Erell[1]
J.Blaustein Institute for Desert Research, Ben-Gurion University of the Negev, Sede-Boker Campus, 84990, Israel

2.1
Introduction

The desert, whose hot and dry climate is generally unsuitable for farming without importing water, was traditionally considered an unfavorable place for human habitation. Although deserts encompass more than one-third of the world's land mass (Maddock 1977), there are few historic and pre-historic urban settlements in desert areas. In recent years, however, the pace of desert urbanization has changed; a number of urban settlements in desert regions across the globe were either established or received a significant growth impetus. Examples include Phoenix and Tucson, Arizona; Be'er Sheva and Eilat in Israel; Pilbara in Western Australia; Ashgabat and Mary in Turkmenistan.

The change in the pace of desert urbanization may be attributed to the following factors:

- The establishment of territory-consuming industrial enterprises, military and research installations. Such installations often require large parcels of vacant land due to technological or security reasons (Clealand 1978; Kneese 1978). Relocation of these functions to desert areas was accelerated by the growing shortage of undeveloped land in non-arid areas, particularly in overpopulated urban regions.
- Exploration of various natural resources (mining, power engineering, etc.).
- Substantial improvements in the means of transportation and communication. The technical and technological advances such as highways, rail, etc. made distant desert areas far less remote in terms of daily commuting, transportation of goods, provision of essential services and utilities for the local population.
- The development of means of pumping fresh water considerable distances from its natural sources. While agricultural production in deserts remains limited due to economic considerations, urban development appears to justify the often large investments required for the provision of drinking water.
- Geo-political and security considerations. In a number of cases, specifically in Israel, urbanization of desert areas stemmed from the necessity to secure the

---

[1] The chapter is based on a shorter article on this topic published by these authors in International Journal of Urban and Regional Research (1998), 22(2): 216-232.

national presence in frontier areas, or from the need to disperse existing population from densely populated central districts (Barkai 1980).
- Accelerating desertification processes. These processes currently affect, by some estimations, about one-sixth of the world's population (UN 1994). A large number of urban settlements originally established in drylands are now faced with severe desert conditions.

While a number of studies were carried out on such aspects of desert development as water, soil, mineral resources, vegetation, building and climatic impact on human behavior (Kamon 1978; Bar-Cohen 1979; Saini 1980), research on regional and urban aspects of desert development was mostly limited to isolated case studies or general reviews of urban planning patterns of desert settlements (Clealand 1978; Golany 1978; Green 1982; Gradus 1992). In particular, there has been very little research on the unique features of long-term development in peripheral desert localities, compared with equivalent processes in core regions.

Understanding these processes, such as the rate and structure of population growth, attractiveness to private developers, etc. seems, however, to be of importance for contemporary urban and regional planning since this knowledge may lead to the development of more sustainable urban planning policies for existing desert regions and the areas affected by desertification.

The present case study of urban settlements in Israel attempts to answer the following research questions:

- Do peripheral desert settlements exhibit any long-term development peculiarities as compared with urban settlements located in central non-desert areas of the country?
- If so, what contributes to the development peculiarities of peripheral desert settlements?
- Finally, which planning strategies and policies can enhance the socio-economic sustainability of existing settlements in peripheral desert areas and facilitate prospective urban development there?

This chapter is primarily focused on physical variables of desert urbanization in Israel (location of urban settlements, population size, etc.) but also includes socio-economic and political aspects of the process, which may represent a legitimate subject for a separate study.

2.2
Desert Urbanization in Israel: Prerequisites and Historical Background

Although the Negev desert contains Israel's largest reservoir of natural resources (copper, phosphates, bromine, magnesia, potash, clays), the quantity and location of these resources (primarily in the Dead Sea area) do not represent a durable economic base for a wide-scale urban development of the entire region (Schechter

1979). The foundation of new towns in the Israeli Negev desert stemmed, therefore, primarily from ideological and geo-political considerations, namely:

- Establishing a national presence and sovereignty over the territory of the Negev, which constitutes nearly two-thirds of the land area of Israel (Gradus and Stern 1980);
- Re-establishing the nation's ties with its land (Drabkin-Darin 1957);
- Redistributing population from congested central regions to sparsely populated peripheral areas, stemmed from security and ecological considerations (Barkai 1980);
- Accelerating the process of new immigrants' integration into the host society in multi-cultural and socially diverse new towns in remote areas (Golany 1979).

Sheffer (1978) points out another factor that may have also stimulated the initial urban development in the Negev. The new immigrants of the 1950s, which constituted a predominant source of the initial growth of the new settlements in the area, largely came to the country from Middle East and North Africa. They did not import capital, were less educated than previous waves of immigrants, and thus were dependent on the state for their absorption. The location of these immigrants in isolated peripheral settlements might thus be perceived as an attempt to control them politically.

Given the above considerations, early urban development in the virtually virgin region did not follow an existing transportation network and led to the establishment of new settlements widely scattered across the area at varying distances from each other.

In this fashion, seven new development towns (Arad, Dimona, Mitzpe-Ramon, Netivot, Ofaqim, Sderot and Yeroham) were established in the Negev in the 1950's and early 1960's, and the existing towns of Be'er Sheva and Eilat received a significant growth impetus.

In the 1950s - the mid-1960s, the growth of these new towns was sustained primarily by the involuntary location of new immigrants and direct government investment in the area's economy. For instance, nearly 30 percent of the immigrants which entered the country between 1962-66, were directed to new towns. By 1969-70, this number, however, diminished rapidly (Kirschenbaum and Comay 1974). Since the early 1970's, the above allocation policy was gradually replaced by various government incentives designed to encourage the growth of the so called "development towns" indirectly. These included such measures as government loan guarantees, tax exemptions, and the provision of public housing (Gerson 1992).

During the same period, the policy of establishing "development towns" affected non-desert regions of the country as well: Several existing rural settlements were transformed into development towns, and a number of new settlements were established. Those included *Karmiel* and *Ma'alot-Tarshiha* in the northern part of the country, and *Bet-Shemesh, Qiryat-Malakhy, Yehud* and *Or-Aqiva* in the central part. The growth of these development towns occurred simultaneously with those in desert localities and thus represents a unique opportunity for a comparative study of their subsequent patterns of development.

2.3
Desert Settlements: Exogenous Factors

Desert settlements may generally be characterized by a number of exogenous factors which inherently constrain their development. Among the most prominent are the following: *remoteness, spatial isolation, harsh climate, and a lack of previous development in the area.*

Remoteness. Since most major urban centers across the globe, including the main population centers of Israel, are located outside desert areas, the distance from these national centers to any desert settlement is likely to be considerable, at least in terms of daily commuting. As Saini (1980) points out, this objective remoteness of desert settlements almost unavoidably increases the cost of infrastructure and of the transportation of goods, especially building materials. Owing to the remoteness of desert localities, businesses in these areas are also likely to experience substantial problems in recruiting skilled labor (Golany 1978). Given these external factors, the economic base of desert settlements tends to lack diversity, since under otherwise equal conditions, any new enterprise would preferably be established in less remote areas (Kneese 1978). It can thus be expected that compared to central (non-desert) localities, peripheral desert settlements should be more vulnerable to any negative changes in the national economy as a whole, such as recession, hyperinflation, etc.

Spatial isolation. The considerable distances often found between the established urban localities in a peripheral desert area are likely to cause a shortage of joint intra-regional educational and recreational structures, and limit the choice of job opportunities (Kneese 1978; Clealand 1978; Green 1982). This is expected to lead to uneven patterns of population growth in peripheral desert settlements, due to considerable outward migration of residents in "bad" economic years.

Harsh climate. The harsh climate of most desert areas (heat stress, temperature fluctuations, blowing dust, a lack of water resources and vegetation, etc.) places limitations on urban amenities and human comfort, making these localities appear to be even less desirable for business activity in comparison with non-desert settlements. These conditions may also have a negative impact on the patterns of population growth, making desert settlements less attractive and desirable (at least theoretically) for both new immigrants and internal migrants.

Lack of previous development in the area. A great number of desert localities were founded in virtually virgin areas that had no previous urban or rural development. Moreover, these settlements do not have, as a rule, an established agricultural hinterland (farms, villages, etc.), such as that which generally surrounds urban centers in other geographic areas. Such isolated towns are thus deprived of their major natural function, namely the role of a regional center providing services to the surrounding rural area. Having these characteristics in mind, it can thus be expected that migration patterns to and from desert settlements should presumably differ from those of any new settlements located in earlier established urban areas. For instance, it can be expected that peripheral desert settlements should be less given to suburban sprawl and other migration phenomena which can be observed in central overpopulated regions. In other words, the effect of general urbanization-suburbanization trends that might manifest themselves elsewhere across the nation should be somewhat less emphasized in peripheral desert areas.

2.4
Development Paradigms

Given the above theoretical considerations, the following analytical models can be suggested to explain development peculiarities of peripheral desert areas.

2.4.1
Economic Development of Peripheral Desert Areas

A model for economic development is proposed which is based on a number of assumptions (Fig. 2.1).

- *The level of infrastructure development* (transport network, utilities, etc.) generally diminishes with distance from the major urban centers of a country. This hypothetical trend is represented by the descending line "infrastructure" in the diagram (Fig. 2.1a);
- *Availability of skilled labor* declines with the distance from major metropolitan centers. This trend line has an inflection point (#1) which supposedly corresponds with the distances practicable for daily commuting.
- *Land availability* (in terms of both the affordability and amount of undeveloped land) increases with the distance from densely populated metropolitan centers, until land resources for development become virtually unrestricted (inflection point #2). No further substantial increase in land availability can be expected beyond this point.
- *Climatic conditions* are generally more extreme in the interior of desert regions than at their fringes. The uncomfortable climate, a lack of environmental amenities, or at least the popular perception of deserts as inhospitable regions, have an adverse impact on regional development. It was initially assumed, however, that the "weight" of the climatic factor is somewhat less significant than that of the factors considered above (land availability, availability of skilled labor, etc.). The corresponding trend is therefore represented by a dashed line in the diagram.

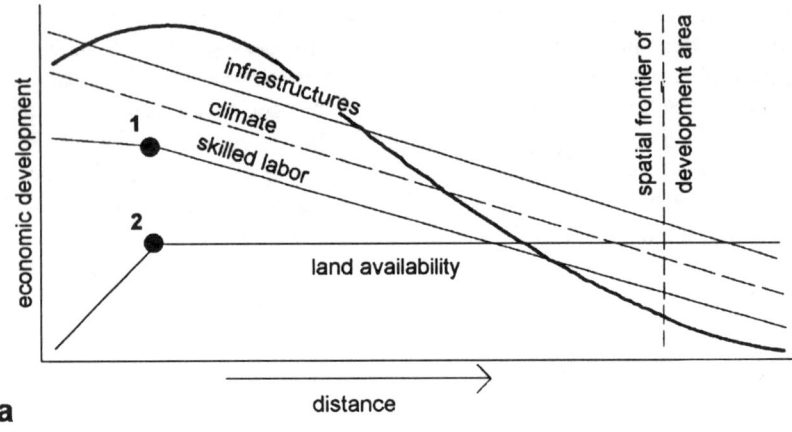

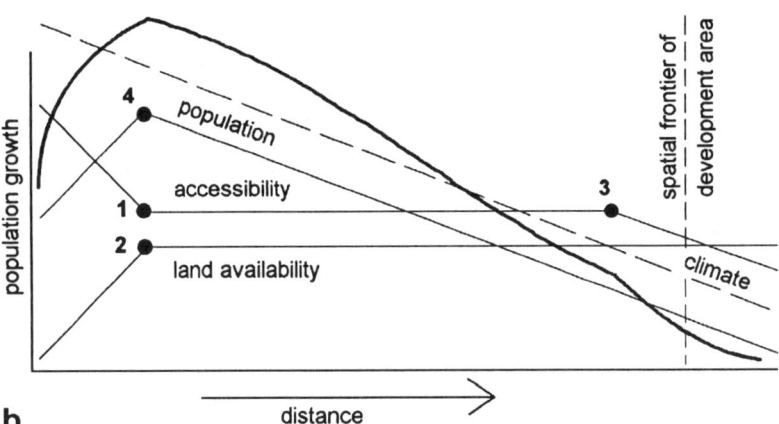

Fig. 2.1. Theoretical models of the factors affecting the rates of economic development (a), and population growth (b) in urban settlements, *vs.* their distance from major population centers

1. The inflection point which corresponds with the external limit of the area practicable for daily commuting;
2. The inflection point that matches the border of the densely populated urban "core" (this point of inflection does not necessarily spatially coincides with that of commuting distance);
3. The inflection point which conforms with the external limit of the area practicable for periodical (weekly) commuting;
4. The inflection point that corresponds with the most populated suburban ring of metropolitan areas

If the above assumptions are correct, their combined effect should lead to an initial peak in the development potential outside metropolitan centers, as a result of increasing availability of land. Beyond a certain distance from the center, no further increase in the availability of land can be expected. Concurrently, a relative decline in the availability of skilled labor, a deterioration of transport infrastructure, and worsening climatic conditions result in a progressive decline of the potential for development. This decline can thus be expected to cause significant differences in growth rates between urban localities in peripheral development areas and urban settlements located in central areas of the country, particularly those near which a peak of economic activity can be expected (see the bold curved line in Fig. 2.1a). An exception to this general rule may, of course, occur where development is based on unique mineral resources or other extraordinary factors.

2.4.2
Population Growth

This hypothetical model for population growth describes the *migration component* of a settlement's population increase (i.e. a settlement's attractiveness to new immigrants and in-country migrants) and rests on the following assumptions (Fig. 2.1b).

- *Land availability* has a positive effect on migration flux to a particular settlement, since the price and availability of land are reflected in the cost of housing. In particular, some types of houses (such as detached or semi-detached single-family dwellings) should become more available in peripheral settlements. As in the model for economic development (see Fig. 2.1a), the trend line in question has a point of inflection (#2, Fig, 2.1b) whose spatial location is determined by the location of hinterland areas in which land resources for development become virtually unrestricted.
- *Impeded access to a major metropolitan center*, where a great number of jobs and services is concentrated, has a negative effect on a settlement's attractiveness to potential migrants since it increases direct and indirect expenses associated with daily and weekly commuting. This general trend may have two inflection points (#1 and #3) which correspond with the distances practicable for daily and periodic commuting, respectively.
- *The level of urbanization* (population size and density) of particular geographic areas generally has a positive impact on the area's attractiveness to potential migrants. New immigrants, in particular, tend to settle in densely populated central regions rather than in relatively undeveloped peripheral areas (Fischer 1976; De Jong & Fawcett 1981; Moore & Rosenberg 1995). The most densely populated area, which apparently corresponds with the external suburban ring around major metropolitan centers (see inflection point #4, Fig.2.1b) might form, therefore, an area of intensive population growth.
- *Climatic harshness* negatively affects the attractiveness of an area to potential in-migrants and immigrants (this trend is represented by the dashed line in Fig. 2.1b).

The composition of the above trend lines forms a relatively sophisticated curve initially ascending within the commuting area of major population centers and then descending with advancing towards more remote peripheral areas (see the bold curve in Fig. 2.1b).

While possible implications of the above models in planning will be considered in Section 2.8, we shall now test these models using available statistical data.

2.5
Research Method

To study development peculiarities of desert settlements, the "sample-control" method was employed in the present case study.

The city of *Be'er Sheva* - the regional center of the Negev desert - whose current population is approximately 150,000 residents was included in the research scope as *Set 1*. Six other cities of similar size located in the central part of the country were selected as a control (*Control 1*): *Holon* (165,000 residents, as of 1994), *Rishon-LeZiyyon* (160,000), *Petah-Tiqwa* (150,000), *Netanya* (145,000), *Bat-Yam* (140,000) and *Ramat-Gan* (120,000) (Fig. 2.2).

Five small desert towns - *Dimona (30,000 residents), Ofaqim (19,000), Arad (19,000), Sderot (17,000) and Netivot (14,000)* - were selected as the second sample *(Set 2)*. Although all the settlements, included in this set, were established as development towns in the 1950s- the early 1960s, some differences between them deserve mention. While *Dimona, Ofaqim, Sderot* and *Netivot* are generally characterized by relatively high levels of unemployment and correspondingly low levels of socio-economic development, the town of *Arad* tends to display a stronger socio-economic base (Newman, et al. 1995). These differences are perceived, however, as a certain advantage for the present comparative analysis since they allowed us to cover a wide range of urban settlements in the area.

As the respective control *(Control 2)*, a group of urban settlements having similar size and located in proximity to major metropolitan centers of the country (either *Tel Aviv* or *Haifa*) were considered. They included: *Karmiel* (33,000 residents), *Qiryat Gat* (43,000), *Yavne* (27,000), *Bet Shemesh* (23,000), *Qiryat Malakhy* (20,000), *Yehud* (18,000), *Ma'alot Tarshikha* (15,000) and *Or Aqiva* (14,000) (Fig. 2.2). All these towns were established within a short period in the 1950's - early 1960's, and, all else being equal, are presumed to be at similar stages in their development as the above desert settlements.

With respect to the selection of the settlements for the present comparative analysis, some important comments should be made. First, none of the settlements included in the samples is located in the Jerusalem metropolitan area, in Judea and Samaria, or in the Gaza area. The development towns in these areas were deliberately excluded from the analysis since it was assumed that the growth of these localities has been affected by government policies and ideological considerations, rather than by more "mundane" factors.

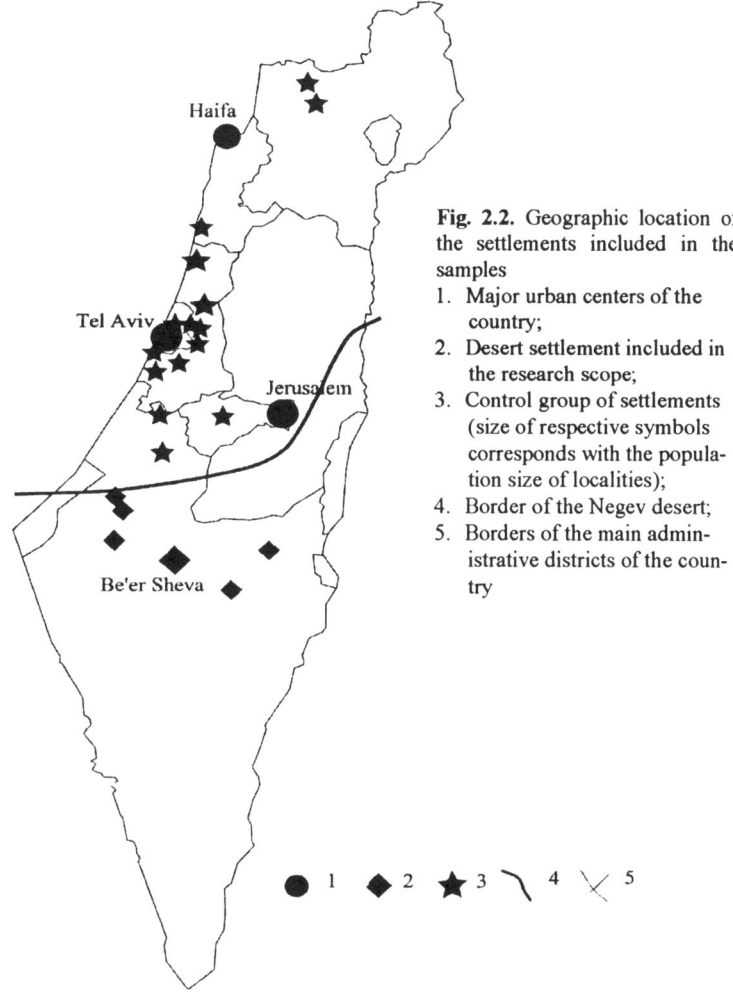

Fig. 2.2. Geographic location of the settlements included in the samples
1. Major urban centers of the country;
2. Desert settlement included in the research scope;
3. Control group of settlements (size of respective symbols corresponds with the population size of localities);
4. Border of the Negev desert;
5. Borders of the main administrative districts of the country

While selecting towns for the present analysis, some social characteristics of the settlements were also considered. These include ethnic makeup of the population (*percentage of Asian-African born*) and the *proportion of large families* (see Appendix). As the table shows, the average differences between small desert settlements (*Set 2*) and their respective control (*Control 2*) with respect to the above indicators are minor. It was thus assumed that these differences should not affect the result of comparative analysis. There is, however, a certain internal heterogeneity of the set (*Arad* and the rest of the set) and its control (*Karmiel* and the rest of the control). This heterogeneity is perceived as a certain advantage for the present comparative analysis since it allowed us to cover a wider range of urban settlements located in the respective geographic areas.

As for *Be'er Sheva (Set 1)*, the differences between the town and its control (*Control 2*) with respect to the above indicators appear to be significant. However, there are no other urban settlements of comparable size which are located in the central district of the country, and which have a closer socio-economic makeup. With due regard for this, the possible impact of the above disparities on the outcome of the analysis will be considered.

The following aspects of the settlements' development were studied:

- Overall population growth, measured as a percentage annual change;
- Structure of annual population growth, defined as the ratio between the migration component of growth (foreign immigration plus in-country migrations) and net natural growth (difference in birth and death rates);
- Annual rate of private construction, in 1,000 m^2.

While the first indicator (overall population growth) is a frequently used development datum (Kneese 1978; Mills 1972; Levy 1985; Vining 1982), the use of the second and third indices calls for some explanations.

The ratio between migration and natural growth may be considered an important indicator of a settlement's development, since an excess of migration over natural growth indicates that the settlement grows mainly because of its ability to attract newcomers and retain existing residents rather than due to natural causes, i.e. birth-death rates (for more details on this indicator of urban growth see Chap. 3 of this book).

The overall rate of construction is widely accepted as a key indicator of socio-economic development (Layton 1972; Levy 1985; McGranahan 1972; Mills 1972; Smith 1975). The *"private"* component of this indicator seems, however, to be even more informative. The rate of public construction at a particular urban locality may reflect government policies rather than genuine attractiveness to investors (for more details on this indicator see Chap. 4 of this volume).

Seven points in time were selected for the present case study using comparable sets of data (Israel Central Bureau of Statistics, "Local Authorities in Israel: Physical Data," 1966-1996. Data on migration and immigration are only available from 1971. The analysis of the structure of annual population growth thus refers only to this period.

2.6
Development Peculiarities of Desert Settlements

The dynamic of the *overall population growth* of urban localities included in the sample (Fig. 2.3) indicates that the population of desert settlements (Sets 1 and 2) grew on the average more slowly in 1970-94 than that of the respective non-desert control. For example, as Fig. 2.3 shows, the overall annual rate of population growth of the small desert settlements sampled (Set 2) was, on the average, 1-2% lower than that of the respective control settlements in the central region (Control 2).

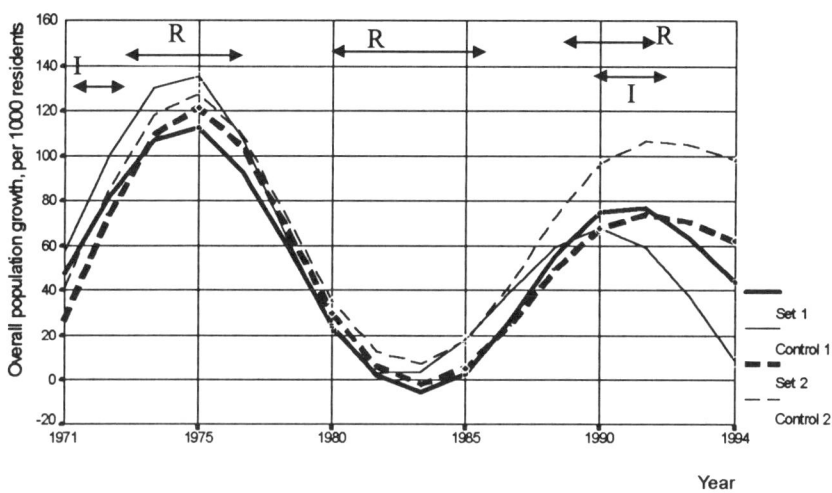

Fig. 2.3. Population growth of the urban localities included in the sample (1971-94)
The curves represent statistical averages computed separately for the respective sets of desert settlements and the controls. The arrows indicate the periods of mass immigration to the country (I), and those of relative economic recession (R)

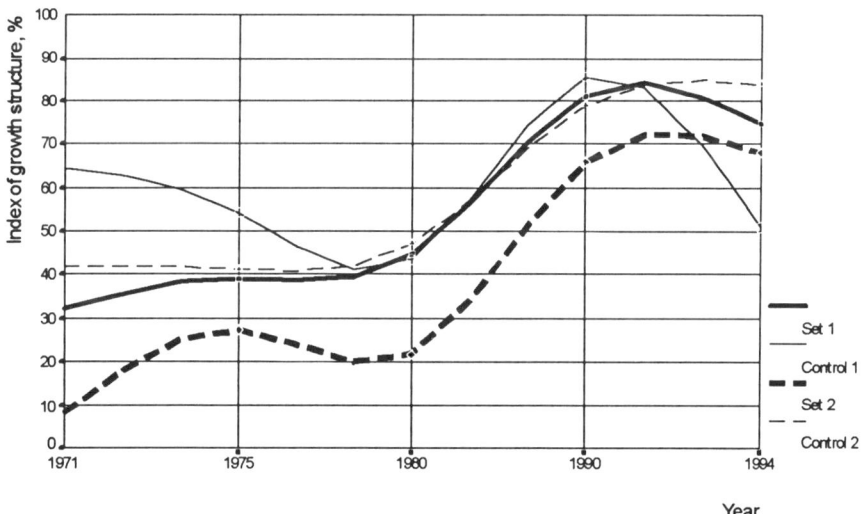

Fig. 2.4. Changes in the structure of population growth of the urban localities included in the sample (1971-1994)

Keeping in mind the above "overrepresentation" of big "oriental" families in *Be'er Sheva* (see the previous section), the analogous difference in the rate of growth between the city and its control (Control 1) could also be expected (under otherwise equal conditions). While this general trend is not entirely unexpected, one particular phase deserves notice. Since the recent peak of mass immigration form the former Soviet Union in 1990-91, large urban localities in the central areas of Israel *(Control 1)* exhibited a substantial decline in the rate of population growth, while smaller urban settlements in these areas *(Control 2)* continued to grow at a relatively high rate.

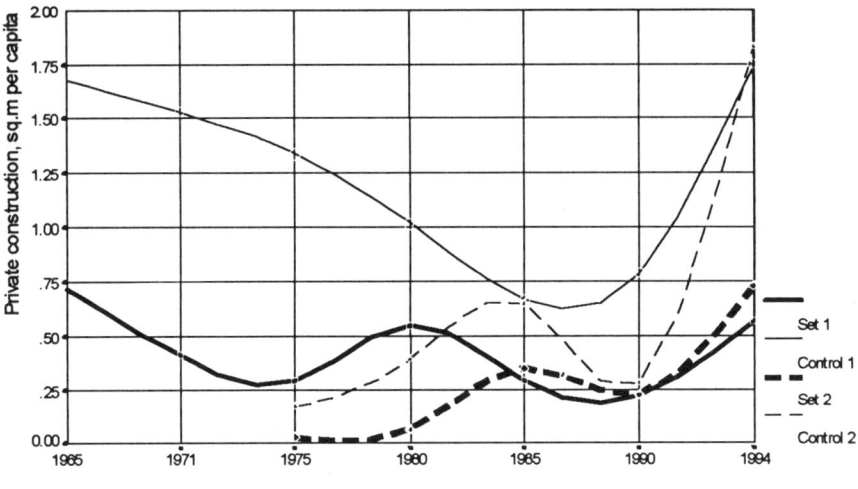

Fig. 2.5. Annual rates of private construction in the urban localities included in the sample (1965-1994)

This phenomenon, which is apparently due to the dispersion of population from major urban centers of the country, was somewhat less evident in peripheral desert areas. The disparities in growth rates between desert localities of different population size (Set 1 and Set 2) are indeed less substantial (Fig. 2.3). These data thus support the hypothesis that the effect of general urbanization and suburbanization trends appears to be delayed and attenuated in peripheral desert areas, compared with the country's central districts.

This attenuation of the urban growth in peripheral desert areas may be attributed to a number of factors. As Portnov and Pearlmutter point out (see Chap. 3 of this book), population movements in Israel appear to represent two simultaneous migration streams which are substantially different in nature: 1) *new immigrants*, being less informed about the country and local conditions, primarily concentrate in big and affluent cities of the country, where they hope to obtain suitable employment in the future or can count on initial help from friends and relatives, while 2) the *existing population* of these centers which is, of course, more economically independent, mobile and informed moves outward from the overpopulated and "expensive" metropolis to smaller localities nearby where suitable housing is more available and affordable. The relocation of internal migrants may be referred to as the *"hinterland-housing" stream*, while the initial distribution of the new immigrants across the country's urban settlements may be described as the *"centripetal"* or *"metropolitan"* stream. Neither of these migration "streams" is thus directed at peripheral settlements. This leads, in turn, to the attenuated patterns of their population growth described above.

While the analysis of the *structure of population growth* (Fig. 2.4) also appears to exhibit the same trend, certain additional insights can be gained from the diagram.

First, *equal rates of migration and natural growth* (corresponding to the value of 50% on the Y-axis in the figure) were achieved by the desert localities (Sets 1 and 2) and their respective controls (Controls 1 and 2) at different time periods. While urban settlements included in *Control 1* began the time-span in question (1971-1994) having a substantial surplus of migration over natural growth, Be'er Sheva (*Set 1*) passed this threshold only in 1982 (see Fig. 2.4).

The trend is similar for small towns. The small towns included in *Control 2* were close to crossing the 50% threshold in the late 1970's, eventually passing it in 1981, while desert localities of similar size approached this threshold only in 1988-89. Thus, peripheral desert localities acquire a similar degree of attractiveness to migrants at a later stage in their development than their non-desert controls, and at a greater population size.

The *rates of private construction* are also substantially different in desert localities from their non-desert controls (Fig. 2.5). The following general trends can be observed:

- The overall rate of private construction is considerably lower in peripheral desert towns than in settlements located in central areas of the country, for all periods,. This conclusion is in line with the research model shown in Fig. 2.1a. According to this model, the overall potential for economic development is expected to be lower in peripheral desert areas than elsewhere due to deficient infrastructure, a lack of skilled labor and climatic harshness (see the bold line in Fig. 2.1a).
- Periods of economic recession and hyperinflation (1973-1977, 1980-1985, 1989-1991) had an especially detrimental effect on the rate of private construction in the peripheral desert localities, while the positive effect of relatively favorable economic years (1992-1994) is less pronounced there.
- Discrepancies between the rates of private construction in large and small urban settlements in central parts of the country (Controls 1 and 2) are rather substantial, while large and small urban localities in peripheral desert areas *(Sets 1*

and 2) are similar in their patterns of construction activity. It appears that peripheral desert settlements and analogous urban localities in central non-desert regions acquire the *same degree of investment attractiveness after reaching different population thresholds*. This phenomenon can be explain using the above theoretical model (Fig. 2.1a). Since the overall development potential of remote desert areas is, as this model suggests, relatively low, *no substantial differences* among particular urban localities in these areas can be expected. The lack of intra-regional discrepancies thus supports the model in question, at least indirectly.

2.7
Influence of the Desert

In the previous subsection, the disparities between peripheral desert localities and their respective non-desert controls were discussed with respect to particular development indicators. It was impossible, however, to affirm whether those disparities were caused by climatic conditions of the desert or by other external factors associated with settlements' location in remote desert areas. The present section attempts to answer this question.

Several exogenous factors presumed to influence development patterns of urban settlements were discussed in Sections 3 and 4. The main factors were: population size, distance from major urban centers of the country (*remoteness*), spatial isolation (*grouping*), and climatic harshness.

The statistical significance of the influence of these factors on the development indicators included in the study area (*overall population growth, growth structure* and *rate of private construction*) may be identified and measured by analysis of variance. It should be noted that in order to ensure the homoscedasticity of errors, a logarithmic transformation was applied to the original variables. Hartley's F-Max test for violations of the homoscedasticity assumption confirmed that the variances are indeed similar.

The following exogenous factors were included as variables in an analysis based on 1994 data covering all 20 towns in the study:
- settlement population ['000s];
- distance [km] to the closest urban metropolis of the country (either Jerusalem, Tel Aviv, or Haifa);
- settlement isolation (grouping), measured for each settlement as the number of other urban settlements located within a practical range for daily commuting from the locality, assumed to be 20 km;
- index of climatic harshness (a so called temperature discomfort index) calculated for each urban locality as the mean annual number of days with heat stress (Bitan and Rubin 1991).

The results of the analysis are presented in Table 2.1. The influence of climatic harshness on all the development indicators in question is not statistically significant. There is, therefore, insufficient evidence to support the assumption that harshness of climate has any substantial influence upon diversifying the patterns of settlement development in Israel. The other variables – population size, distance to major urban centers of the country, and particularly spatial isolation of urban settlements (grouping) – do, however, have a significant effect.

Table 2.1. Analysis of variance (F values)

Dependent variable	Factor			
	Population	Distance	Grouping	Climate
Overall growth (OG)	9.40^a	2.94^b	7.92^a	1.05
Indicator of growth structure (I+M)/NG	4.18^b	2.16	4.35^b	1.76
Private construction	0.22	0.62	4.18^b	0.21

^{a} indicates a 0.01 confidence level; ^{b} indicates a 0.05 confidence level;

2.8
Conclusions and Policy Implications

Since the research indicates that scattered patterns of urban development, and remoteness of existing settlements from the major urban centers of the country are the major forces which attenuate population growth and economic development of desert settlements, a number of possible planning policies and strategies, stemming from the results of the present analysis, can be suggested. These strategies are aimed at enhancing the socio-economic sustainability of existing settlements in peripheral desert areas and at facilitating prospective urban development there.

In order to reduce *spatial isolation* of peripheral desert settlements, the following patterns of urbanization in peripheral desert areas could be employed (Fig. 2.6):

- *Development clusters with a clearly expressed urban core*: This urbanization pattern may be relevant to desert areas which already have existing regional centers represented by relatively big urban localities. The process of urban development in this case may represent the consecutive formation of a group of satellite settlements situated within the distance practicable for daily commuting, and which may form a single economic unit with the regional center (Fig. 2.6a).
- *Development clusters of small urban settlements having no dominant urban core*: This pattern of urbanization may be relevant to desert areas whose current settlement patterns are less intensive, and where existing small settlements are widely scattered across the area. Under such circumstances, urban development of the region may lead to the establishment of new settlements so as to form development clusters with existing small urban localities (Fig. 2.6b). The settlements in such clusters are expected to share some essential functions

(employment, educational, cultural, recreational services, etc.), which each of the small localities cannot individually sustain.

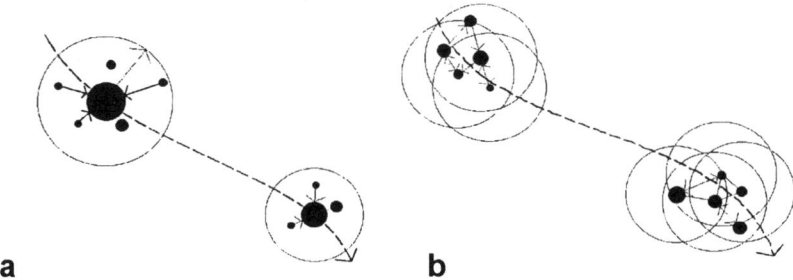

Fig. 2.6. Planning strategies suggested for accelerating urban development in peripheral desert areas
a - development clusters with clearly expressed urban core; b - development clusters formed by a group of small urban settlements located within a distance practicable for daily commuting which have no clearly expressed urban core. (The dashed lines represent the general direction of urban expansion in a region: from earlier established urban areas to developing desert frontiers)

Unlike small population size and spatial isolation, the *remoteness of desert settlements* with respect to a major urban metropolis of the country is a geographical fact. The above analytical models (Section 4) make it possible, however, to consider some solutions to this problem (Fig. 2.7).

For instance, an increase of the economic potential of peripheral desert areas can be achieved by the combination of the following development policies (Fig. 2.7a):

1. *Stricter land use regulation in central non-desert areas.* The reduction in available land in central non-desert areas is a natural process, which may be accelerated by progressive land taxation, and strict zoning. In turn, this would increase the relative attractiveness of undeveloped land in peripheral regions by "pushing" the land availability threshold towards the spatial frontier of development areas (see Figs. 2.1a and 2.7a); Further improvement in the means of transportation and expansion of existing transport networks. These measures can contribute not only to increasing the competitiveness of the local enterprises in terms of production expenses, but also to making skilled labor more available. These changes are represented by the "flattened" infrastructure and labor force lines in Fig. 2.7a;
2. *Development of a progressive system of investment incentives* (tax exemptions, favorable loans, etc.) which are to increase with distance from existing urban

centers (see the ascending "investment incentives" line in Fig. 2.7a. This line, as we may recall, was not represented in the "original" model in Fig. 2.1a).

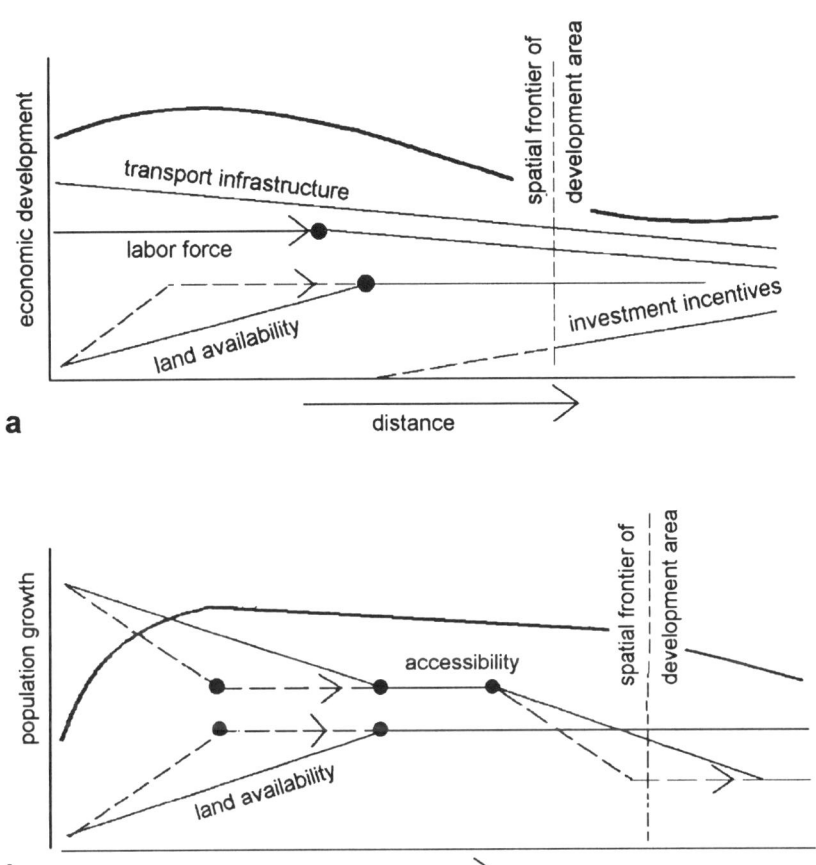

Fig. 2.7. Expected impact of various planning strategies on (a) economic growth, and (b) population growth of development areas

In addition to the expected effect of the aforementioned measures on the economic development of peripheral desert regions, some positive changes in the *population growth* of these areas can also be expected. As the transportation network develops, improving accessibility to the country's central regions, and moving the threshold of land availability deeper into the peripheral desert region, the population growth of these areas may become more intensive and sustainable. These positive changes are represented in Fig. 2.7b. As this diagram shows, the adjusted thresholds of *"accessibility"* and *"land availability"* can lead to the "equalization" of the rates of population growth in different geographic areas (see the bold lines in Figs. 2.1b and 2.7b).

References

Bar-Cohen A (1979) Space cooling and thermal control by skyward radiation. In: Golany G (ed) Arid zone settlement planning: the Israeli experience. Pergamon Press, New York, pp 164-196

Barkai Z (1980) Urbanization processes in the central coastal plain and their effect on a national economic planning and on policy of population dispersal. In: Israel builds 1977. Ministry of Construction and Housing, Jerusalem, pp 8-13

Bitan A, Rubin S (1991) Climatic atlas of Israel for physical and environmental planning and design. Ramot Publishing Co, Tel-Aviv

Clealand CB (1978) Community organization and problems in the American arid-zone city. In: Golany G (ed) Urban planning for arid zones, John Wiley & Sons, New York, pp 89-99

De Jong GF, Fawcett JT (1981) Motivation for migration: an assessment and a value-expectancy research model. In: De-Long GF, Gardner RW (eds) Migration decision making: multidisciplinary approaches to microlevel studies in developed and developing countries, Pergamon Press, London, pp 13-53

Drabkin-Darin H (1957) Housing in Israel: Economic & sociological aspects. Gadish Books, Tel Aviv

Evenari M, Shanan L, Tadmor N (1971) The Negev - The Challenge of a Desert. Harvard University Press, Cambridge

Fischer CS (1976) The Urban Experience. Harcourt Brace Jovanovich, New York

Gerson M (1992) The programs of aid for housing. In: Golany Y, Eldor S and Garon M (eds) Planning and Housing in Israel in the Wake of Rapid Changes. Ministries of the Interior and of Construction and Housing, Jerusalem, pp 201-220

Golany G (1978) Planning urban sites in arid zones: the basic considerations. In: Golany G (ed) Urban planning for arid zones. John Wiley & Sons, New York, pp 3-21

Golany G (1979) Israeli development policies and strategies in arid zone planning. In: Golany G (ed) Arid zone settlement planning: the Israeli experience. Pergamon Press, New York, pp 3-42

Gradus Y (1992) Beer-Sheva - capital of the Negev desert. In: Golany Y, Eldor S and Garon M (eds) Planning and Housing in Israel in the Wake of Rapid Changes. Ministries of the Interior and of Construction and Housing, Jerusalem, pp 251-266

Gradus Y, Stern E (1980) Changing strategies of development: toward a regiopolis in the Negev desert. J of American Planning Association 46(4): 410-423

Green HC (1982) Town design in the arid Pilbara of Western Australia. In: Golany G (ed) Desert planning: international lessons. The Architectural Press, London, pp 15-30

ICBS (1966-1996) Local Authorities in Israel: Physical Data. Israeli Central Bureau of Statistics, Jerusalem (in Hebrew)

Kamon E (1978) Physiological and behavioral responses to the stresses of desert climate. In: Golany G (ed) Urban planning for arid zones. John Wiley & Sons, New York, pp 41-60

Kirschenbaum A, Comay Y (1974) Dynamics of Population Attraction to New Towns - the Case of Israel. In: Dialogue in development - natural and human resources. Proceedings of the 3rd World Congress of Engineers and Architects. Jerusalem, pp 18 - 30

Kneese AV (1978) The economic and economically related aspects of new towns in arid areas. In: Golany G (ed) Urban planning for arid zones. John Wiley & Sons, New York, pp 123-138

Layton AP (1972) Some Australian experience with leading economic indicators. In: Lahiri K, Moore J (eds) Leading economic indicators: New approaches and forecasting records. Cambridge University Press, Cambridge, pp 211-230

Levy JM (1985) Urban and metropolitan economics. McGraw-Hill Book Company, New York

Maddock T (1977) Drought is the dilemma: the philosophy of water management. In: Arid zone development. Ballinger Publishing Company, Cambridge, pp 21-31

McGranahan D (1972) Development indicators and development models. In: Baster N (ed) Measuring development, Frank Cass, London, pp 91-102

Mills ES (1972) Urban economics. Scott, Foresman and Co, London

Moore EG, Rosenberg MW (1995) Modeling migration flows of immigrants groups in Canada. Environment and Planning A 27: 699-714

Newman D, Gradus Y, Levinson E (1995) The impact of mass immigration on urban settlements in the Negev 1989-1991. Working Paper No 3. The Negev Center for Regional Development, Be'er Sheva, Israel

Saini BS (1980) Building in hot dry climates. John Wiley & Sons, Chichester

Schechter J (1979) Research challenges of arid zones: the Israeli example. In: Golany G (ed) Arid zone settlement planning: the Israeli experience. Pergamon Press, New York, pp 43-66

Sheffer G (1978) Elite cartel, vertical domination, and grassroots discontent in Israel. In Tarrow S, Katzenstein PJ, Graziano L (eds) Territorial politics in industrial nations. Praeger, New York, pp 64-96

Smith WF (1975) Urban development. The process and the problems. University of California Press, Berkeley

UN (1994) United Nations convention to combat desertification in countries experiencing serious drought and/or desertification, particularly in Africa. UN, New York

Vining DR Jr (1982) Migration between the core and the periphery. Scientific American 247-6: 37-45

Appendix
Population composition of the samples

	Percentage of Asia and Africa born		Percentage of the population belonging to large families	
	1967[a]	1972[a]	1973[b]	1977[b]
Set 1 (Be'er Sheva)	42.3	37.2	32.3	27.9
Control 1:				
Holon	24.0	22.4	10.6	11.5
Rishon-LeZiyyon	26.0	24.0	16.0	15.8
Petah-Tiqwa	26.1	24.3	16.2	15.7
Netanya	27.9	27.5	20.0	18.6
Bat-Yam	24.8	25.4	10.2	9.6
Ramat Gan	20.3	20.2	6.3	6.9
Average for the set:	24.9	24.0	13.2	13.0
Set 2:				
Dimona	60.2	52.9	41.2	39.3
Ofaqim	64.2	55.9	50.2	50.6
Arad	-	15.2	13.2	15.7
Sderot	57.7	48.7	44.6	42.0
Netivot	60.2	53.0	62.0	61.4
Average for the set:	60.6	45.0	42.2	41.8
Control 2:				
Karmiel	-	20.9	12.6	13.7
Qiryat Gat	53.2	47.1	42.1	36.4
Yavne	57.2	51.7	45.4	40.9
Bet Shemesh	57.7	48.5	46.3	40.3
Qiryat Malakhy	53.5	42.8	43.1	39.5
Yehud	39.8	36.2	18.6	15.9
Ma'alot Tarshikha	69.7	52.3	44.3	46.2
Or Aqiva	53.6	49.9	40.4	38.2
Average for the set:	55.0	43.7	36.6	33.9

[a]Source: Berman Y (1977) Social profile of Israeli towns. The Ministry of Welfare, Jerusalem (in Hebrew). [b]Source: Berman Y (1978) Social profile of Cities and Town in Israel. The Ministry of Labor and Social Affairs, Jerusalem

3 Sustainable Population Growth of Urban Settlements: Preconditions and Criteria[1]

Boris A. Portnov and David Pearlmutter
J. Blaustein Institute for Desert Research, Ben-Gurion University of the Negev, Sede-Boker Campus, 84990, Israel

3.1
Introduction

In the previous chapter, two criteria of urban development – the *rate of private construction* and the *migration balance/natural growth index* (the MB/NG index) – were used to gauge the performance of desert localities in comparison with urban settlements located elsewhere. In the present chapter, we shall discuss in detail the derivation of the index (MB/NG) and a manner in which it may serve as a general indicator of sustainable population growth. The performance and determinants of this indicator will also be analyzed using a broader settlement sample.

3.2
Defining Sustainability

In the wake of rising environmental awareness, various aspects of *sustainable development* (SD) have become the focus of academic study. Given the multifaceted nature of SD, it is not surprising that definitions of the phenomenon vary widely: as Haughton and Hunter (1994) point out, more than twenty distinctive definitions of SD are currently in use. The most widely cited definition of SD is, however, that of the World Commission on Environment and Development (the Brundtland Commission), which describes SD as "development that meets the needs of the present without compromising the ability of future generations to meet their own needs" (Turner 1993).

Defined in such general terms, SD is therefore open to a variety of interpretations, which encompass at least three major dimensions of the issue - environmental, economic, and socio-demographic.

- *The environmental dimension* of SD embraces the multiplicative interaction of three conditionally independent variables: population - consumption - technology. In other words, it considers SD in terms of upward and downward pres-

---

[1] The chapter is based on a shorter article on this topic published in Review of Urban and Regional Development Studies (1997), 9(2): 129-145

sure of the population on existing environmental resources, such as food, energy sources, water, etc. (Sage 1994).
- *Economic aspects* of SD are those conditions that insure that economic development can endure over time. It has been proposed that economic development, rather than being narrowly defined in terms of gross national product per capita or real consumption per capita, based on a more comprehensive set of welfare indicators, including education, health, and quality of life (Turner 1993).
- *Socio-demographic aspects* of SD are those that provide overall stability of population growth in particular geographic areas, and deal with specific aspects of the issue such as the "optimal size" of a settlement and the rural-urban balance of a region.

The formal criteria for gauging the degree of sustainability of population growth in urban settlements will be discussed, in some detail, in the following section.

3.3
Measuring Population Growth

Five criteria are often used to measure the population growth of urban settlements: the overall rate of population growth; the relative rate of population growth; percentage change of urban population in the area; the average rate of net migration and the natural growth rate.

The overall rate of population growth. While this criterion is the most obvious and direct measure for gauging the process of urban growth, its reliability and usefulness are fundamentally limited. For example, two urban settlements may have the same overall rate of population growth, 4% per annum. Whereas the first settlement has a positive natural growth (+5.0 % per annum) and a negative migration balance (-1.0 %), the respective components of growth in the second locality are both equal to +2.0% annually. From the standpoint of regional and urban planning, these two settlements clearly represent different cases, while the commonly used criterion of population growth treats them identically.

The relative rate of population growth. This index measures population growth in a locality relative to the national or regional growth rate, taking into account the socio-demographic "context" of the settlement. For example, a local population growth rate of 1.0% per annum may be considered high in Sweden, whose national population growth does not exceed 0.6%, and low in Egypt, where annual growth is close to 2.0%. This indicator, like the overall rate of growth, does not take into account the structure of population growth (i.e. the ratio between migration balance and natural growth).

Percentage change of urban population in the area. This measure is often used to trace the progress of urban growth in a particular geographic area, and is an important component of national statistics worldwide (Brown and Jacobson 1987). However, it should be kept in mind that a region's urban population growth does not necessarily indicate whether this growth occurs evenly across all urban localities, or is concentrated only in large metropolitan centers, for instance. While this indicator seems to be a basic tool for evaluating the process of urban development in entire countries or regions, it may have only limited application in planning.

The *average rate of net migration*. Unlike the previous index, this indicator is applicable to single urban settlements as well as to entire geographic regions. However, this indicator reflects only one component of population growth -- migration -- neglecting the effect of natural growth.

The *natural growth rate* is essential for the proper understanding of the processes of urban growth in particular geographic areas since this indicator reflects both socio-demographic makeup and, to a some extent, environmental and location preferences of the local population. As De Jong and Fawcett (1981) justly argue, young couples with children are often attracted to suburbs in which single-family housing is more available and affordable, while middle and older age migrants tend to settle in "quiet" peripheral communities. The rate of natural growth thus reflects, directly and indirectly, various parameters of socio-economic development of urban settlements. Nevertheless, the indicator in question cannot substitute for migration indices of population growth.

It thus appears that none of the above measures represents a sufficiently accurate tool for gauging the degree of sustainability of population growth in a particular urban settlement. The following sections will propose an alternative approach, based on a *compound indicator* incorporating the comparative advantages of two of the above direct measures of population growth -- migration balance and natural growth.

3.4
Urban Settlements in Israel: Inequalities of Population Growth

Statistical data on the population growth of urban localities in Israel (Local Authorities in Israel: Physical Data 1994-96; Statistical Abstract of Israel 1993-96) indicate that, for the period of 1992-93, this growth varied considerably across settlements of different sizes. In particular, the highest rates of annual population increase were exhibited by urban localities of 10,000-25,000 residents (Fig. 3.1).

To understand the nature of this phenomenon and its implications for planning, a case study was undertaken with several major objectives: 1) to explain why localities of the above size tend to grow faster than other urban settlements; 2) to determine the main factors and prerequisites which contribute to the population growth patterns of the country's urban settlements in general, and 3) to develop planning strategies which are conducive to the sustainable growth of urban settlements in sparsely populated regions of the country.

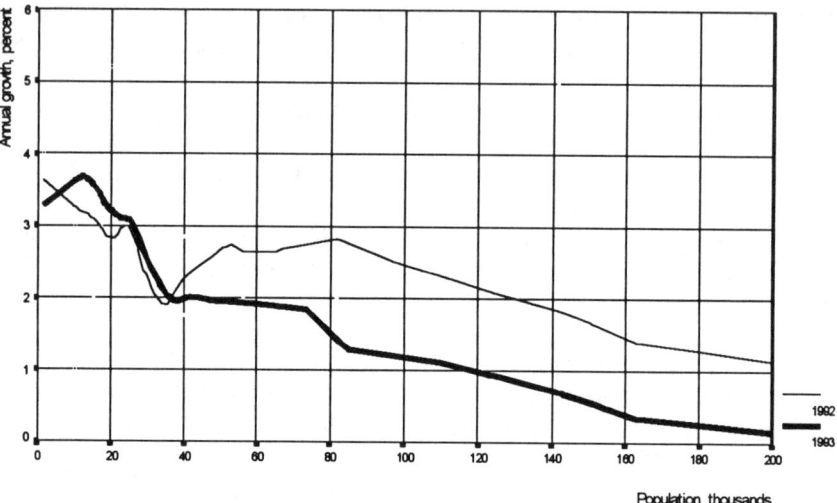

Fig. 3.1. Annual population growth across Israeli settlements of different size in 1992-93
The diagram shows that in 1992-93, the most intensive population growth occurred mainly in settlements of 10,000-25,000 residents

Since it was assumed that diversities in population growth are unlikely to occur because of *natural growth* (fertility-mortality rates) alone, the attention was *primarily* concentrated on population growth due *in-country migrations* and *foreign immigration,* which can be affected by urban and regional development planning more than natural growth rates. [2]

3.5
Patterns of Urbanization in Israel and General Development Policies

The vast majority of urban settlements in Israel are concentrated in and around the Mediterranean coastal plain, and in close proximity to its major metropolitan centers - Tel Aviv, Jerusalem, and Haifa (see Fig. 3.2). The overall population of these urban centers along with their immediate hinterland (i.e. the Tel Aviv, the

---

[2] During 1992-94, foreign immigration to Israel amounted to 50-60 percent of the country's annual population growth.

Central, the Jerusalem, and the Haifa districts) amounts to over 3.5 million residents, or nearly 70% of the country's population.

As Lerman and Lerman (1992) point out, the centralized concentration of the country's population and lack of development in outlying regions create significant problems for national security as well as for maintaining the future capacity for immigrant-absorption. Realizing the extent of these problems, Israeli government and planning officials have favored at various times the implementation of a policy promoting *dispersal of the population* to the peripheral regions of the country (Newman *et al.* 1995).

From the early 1950's to late 1960's, this policy was implemented by the direct settlement of new immigrants in sparsely populated areas of the country. Since the early 1970s, this approach was gradually replaced by geographically selective "aid programs," which provided financial incentives to builders and buyers in specified development areas (Fialkoff 1992).

The subsequent population growth of settlements in these development areas was generally lower than initially expected (Newman et al. 1995). Moreover, the pattern of growth did not exhibit, at least at first glance, any direct relationship with the settlements' geographic location either in the peripheral areas or in the central districts of the country (Figs. 3.2 and 3.3). This irregularity prompted considerable discussion among planners and decision-makers about the socio-economic nature of the factors and premises involved in determining population growth patterns of various urban settlements located in particular regions of the country.

One early survey (Kirschenbaum and Comay 1974) examined the relationship between social and economic conditions in new settlements and their differential influence in attracting migrants, during the 1960's. Key factors found to influence the flow of *in-country migrants* to a particular settlement included 1) employment; 2) welfare conditions; 3) housing availability, and 4) proximity to large urban centers of the country.

A more recent study of in-country migrations in Israel (Anson 1993) indicated that the main patterns of the process are interchanges between the major urban centers of the country and their peripheral districts, for example, between the Tel-Aviv and the Central district, and between the Haifa district and the Northern one. While the importance of social and economic premises in these processes was clearly noted (income factors, employment, unemployment, social composition of the population, etc.), any specific role of settlement size and location in these interchanges was not revealed.

In contrast, Newman *et al.* (1995), in their study of the Negev region, concluded that the small size of urban settlements in development areas of the country, when complicated by the lack of employment opportunities and poor housing conditions, could result in a major outflow of population and in a sequential settlement decline. However, the local area of the study (limited to eight urban settlements in the Negev) did not allow the authors to reveal any specifics of that trend.

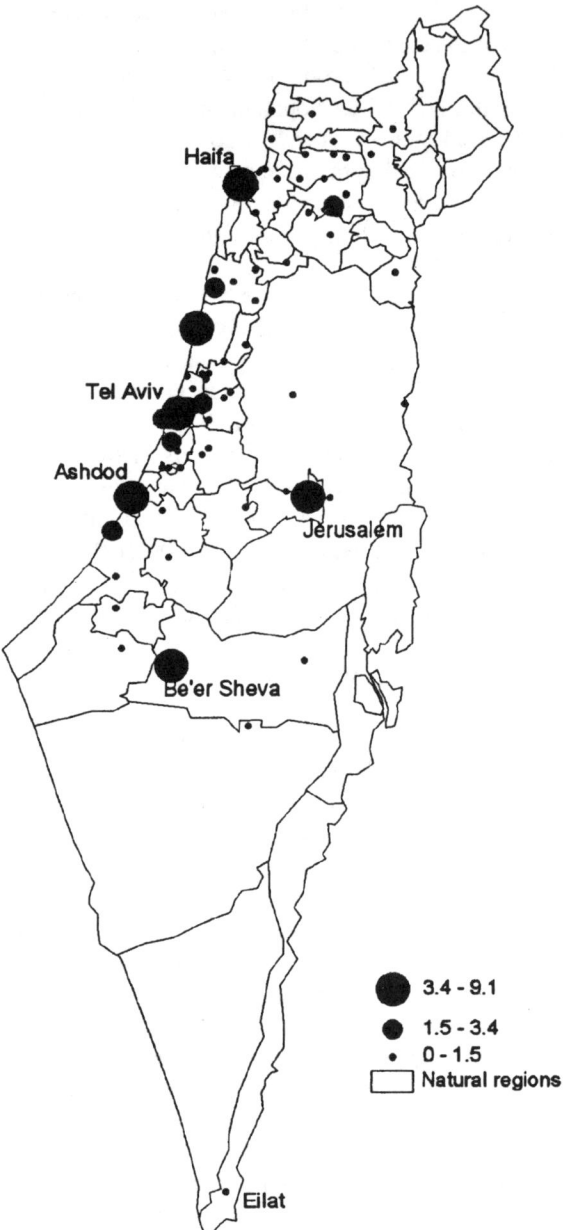

Fig. 3.2. Immigration in 1994, thousands of immigrants

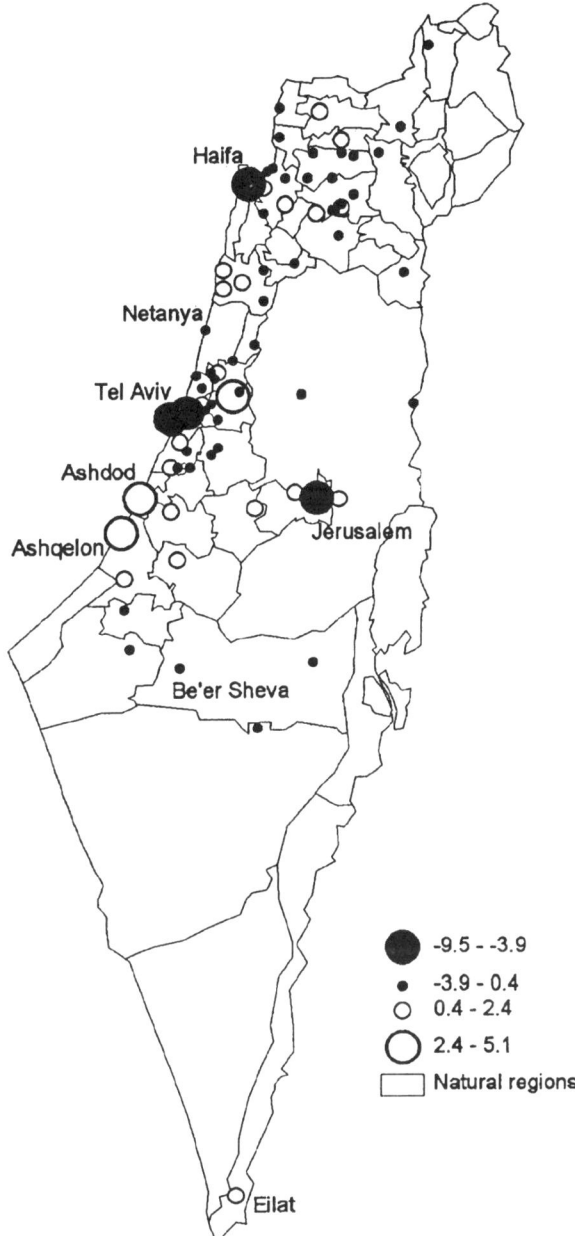

Fig. 3.3. Net balance of in-country migration, thousands of migrants

3.6
A Generalized Model of In-country Migrations

The following generalized model can be suggested as a description of the major directions of population flow between urban settlements in Israel (Fig. 3.4).

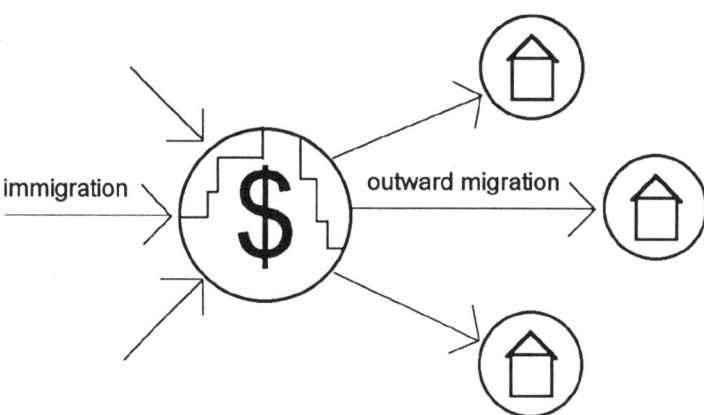

Fig. 3.4. A generalized model of inter-urban population flows in Israel

The model reflects the tendency of foreign immigrants to settle in the large, economically developed urban centers of the country (Fig. 3.5). Possible motives and driving forces of this trend include the cultural preference of living in big cities, hopes of obtaining suitable employment in the most prosperous centers of the country, affiliation to friend and relatives, and a lack of information about small urban localities.

The model also shows the trend of outward migration (by both recent immigrants and "veteran" population) from large urban centers to localities of smaller size in which housing is more widely available, assuming that *an influx of new immigrants into a particular settlement might "push" its current residents to move outward*. Though indirectly, these parallel processes may be correlated with housing prices: the mass influx of new immigrants to a particular locality boosts housing demand and increases housing prices, which through a spiral process can eventually cause the outflow of current residents who cannot afford decent housing in the metropolitan centers to areas of the country where housing is more widely available and more affordable.

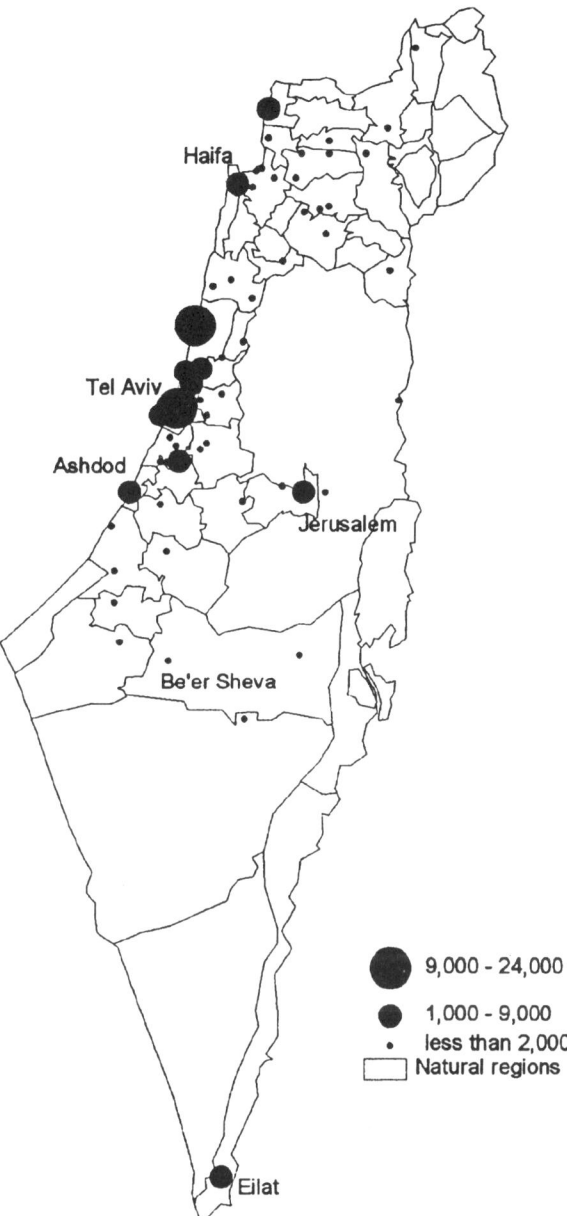

Fig. 3.5. Personal bank savings in 1994, $US per capita

3.7
Research Method

To test this model, data were obtained for 80 major urban settlements of the country with a population of over 10,000. The sample included three major urban municipalities of the country (Jerusalem, 600,000 residents; Tel Aviv, 357,000 residents, and Haifa, 247,000), together with nine urban localities of 100,000-199,999 residents, eight settlements of 50,000-99,999 residents; 27 localities of 20,000-49,999 residents, and 33 small towns of 10,000-19,999 dwellers.[3]

Three points in time - *1992, 1993, and 1994* - were chosen to observe trends of settlement population growth using comparable and complete sets of data. Before 1992, population migration in the country was largely affected either by governmental "housing interventions" or by the extremely high number of immigrants in 1989-1991, which could mask the normal trends.

According to a suggestion that the "sustainability of population growth" of an urban settlement is primarily determined by the *ability of the settlement to maintain relatively high rates of annual population increase by attracting newcomers and retaining existing population*, the case study addressed the following growth constituencies used in the following analysis to establish the integrated sustainability index (see Section 3.9):

- The annual net balance of internal (in-country) migration to a settlement, drawn from the Central Population Register of the Israeli Ministry of Interior (These data do not include the information on the initial settling of new immigrants, but do include sequential changes of their address);
- The annual immigration rate to a settlement, reflecting initial distribution of the new immigrants to the country, i.e. a first place of residence;
- The natural population growth of a settlement computed as the difference between number of births and deaths per annum.

In order to determine the factors influencing population growth, the following selected socio-economic and physical indicators were considered:

- indices of climatic harshness (a so called temperature discomfort index, and a heating degree days index) calculated for each urban locality respectively as the mean annual number of days with heat stress, and the average number of days during which space heating is required (see for definition Bitan and Rubin 1991);

---

[3] The data for the present analysis were obtained from the following main sources: Statistical Abstract of Israel (Central Bureau of Statistics, 1984-1995), Bulletin of Statistics and Supplement (Central Bureau of Statistics, Monthly), and Local Authorities in Israel: Physical Data (ICBS 1994-95). Climatic indices for the study were drawn from the Climatic Atlas of Israel (Bitan and Rubin 1991). The settlements selected for the sample did not included those located within Judea, Samaria and Gaza Areas which are treated by Israeli Central Bureau of Statistics as separate statistical districts of the country.

- population of a settlement, thousands of residents;
- number of unemployed in a particular settlement, measured separately in absolute numbers (thousands) and in percentages;
- annual change in unemployment rate, percent relative to the previous year;
- annual rates of housing, industrial and commercial building completions (both public or private investments), measured in sq.m of gross building area;
- annual rate of road construction in a particular settlement, km;
- private bank savings, $US per capita;
- car ownership level, measured as the average number of private cars per 1000 residents;
- distance to the closest urban metropolis of the country (either Jerusalem, Tel Aviv, or Haifa), km;
- the average number of pupils per class in elementary and secondary schools taken as an indicator of educational conditions.

The list of variables was, therefore, composed to include the main factors that a hypothetical immigrant or an in-country migrant could be expected to employ in the migration decision-making process: *housing, employment, education, and the socio-economic prosperity of the area*. The "standard" list of possible motivations (see inter alia Ehrlich, Ehrlich and Holdren 1972; De Long and Fawcett 1981; Moore and Rosenberg 1995) was in this case supplemented with indicators of climatic harshness, and with the data on construction activity in particular settlements (road, industrial, commercial construction, etc.).

The group of indices for *climatic harshness* was included to test an initial assumption that considerable differences in climatic conditions among various geographic regions of the country should have some affect on in-country migrations, making areas of harsh climate less desirable for newcomers.[4] The group of indicators for *construction activity* was added to trace the potential of local physical planning in influencing ongoing processes of in-country migrations. Some "traditional" components of the motivation list (level of social and health services, the family and friend affiliation, earning differentials, and size of existing immigrant community) were deliberately dropped due to restrictions on the availability and comparability of data.

---

[4] Although the land area of Israel is relatively small (about 21.5 thousand km^2), the country contains 14 different climatic zones. For example, the average annual number of days with heat stress on a summer day ranges from 4 in Jerusalem to 24 near the Dead Sea.

3.8
Components of Population Growth

The correlation between the sources of the settlements' annual population growth (natural increase, internal migration, and immigration), and the size of localities (population) is diagrammed in Fig. 3.6.[5]

The following general trends can be observed. The attractiveness of urban settlements for new immigrants generally grows with settlement size, while with respect to in-country migrations the reverse trend seems to be clear: the larger a locality, the lower its rates of growth due to in-migrations. After reaching a particular population size (about 120,000 inhabitants), the balance of in-country migration in localities drops below the zero level. The rates of natural population growth also tend to decline as overall settlement size increases.

One aspect of the above trends seems to have particular importance for the proper understanding of the issue in question. This is the drastic change in components of population growth exhibited by settlements whose overall population ranges from approximately 10,000 to 50,000 residents (Fig. 3.6). While average rates of natural population growth in these settlements abruptly drops from 2.8 to about 0.8 percent, their attractiveness for new immigrants gradually increases and almost reaches the respective levels of attractiveness exhibited by large urban localities of 100,000-200,000 dwellers, i.e. 1.5 percent per annum. In addition, this particular group of settlements has a relatively high positive balance of internal migration which varies between 0.5-0.6 percent per annum.

The structure of growth components reverses within this settlement group (10,000-50,000 residents) *from the predominance of the natural growth to that of migration and immigration constituencies.* While for settlements of 10,000 residents, the annual natural growth constitutes on the average 65 percent of the overall population increase, for the settlements of 40,000 residents it does not exceed 35 percent.

Within the group of settlements of 20,000-30,000 residents, a 50 percent *breaking point* is observed: that is after reaching the above population size, the population growth of urban settlements in the country tends to become *less dependent on natural causes* (fertility and mortality rates) *than on the settlements' ability to attract newcomers and retain current residents.* This, in turn, is clearly indicative of fundamentally new growth capabilities, at least in terms of urban

---

[5] The diagram represents average rates of growth for 1992-94. While computing and diagramming averages, it was taken into consideration that deviations between corresponding years were minor. It should also be noted that the diagram portrays the population growth trends for the settlements, which fall within the range of 10,000-200,000 residents. Although the major urban centers of the country (Jerusalem, Tel Aviv, and Haifa) were also included in the analysis, their corresponding values are not represented in the present diagram. This has been done to emphasize the patterns of growth in the settlements of smaller size, which constitute over 95 percent of the sample.

and regional planning. Such growth capabilities signal clearly *a turning point in the sustainability of a settlement's future population growth.*

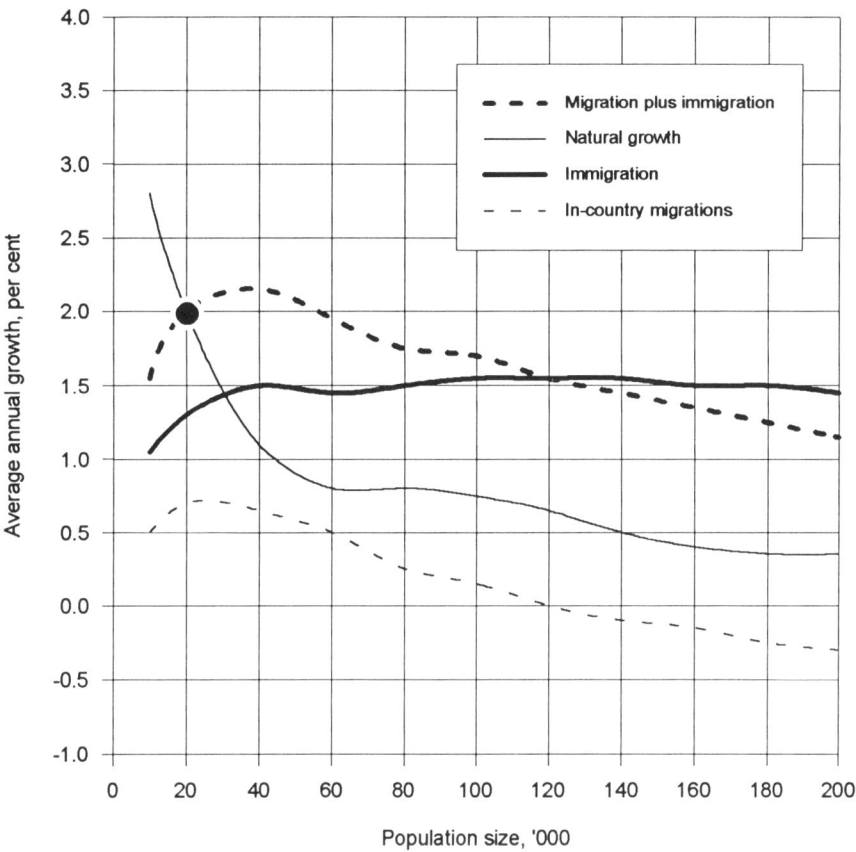

Fig. 3.6. Sources of population growth of Israeli urban settlements in 1992-94

The diagram represents respective fit lines computed by the Lowess fit method for the 80 urban settlements included in the study area. The filled circle indicates the breaking point in the settlement growth after which migration becomes the predominant source of a settlement's annual population increase.

To explain why a settlement becomes more attractive to migrants upon reaching a 20,000-30,000-population threshold, a number of factors should be taken into consideration. After reaching this threshold, a settlement achieves a certain level of socio-economic development. In particular, a settlement of this size may have a relatively large shopping center, a number of schools, including a few secondary ones or even a college, developed medical and sport facilities, and other social services which a smaller locality can not, in most cases, sustain.

Also, the patterns of physical development in settlements of the above size begin to approach those of large urban localities in terms of planning layout, density, types of building, etc.

3.9
The MB/NG Ratio as an Integrated Indicator for Measuring the Sustainability of Population Growth

Following the empirical findings on changes in the components of population growth (Fig. 3.6), a more general model can be suggested (Fig. 3.7). The model considers two major components of overall population growth (OG): *migration balance* (MB) - the net balance of internal migration and foreign immigration, and *natural growth* (NG) - the net balance of births and deaths. (Both components are measured as the annual percentage change in the local population size).

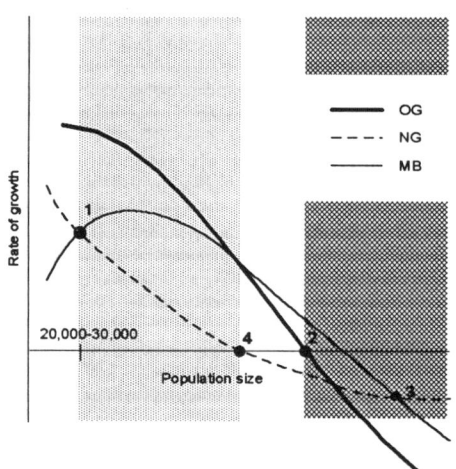

Fig. 3.7. Changes of the components of population growth

OG – overall growth; NG – natural growth; MB – migration balance

This model reflects two general trends:

- The *net balance of migration* (MB) tends to be relatively low in small urban communities (10,000-15,000 residents). In larger urban settlements (above 20,000-30,000 residents), the net balance of migration becomes the main source of population growth. Finally, after peaking, the migration balance starts to decrease until outward migration exceeds the inward flow of newcomers. Such a negative MB can, for instance, be found in long established urban centers that experience suburban sprawl and other forms of outward migration of current residents.

- The *rate of natural growth* (NG) is relatively high in small towns, and gradually decreases as population size of the settlement grows (Fig. 3.7). Eventually, when the settlement reaches a certain population threshold, the rate of NG may become negative. This trend is primarily attributed to the overall aging of the settlement's existing population, while potential new residents of child-bearing age are attracted to other urban areas where appropriate housing is considered more available and affordable.

The model shows three major thresholds delimiting various stages of a settlement's population growth. Each of these thresholds can be expressed in terms of the following generic equation:

$$|MB| = |NG|,$$

where |MB| and |NG| are the absolute values of migration balance and natural growth, respectively.

These thresholds are:

- Point #1 (Fig. 3.7): NG = MB where both NG and MB are positive. This threshold corresponds to the inflection point after which MB becomes major source of a settlement's population growth (see Fig. 3.6).
- Point #2 (Fig. 3.7): |NG| = MB where NG is negative and MB is positive. If these conditions are met, the overall population growth of a settlement equals zero.
- Point #3 (Fig. 3.7): |MB| = |NG| where both MB and NG are negative. These conditions correspond to the absolute population decline.

An additional point of inflection (Point #4 in Fig. 3.7) is also important. After reaching this inflection point, NG drops below a zero level and positive OG in a locality becomes totally dependent on the influx of migrants.

Given these thresholds, different values of the MB/NG ratio can be interpreted according to a settlement's ability to sustain its population growth.

- NG > MB, while both NG and MB are positive (0<MB/NG<1, MB>0 & NG>0): If these conditions are met, the population of a settlement grows mainly due to natural causes (fertility-mortality rates), rather than due to the settlement's attractiveness to newcomers. This phase of growth is bounded by Point #1 (Fig. 3.7) and may be defined as *transitional* (or incipient) growth.
- MB > NG and both NG and MB are positive (MB/NG>1, MB>0 & NG>0): Under these conditions, the population of a *settlement primarily grows because the locality is attractive to newcomers* (internal migrants and foreign immigrants). It is suggested that this phase of growth may be defined as *sustainable*. The right margin of this phase is bounded by Point #4 (Fig. 3.7).
- MB > |NG| while MB is positive and NG is negative (MB/NG<-1, MB>0 & NG<0): The population of the settlement is almost constant, owing to positive MB which is offset by negative NG. This phase of growth is limited by Points #4 an #2, and may be defined as *transforming growth*, since the overall growth

of the locality can easily become negative if the settlement lacks a "migration feedback".
- MB < |NG| when NG is negative (-1<MB/NG<0, MB>0 & NG<0). These conditions, when met, clearly indicates the absence of any growth or, in other words, the phase of *absolute population decline*. In Fig. 3.7, the onset of this phase is marked by Point #2.

This list does not, however, encompass the whole range of possible values of the MB/NG ratio. A more detailed list of these ratios is represented in Table 3.1.

Table 3.1. Categories suggested for grouping of urban settlements according to the MB/NG ratio

MB/NG ratio	Natural growth	Suggested description of the state of growth	Comment
Above 1.0	Positive	Sustainable growth	The population of a settlement primarily grows due to its attractiveness to migrants
From 0 to 1.0	Positive	Transitional growth	The population of a settlement primarily grows due to natural causes (fertility/mortality rates). MB is positive but is lower than NG.
From -1.0 to 0	Positive	Transforming growth	The population of a settlement grows due to natural causes while MB is negative.
Less than -1.0	Positive	Decline	The population of a settlement declines due to negative MB which exceeds positive NG
Less than -1.0	Negative	Transforming growth	The population of a settlement rises because positive MB exceeds negative NG.
From -1.0 to 0	Negative	Decline	The population of a settlement declines due to negative NG whose absolute value exceeds positive MB
From 0 to1.0	Negative	Decline	The population of a settlement declines due to negative values of both MB and NG
Above 1.0	Negative	Decline	As above

The importance of the *MB/NG index* clearly surpasses its direct use as a simple measure of population growth. Indeed, the ongoing *migration attractiveness* of a particular settlement is interconnected with "sound" economic development, a favorable physical environment, and other preconditions, which are essential for "sustainable urban growth" in general. It is also important that the indicator in question allows normalization for natural growth rates that may vary substantially from one settlement to another. The proposed index (the MB/NG ratio), then, may serve as a valuable tool for urban and regional planning, in particular due to its potential use in gauging the impact of public policies, which are aimed at sustaining population growth and economic development in lagging urban areas.

3.10 Factors Influencing Migration Attractiveness of Urban Areas

The importance of the migration component of urban growth argued in the previous section requires a clarification of the factors and forces affecting migration attractiveness of an urban settlement. In the section on research methodology, a list of factors was introduced which are likely to be employed in migration decision-making. Such a correlation may be identified and measured through the procedure of multiple regression analysis.

The linear regression model used for the analysis can be expressed in terms of the following formula:

$$SA_i = B_o + B_1 \times F_1 + ... + B_n \times F_n + \varepsilon,$$

where SA_i is an indicator of a settlement's attractiveness which represents either the annual number of new immigrants who opted to settle in the locality (SA_1), or the annual balance of its immigration (SA_2); $B_o, B_1, ..., B_n$ are the respective regression coefficients; $F_1, ..., F_n$ are the above listed research variables (see section on research method), and ε is a random error term.

Table 3.2 portrays the most statistically significant factors of a settlement's migration attractiveness selected by the *stepwise multiple regression* (SMR) procedure. Each of the factors is significant at a .05 confidence level.

The 1993 and 1994 hierarchies exhibit relatively little variation. It appears that alterations in the factor hierarchy from year to year should not be treated as crucial, since absolute changes in the corresponding factors' statistical significance in 1993 and 1994 are not substantial (Table 3.2) and could simply derive from the factors' partial intercollinearity (Appendix 2). Keeping this in mind, we shall discuss the *integral* effect of the above selected factors, rather than the effect of each of them upon the independent variables in question.

As Table 3.1 shows, *population size* and *per capita bank savings* are the main factors that increase *immigration* to a settlement. For *in-migrants*, *population size of "target" localities* has a significantly negative effect, *housing construction* has a significantly positive effect, and, lastly, the factor of *bank savings,* that emerged only in 1994 as having statistical significance, also has a negative effect on in-country migration (Table 3.2). These data are thus in line with the theoretical model describing the forces that govern in-country migrations (Fig. 3.4).

The opposite effect of the variables (population size, bank savings and housing construction) on in-country migrations and immigration calls for some additional explanation. This effect indicates that the two migration streams are substantially different in nature: 1) *new immigrants*, being less informed about the country and local conditions, primarily concentrate in big and affluent cities of the country, where they hope to obtain suitable employment in the future or can count on initial help from friends and relatives, while 2) the *existing population* of these centers which is, of course, more economically independent, mobile and informed moves outward from the overpopulated and "expensive" metropolis to smaller localities where suitable housing is more available and affordable. This migration "stream" can, thus, be conditionally named the *"hinterland-housing" stream*, while the initial distribution of the new immigrants across the country's urban settlements can be rather described as the *"centripetal" or "metropolitan" stream* (Portnov 1998).

Some factors included in the initial "motivation list" (see Section 3.10), and which *are not included* in the above equations as statistically significant (Table 3.3) deserve also to be mentioned. Among them are the *"distance to major urban centers of the country,"* and the *"climatic harshness"* of a geographic area. The limited role of these factors in determining the country's migration patterns seems to be useful in elaborating the planning policies for development of the peripheral regions of the country since remoteness and climatic harshness are unlikely to be substantial obstacles *per se* for potential newcomers.

3.11
Conclusions and Policy Implications

The national goal of population dispersion from densely populated urban centers of Israel has been traditionally pursued by means of the development of a broad network of new towns in the peripheral districts of the country. Similar regional development policies are found in other countries (Sweden, Norway, Brazil, Japan, Egypt, etc.), which face problems of peripheral underdevelopment. As Gradus and Stern (1980) pointed out, the major reason for the subsequent failure of this and similar strategies in a vast number of cases stemmed from the old problem of "too small, too many and too near."

The tendency revealed in the present analysis - that urban settlements in Israel tend to become more sustainable in their growth after reaching a particular size (20,000-30,000 residents) - seems to be useful in determining an alternative strategy for developing peripheral regions which are currently sparsely populated. This strategy can be conditionally defined as the strategy of *"redirecting priorities,"* and its main idea is represented by the following diagram (Fig. 3.8).

The strategy assumes, in particular, that development resources should be primarily concentrated on a limited number of selected settlements in the frontier areas of the country until they reach the above threshold of sustainable population growth and become considerably attractive to both new immigrants and in-country migrants. The localities selected during the *first phase* of the development process can be located in close vicinity to existing urban centers, which *have already achieved the above threshold of sustainable population growth.*

Table 3.2. Key factors affecting components of population growth of urban settlements in Israel

Factors	Immigration, thousands						Balance of Internal Migration, thousands					
	1993			1994			1993			1994		
	B	Beta	t	B	Beta	t	B	Beta	t	B	Beta	t
Settlement's population, thousands	4.24	0.52	3.3[a]	3.47	0.72	7.2[a]	-9.65	-1.26	-7.8[a]	-5.87	-0.98	-5.3[a]
Housing construction, thousand m²	-	-	-	-	-	-	5.36	0.89	5.5[a]	3.97	0.77	4.7[a]
Bank savings, $US per capita	1.87	0.34	2.1[b]	0.70	0.18	1.9[b]	-	-	-	-1.19	-0.25	-2.0[b]
Constant (B_o)	-11.36		-4.9[a]	-6.47		-1.0	6.87		4.4[a]	6.68		4.9[a]
F-statistic	17.87			61.36			30.31			18.43		
Standard Error (SE)	1.83			0.987			1.44			1.57		
R Square	0.598			0.714			0.716			0.535		

[a] indicates a two-tailed .01 significance level; [b] indicates a two-tailed .05 significance level

Table 3.3. Exogenous factors which are not included in the equations as statistically significant

Factors	Immigration, thousands						Balance of Internal Migration, thousands					
	1993			1994			1993			1994		
	Beta	t	Sign.	Beta	t	Sign.	Beta	t	Sign.	Beta	t	Sign.
Temperature discomfort index[a]	-0.053	-0.37	0.71	-0.066	-0.80	0.43	0.130	0.81	0.42	0.095	0.92	0.36
Heating degree days index[a]	0.056	0.37	0.71	0.050	0.58	0.56	-0.030	-0.18	0.86	0.011	0.11	0.92
Number of unemployed, thousands[a]	0.101	0.37	0.71	0.072	0.42	0.67	0.232	0.69	0.50	0.196	0.87	0.39
Number of unemployed, %[b]	0.055	0.38	0.71	0.027	0.32	0.75	0.130	0.74	0.47	0.100	0.87	0.39
Annual change in unemployment, %	-0.019	-0.13	0.89	-0.016	-0.19	0.85	0.071	0.44	0.67	-0.028	-0.25	0.80
Housing construction, thousand m²[a]	0.191	1.02	0.32	0.038	0.30	0.77	-	-	-	-	-	-
Industrial and commercial construction, thousand m²[a]	-0.010	-0.05	0.96	0.135	1.15	0.26	-0.022	-0.09	0.93	0.036	0.22	0.83
Road construction, km[a]	0.065	0.43	0.67	-0.024	-0.23	0.82	0.196	1.19	0.25	0.177	10.55	0.13
Bank savings, $US per capita[a]	-	-	-	-	-	-	-0.185	-0.85	0.40	-	-	-
Car ownership level, cars per 1000 residents[a]	-0.069	-0.44	0.67	0.00	0.04	0.96	0.073	0.42	0.67	0.002	0.02	0.99
Distance to metropolis, km[a]	0.174	1.18	0.25	-0.080	-0.83	0.41	0.059	0.32	0.75	0.073	0.38	0.69
Educational conditions, pupils per class[a]	-0.085	-0.59	0.55	-0.079	-0.95	0.35	0.101	0.62	0.54	0.195	10.90	0.06

[a] Indicates factors which values represent the logarithm of respective original values

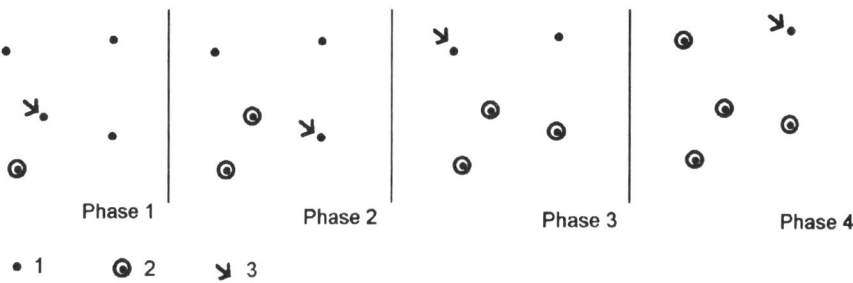

Fig. 3.8. The diagram illustrates the planning strategy of redirecting priorities suggested for developing peripheral districts of Israel

1 - a small urban locality; 2 - an urban locality achieved the above threshold of sustainable growth; 3 - concentration of development resources

Support of the selected localities should of course provide a balanced investment both in housing development and employment generating economic sectors. In addition to direct government intervention, various forms of indirect involvement such as incentives for private investors and tax exemptions should also be applied. As the annual population growth of the initially selected localities becomes more *sustainable*, the support may be redirected on a stage by stage basis to other adjacent urban settlements. This process of the temporary and hierarchical concentration of resources can thereby be moved deeper and deeper into the frontier areas of the country.

The strategy of "redirecting priorities" suggested here thus has certain parallels with Perroux's "growth pole" theory subsequently developed by Friedmann (1973) and Hansen (1975), and implemented in various empirical planning studies across the globe (see *inter alia* Kuklinski 1978). The differences between these two development strategies are, however, substantial. First, the strategy of *redirecting priority* does not place a particular emphasis on the fact that the selected development settlements will "radiate development impulses" into their hinterland, but it rather assumes that these settlements will sustain their own population growth due to their relative *attractiveness to newcomers*. Second, in contrast to the "growth pole" strategy, the strategy in question defines a certain population threshold upon reaching which development assistance can be switched to adjacent locations.

Although the concrete findings of this study - the population threshold conducive to migration attractiveness of a settlement and the intensity of various influencing factors - are definitely specific to the conditions in Israel at this time, the mode of analysis and its applications for planning policy may be applicable to regional and urban planning elsewhere. In particular, the strong interrelationship found between population size, housing construction, on the one hand, and migration, on the other, is probably characteristic of development processes elsewhere. The planning strategy of *"redirecting priorities"* suggested here may thus be of value to planners and decision-makers in any country which experiences acute problems of inter-regional and inter-urban inequalities in population growth and socio-economic development.

References

Anson J (1993) Geographical regions as ethnic groups? Social geography as the sociology of space. In: Anson J, Todorova E, Kressel G (eds) Ethnicity and politics in Bulgaria and Israel. Aldershot, Avebury, pp 13-28

Bitan A, Rubin S (1991) Climatic atlas of Israel for physical and environmental planning and design. Ramot Publishing Co, Tel-Aviv

Brown LR, Jacobson J (1987) Assessing the future of urbanization. State of the World 1987. WW Norton & Company, New York and London

De Long GF, Fawcett JT (1981) Motivation for migration: an assessment and a value-expectancy research model. In: De-Long GF, Gardner RW (eds) Migration decision-making. Multi-disciplinary approaches to microlevel studies in developed and developing countries. Pergamon Press, London, pp 13-53

Ehrlich PR, Ehrlich AH, Holdren JP (1972) Ecosience: population, resources, environment. W.H.Freeman and Company, San Francisco

Fialkoff C (1992) Israel's housing policy during a period of massive immigration. In: Golani Y, Eldor S, Garon M (eds) Planning and housing in Israel in the wake of rapid changes. Ministries of the Interior and of Construction and Housing, Jerusalem, pp 169-177

Friedmann J (1973) Urbanization, planning, and national development. SAGE Publications, London and Beverly Hills

Gradus Y, Stern E (1980) Changing strategies of development: toward a regiopolis in the Negev desert. J of the American Planning Association 46(4): 410-423

Hansen NM (1975) Criteria for a growth center policy. In: Friedmann J, Alonso W (eds) Regional policy. The MIT Press, Cambridge, pp 566-587

Haughton G, Hunter C (1994) Sustainable cities. Jessica Kingsley Publishers, London

ICBS (1984-1995) Statistical abstract of Israel. Israeli Central Bureau of Statistic, Jerusalem

ICBS (1994-95) Local authorities in Israel: physical data. Israeli Central Bureau of Statistics, Jerusalem

Kirschenbaum A, Comay Y (1974) Dynamics of population attraction to new towns - the case of Israel. In: Pruschansky J (ed) Dialogue in development - natural and human resources. Proceedings of the 3rd World Congress of Engineers and Architects, Jerusalem.

Kuklinski A (1978) Regional policies in Nigeria, India, and Brazil. Mouton Publishers, Hague

Lerman R, Lerman E (1992) A comprehensive national outline plan for construction, development, and absorption of immigrants - N.O.S # 31. In: Golani Y, Eldor S, Garon M (eds) Planning and housing in Israel in the wake of rapid changes. Ministries of the Interior and of Construction and Housing, Jerusalem, pp 29-47

Moore EG, Rosenberg MW (1995) Modeling migration flows of immigrants groups in Canada. Environment and Planning A 27: 699-714

Newman D, Gradus Y, Levinson E (1995) The impact of mass immigration on urban settlements in the Negev 1989-1991. Working Paper No 3. Negev Center for Regional Development, Be'er Sheva

Portnov B (1998) The effect of housing on migrations in Israel; 1988-1994, J of Population Economics, 11: 379-394

Sage C (1994) Population, consumption and sustainable development. In: Redclift M, Sage C (eds) Strategies for sustainable development: local agendas for the Southern Hemisphere. John Wiley & Sons, Chichester, pp 35-60

Turner RK (1993) Sustainability: principles and practice. In: Turner RK (ed) Sustainable environmental economics and management: principles and practice. John Wiley & Sons, Chichester, pp 3-36

Appendix 1

Homogeneity-of-variance test for violation of the equal variance (homoscedasticity) assumption using Hartley's F-Max Method

Factor	Variance	
	Before logarithmic transformation	After logarithmic transformation
HOUSING	3411.45	0.26
SAVINGS	5834367.91	0.23
POPULATION	6690.28	0.16
F-Max	1710.23	1.63
F-Critical ($\alpha=.05$)	1.85	1.85

Note: While the logarithmic transformation was applied to nearly all research variables (see Table 3.2), the appendix reports the results of the test for the variables selected as statistically significant by the stepwise multiple regression procedure

Appendix 2

Pearson correlation coefficients for the variables included into regression equations as statistically significant

Factor	HOUSING	SAVINGS	POPULATION
HOUSING	1.0000	.4020[a]	.6782[a]
SAVINGS	.4020[a]	1.0000	.5863[a]
POPULATION	.6782[a]	.5863[a]	1.0000

[a] Indicates a 2-tailed 0.01 significance level

4 Private Construction as a General Indicator of Urban Development[1]

Boris A. Portnov and David Pearlmutter
J. Blaustein Institute for Desert Research, Ben-Gurion University of the Negev, Sede-Boker Campus, 84990, Israel

4.1 Introduction

Overcoming inequalities in the level of socio-economic development among regions is, undoubtedly, a key issue for urban and regional planning worldwide. The importance of this issue has grown in recent years, due to the fact that differentials in development potential between areas have tended to increase rather than narrow (Wong 1995).

In light of the increasing gap between regions, the goal of sustaining development in economically lagging, particularly peripheral, areas becomes essential. Achieving this goal can contribute to the effective redistribution of wealth and opportunities within a nation, while, if not achieved, can lead to serious social divisions (Bourne 1975; Armstrong and Taylor 1993).

To assess socio-economic disparities in regional development and gauge the progress of regional development programs, effective monitoring indicators are needed. A number of such criteria are currently in use; examples include employment growth, unemployment, overall population growth, net migration balance, and the ratio of local to external daily journeys (Markusen 1996; Diamond and Spence 1983; Clawson and Hall 1973). One additional measure, whose potential for urban and regional planning seems to be somewhat underestimated, is the rate *of private construction in a settlement*. While the *overall rate of construction* is widely accepted as a key indicator of socio-economic development (Layton 1972; de Leeuw 1992; Levy 1985; McGranahan 1972; Mills 1972; Smith 1975), the "private" component of this indicator has until now received relatively little attention as being a distinct measure on its own.

The goal of the present chapter is threefold: 1) to define the role of the rate of private construction as a general indicator of urban development; 2) to analyze the factors which affect the rate of private construction as it varies across urban settlements in Israel, and 3) to determine the policy premises for enhancing private construction in underdeveloped, primarily peripheral, urban areas.

---

[1] The chapter in partly based on: Portnov BA, Pearlmutter D (1999) Private construction as a general indicator of urban development: the case of Israel. *International Planning Studies* 4(1): 133-161.

4.2
Private Construction as a Development Indicator

Development indicators in urban and regional planning are commonly used in three distinctive ways. These include: 1) measuring the needs or opportunities of each region as a basis for resource allocation; 2) setting up the contextual "baseline" of an area's conditions in order to measure the improvement brought about by public policy intervention, and 3) identifying the opportunities or problems that are most important for each area as a basis for defining and prioritizing policy targets (Wong 1995).

In the framework of the Global Urban Observatory Program (GUOP), the United Nations Center for Human Settlements (HABITAT) developed a set of major urban indicators. Each of these groups of indicators covers a specific aspect of development:

1. *Socio-economic development* (number of households below the poverty line, employment rate, number of hospital beds, child mortality rate, life expectancy at birth, adult literacy rate, school enrolment rates, etc.).
2. *Environmental management* (wastewater treated, solid waste generated, disposal methods for solid waste, and regular solid-waste collection).
3. *Infrastructure* (household connection levels, access to potable water, consumption of water, and median water price).
4. *Local government* (major sources of income, per-capita capital expenditure, debt service charge, local government employees, wages in the budget, contracted recurrent expenditure ratio, etc.).
5. *Transport* (modal split, average travel time, expenditure on road infrastructure, and car ownership).
6. *Housing* (house price to income ratio, house rent to income ratio, floor area per person, permanent structures, land development multiplier, infrastructure expenditure, mortgage to credit ratio, housing production and housing investments) (HABITAT 1997).

While this extensive list includes some construction indicators (housing production and housing investment data), it does not include *private building* as a measure on its own. However, the potential importance of this indicator (private construction) is due to a number of considerations.

First, the effect of private construction on socio-economic development in urban localities is typically time-lagged. For instance, the number of housing units started in a locality in a particular year may directly and indirectly affect the prospective population growth of the settlement through the influx of migrants. The latter assumption appears to be fully justified in the case of Israel (Fig. 4.1).

4 Private Construction as a General Indicator of Urban Development

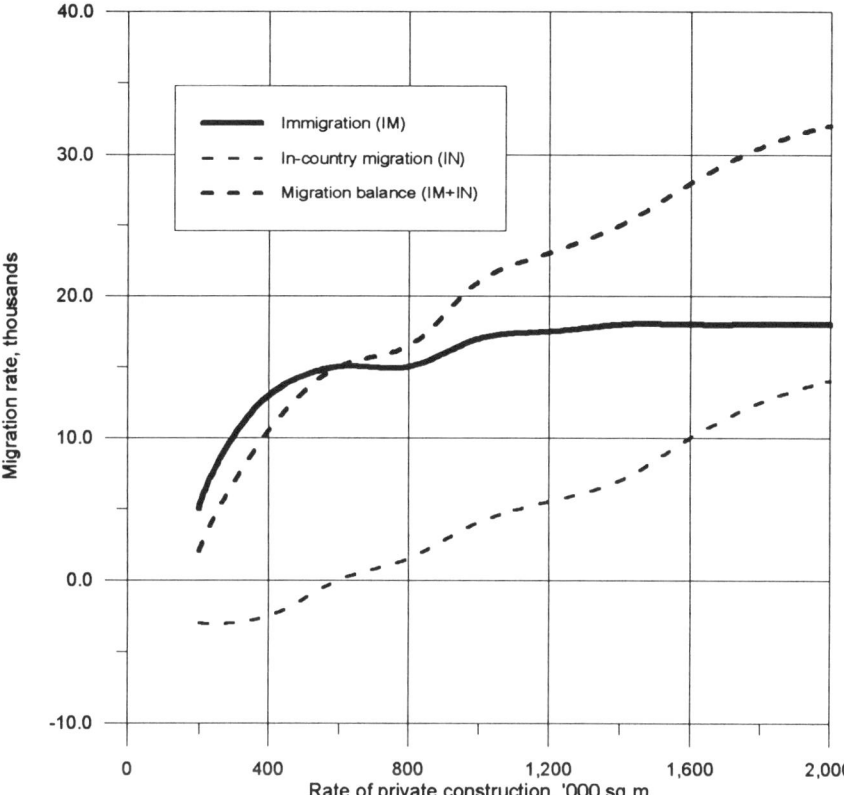

Fig. 4.1. Migration balance by district of Israel vs. annual rate of private residential construction in 1985-95

Source: Diagrammed using annual publications of the Israel Central Bureau of Statistics, "Statistical Abstract of Israel." The trend lines are computed using the Lowess fit method for respective components of population growth. Construction rates are lagged by one year to reflect a lagging response of migration to local housing development (Portnov 1998)

As Fig. 4.1 shows, the influx of migrants to a geographic area appears to grow as the annual rate of private housing construction in the area increases. This trend is in line with the findings of numerous migration studies (Lipshitz 1997; Portnov and Pearlmutter 1997; Portnov 1998) which indicate the particular importance of housing availability on the patterns of in-country migration. On the other hand, private construction rates in the non-residential sector can be even more important, as commercial development, for instance, often reflects the potential for future micro-economic performance and growth. The indicator in question can thus be considered as a "leading" indicator of a settlement's future socio-economic development, as opposed to other more "immediate" measures such as existing income distribution or the current level of unemployment.

Second, from the micro-economic point of view, the annual rate of private construction in a given urban area can be considered as a good indicator of its overall "investment climate." It may be assumed that such an indicator is, for instance, less forthcoming for *public construction*, whose rate may be based more upon social or political considerations than on attractiveness for investment. To justify this assumption, the patterns of distribution of public and private housing construction across administrative districts of Israel may be considered (Table 4.1). As seen, in both 1985 and 1995, private construction was "naturally" concentrated in the Central, Tel Aviv, and Northern districts, which are in fact the most populated areas of the country. At the same time, public construction was greatly "skewed" toward the Jerusalem and Southern districts, Judea, Samaria and the Gaza Area. This trend is clearly due to the governmental policy of population dispersal from the country's overpopulated core to the underpopulated periphery and other "strategically sensitive" areas (Lipshitz 1997; Portnov and Erell 1998 and Chap. 7 of this book).

Table 4.1. Percentage of distribution of population and of public and private construction of residential building by administrative district of Israel in 1985 and 1995

District	1985			1995		
	Population	Public construction	Private construction	Population	Public construction	Private construction
Northern	16.6	7.4	30.0	16.9	3.1	20.7
Haifa	13.7	4.3	11.1	13.2	9.9	9.3
Central	21.0	17.9	28.2	21.6	22.9	37.5
Tel Aviv	23.5	4.5	15.9	20.3	2.7	13.1
Jerusalem	12.0	28.1	8.6	11.8	22.1	5.6
Southern	12.0	6.7	5.9	13.7	31.9	10.7
Judea, Samaria and Gaza Area[a]	1.2	31.1	.3	2.5	7.4	3.1
Total:	100.0	100.0	100.0	100.0	100.0	100.0

Source: ICBS 1987-1996; [a] Jewish localities.

4.3
Private Construction in Israel: Historical Background and Spatial Trends

The following analysis of the geographic distribution of private construction will involve statistical data for urban settlements in Israel. Therefore, we will provide, as an introduction, a short history of private construction and the general patterns of its spatial distribution across different geographic areas of the country.

4.3.1
History of Private Construction in Israel

According to the official definition, *private construction* includes all building that *is not* financed by "the government, the national institutions, the local authorities and companies entirely controlled by these institutions" (ICBS 1997: 66). Private construction also includes building initiated and financed by the National Labor Federation (*Histadrut*). Private construction is thus distinguished from public building by its *source of financing* rather than by building purpose and the form of potential ownership. For instance, buildings constructed privately can be subsequently purchased by the government or local authorities for public housing, while a part of public building, upon completion or in the process of construction, can be offered for sale on the open market for private ownership.[2]

Since the establishment of the State of Israel in 1948, the governmental policy toward private construction has undergone a number of changes.

In the late 1940s – 1950s, in the wake of large-scale foreign immigration from the post-war Europe and Northern Africa, a severe shortage of housing became clearly apparent. Since expensive private building was not able to solve these acute problems, the role of public building increased significantly. Its share in the overall amount of building grew from less than *ten per cent* in the mid-1930s to about 45 percent in 1945-49 (Drabkin-Darin 1957).

While public construction in these years was primarily concentrated in peripheral areas of the country, private and semi-private construction companies (Afridar, Solel Boneh, Palroad, Ramet Ltd., Svirsky and Partners) concentrated their efforts in or near existing urban centers where demand was greater and construction therefore was more profitable.

In 1954, a new governmental policy was introduced. It was based on the assumption that the state's resources should be allocated only for sparsely inhabited development regions lying in the north and south of the country - the *Galilee* and the *Negev* (Drabkin-Darin 1957).

The 1970s - 1980s were marked by a sharp reduction in the central government's role in construction. During these years, the share of public construction dropped from 28 percent of the national total in 1975 to two percent in 1989 (Fialkoff 1992).

With renewed mass immigration in 1989-91, the State became again involved in housing construction and provision. During these years, the number of housing units initiated by the Israel Ministry of Construction and Housing jumped by

---

[2] While the government and municipal companies rent flats to families with very low incomes at subsidized prices, private developers in Israel are typically reluctant to build for rent. The prices of housing in Israel are high while rental prices are relatively low and are not considered as sufficient to compensate investments. Although the government has made efforts to promote renting by offering contractors a tax holiday or subsidized loans, these generous incentives often fail to produce results (Benzaquen 1998).

3,900 percent: from an annual average of 3,000 in 1989 to a total of 120,000 housing units during this two-year-period. In addition to direct state intervention in the housing market, indirect measures were also employed. These included: 1) bonuses granted to private developers for accelerating construction: 2) purchase commitments, e.g. an increase of buy back guarantees of up to 100 percent in the peripheral regions and up to 50 percent in the central areas; 3) inducements for early work completion, and 4) the lowering of land and infrastructure prices, especially in peripheral areas.

Since 1992 the above incentives to private developers have been, however, reduced as foreign immigration to the country has declined. In addition, the direct participation of the government in construction has also decreased. Thus, between 1992 and 1997, the share of public building of housing in the national total dropped almost threefold: from 74 percent in 1992 to less than 27 percent in the first quarter of 1997 (MCH 1997).

4.3.2
Geographic Distribution of Private Construction in Israel: General Trends

The distribution of private residential construction (Housing) and non-residential construction (Non-housing) in 1992-94 as a function of a settlement's distance to the closest major urban center of the country (Jerusalem, Tel Aviv, or Haifa) is diagrammed in Fig. 4.2.

As Fig. 4.2 shows, the rate of *private housing construction* steadily declines as distance to the urban centers increases. This trend is in agreement with the well-known Alonso-Muth-Clark model of urban areas (Clark 1982) which suggests the overall reduction of socio-economic activity with increasing distance from the urban core.

As for *private non-residential construction* (non-housing), the distance-related changes in construction activity are, however, somewhat less clear. In 1993-94, for instance, the rates of non-residential construction initially increased with the distance from the core, then these rates appeared to decline. In 1993, this decline was altered by an increase where the distance reached a 40-km threshold, while in 1994 no such an increase was observed (Fig. 4.2). While the factors that may contribute to this unusual "behavior" of non-residential construction will be discussed later in the chapter, we shall turn now to investigating the distribution of construction rates across urban settlements of different size.

In order to clarify the relationship between *private construction rates* and *settlement size, per capita* rates of construction are considered (Fig. 4.3).

As the diagram shows, with an increase of a settlement's size, the rate of private building steadily grows until reaching the inflection point of *70-80,000 residents*. After this point of inflection, the per capita rate of construction appears to level off (non-residential construction) or even drop (housing).

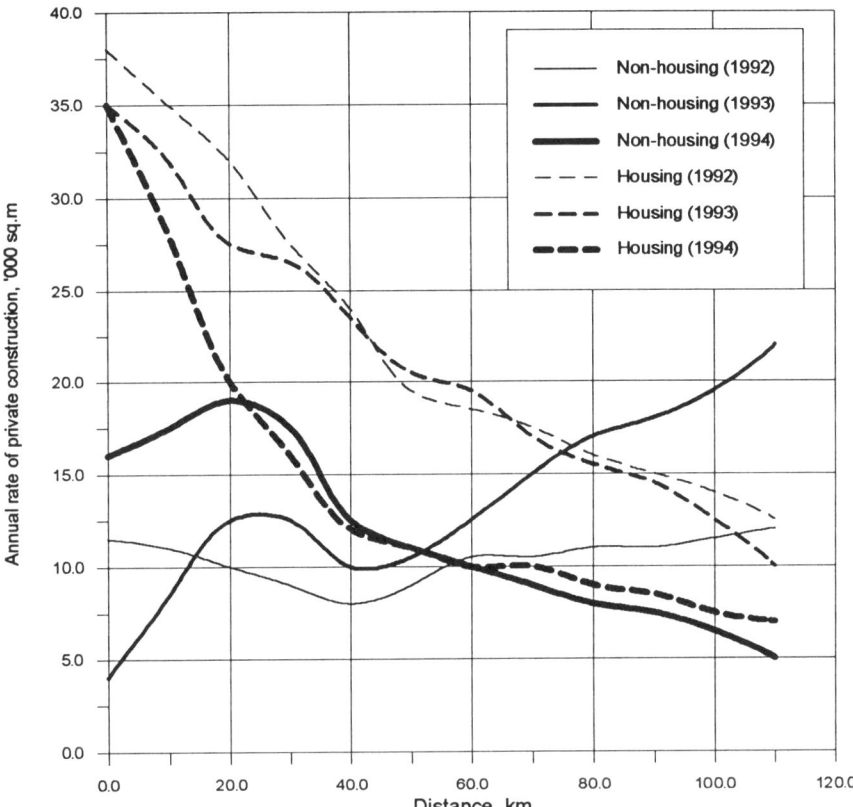

Fig. 4.2. The annual rate of private construction as a function of a settlement's remoteness from the major urban centers of the country

The Lowess fit lines are computed separately for residential (Housing) and non-residential (Non-housing) construction using the 1992-94 data on 80 urban settlements whose population exceeds 10,000 residents

Since the aforementioned inflection point is important for the following discussion, we tested its statistical accuracy. Since the Lowess fit method, used for calculating the trend lines (see Fig. 4.3), does not provide any statistical parameters of the fit but rather helps to illustrate the general patterns of the relationship between variables, an alternative approach was used. In order to test the presence of the aforementioned inflection point, the original set of data was disaggregated into two subsets: a) settlements of less than or equal to 80,000 residents, i.e. settlements located before the above point of inflection, and b) settlements of above 80,000 residents that fall beyond this inflection point (Fig. 4.3).

For each of these two groups of settlements, linear regression models were fitted using per capita rate of private housing construction as the dependent variable and population size (one year lagged) as the predictor.

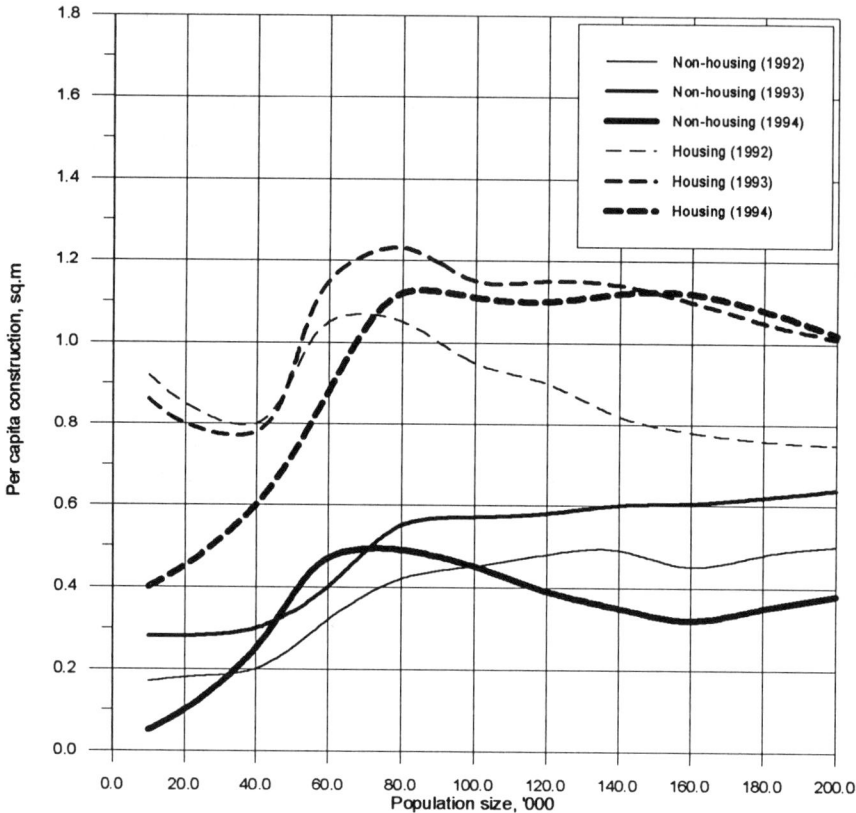

Fig. 4.3. Per capita private construction in urban settlements of different population size
See comments to Fig.4.2

In interpreting the results of the analysis (see Appendix 1), we shall focus on the signs of the second regression coefficient in each equation (B_1), which express the change in the private construction rate for each group (it is to be noted these coefficients are statistically significant: $t>3.0$; $P<0.01$). While in the first subset (population $\leq 80,000$), this sign is positive for all the years covered, in the second group of localities (population $> 80,000$), this sign is negative. This indicates that within the first settlement group, per capita private construction steadily increases as fast as population size of a settlement grows, while in the larger localities (population $> 80,000$), the opposite trend is observed: per capita construction tends to decline as population size of a locality increases. The opposite slope of the regression lines thus helps to support the above conclusion concerning the alteration of "population-rate" trends that corresponds to a 70,000-80,000 inflection point.

To understand why the settlements of 70,000-80,000 residents are especially attractive for private developers, a number of considerations should be taken into account. As mentioned in the previous chapter, after reaching a certain population size, urban settlements tend to become considerably attractive to both in-migrants and new immigrants. At the same time, land for development in small urban communities is still more available than in large population centers. These two groups of factors (migration attractiveness and relative land availability) thus may make the settlements of this size especially attractive to private developers.

4.4
Private Construction in the Hierarchy of Development Data

To determine the relative significance of private construction in the hierarchy of development data, the technique proposed by the United Nations Research Institute for Social Development (McGranahan 1972) can be used. This technique suggests that among a group of quantitative indicators, *that which best correlates with the rest of the collection,* or, more precisely, has the highest average correlation with the others, may be considered as the most informative indicator of the group. This approach is based on the assumption that since development tends to be an interdependent process, *a good development indicator should reflect many more things than it directly measures.*

Following this approach, the present analysis included, in addition to two indicators of private construction (PRHOUS: the annual rate of private housing construction, and PRNHOUS: the annual rate of private non-residential construction in a settlement, thousands of m^2), eleven other commonly used development-related data series covered by the Israeli national statistics in 1993 for 80 urban settlements of the country (ICBS 1994-1996). The description of these indicators is as follows:

- POPUL: population of a settlement, thousands of residents;
- UNEMPL: the absolute number of unemployed in a locality, thousands;
- SAVINGS: average personal bank savings, $US per capita;
- PUBHOUS: the overall annual rate of public residential construction in a settlement, thousands of m^2;
- PUBNHOUS: the same for non-residential public construction, thousands of m^2;
- INFR: the annual rate of road construction (as a proxy for infrastructure) in a settlement, km;
- IMMIGR: foreign immigration to a settlement, thousands of immigrants per annum;
- MIGR: the net balance of in-country migrations to and from a locality, thousands of migrants;

- NGROW: the natural increase of population (i.e. the difference between fertility and mortality rates), thousands of people;
- CAROWN: the number of private vehicles per 1,000 residents, and
- SCHOOLS: school enrolment, the number of pupils per class.

The indicators included in the analysis thus cover all the major groups of development data discussed in the section on general development indices: socio-economic development (migration, immigration, natural growth, school enrolment), infrastructure (road construction), transport (car ownership), and housing (public and private construction). In this case, the list of development indicators was also supplemented with data on non-residential public and private construction and with per capita bank savings. (The latter datum is considered a good indicator of the local population's affluence; see *inter alia* Portnov 1998; Portnov and Pearlmutter 1997). At the same time, some other possible indicators (local government expenditures, environmental conditions, etc.) were excluded from the analysis due to restrictions on data availability and comparability.

To test the interrelationship of the above development data, two different approaches were used: a) hierarchical cluster analysis, and b) the analysis of correlation.

The first approach allowed us to identify relatively homogenous groups of indicators based on the data available for the settlement sample (see Appendix 2), and to calculate standardized distances (proximity matrix) between the indicators at hand (see Appendix 3). At the same time, the second approach (correlation analysis) made it possible to calculate the average bivariate correlation coefficients between each of the above development indices and the rest of the collection (see Appendix 3).

As Appendix 2 shows, the indicators of private construction (PRHOUS and PRNHOUS) are part of a large cluster of interrelated development data that also includes such "traditional" indices of urban development as population size, unemployment, savings and public building. At the same time, other indicators included in the analysis (infrastructure, immigration, migration, natural growth and school enrolment) form separate, relatively isolated, clusters that "joined" the main "core" cluster at the later phases of the analysis. This implies that these "remote" indicators reflect the more "narrow" aspects of development that they directly measure, while the indicators included in the main cluster can be considered as more general measurements.

The indicators of private building (PRHOUS and PRNHOUS) have relatively small mean distances and high average correlations (Appendix 3) with the rest of the set. Thus, according to both approaches of analysis, the construction indicators in question (PRHHOUS and PRNHOUS) reflect, although indirectly, various distinctive aspects of urban development whose scope lies beyond the phenomena these indices directly measure. In other words, the indices in question appear to be general (integrate) measurements of urban development in particular settlements. These indicators may, therefore, represent a valuable tool for urban and regional policy analysis by complementing other traditional indicators of urban development such as population size, unemployment rates, etc.

While the potential implications of private construction as a general indicator of urban development will be discussed further in Section 4.8, we shall turn now to a description of the factors and forces which may affect the rate of private construction in various geographic areas.

4.5
Location Paradigm

To explain the geographic distribution of private construction across various urban areas, a theoretical model is suggested, as shown in Fig. 4.4. This model considers the phenomenon in terms of *profit maximization* and includes both a) *cost factors* (land availability, infrastructure, government incentives and construction costs) and b) *benefit factors* (natural amenities, population/migration, accessibility, and buying power of the local population). Each of these groups of location factors presumably affects the decision-making of private developers and will be discussed in some detail.

Land availability. A shortage of land for new development normally increases land prices and may thus limit the rate of private construction in a given urban area. A decrease in land prices with distance from the city core is evidenced by numerous theoretical and empirical planning studies (Alonso 1991; Muth 1969; Clark 1982; Andoh and Ohta 1997). In Japan, for instance, the impact of Tokyo on land prices in surrounding areas is felt within a 200-km radius from the city center (Andoh and Ohta 1997). In Israel, most urban settlements are concentrated within commuting distances from three major metropolitan centers (Tel Aviv, Jerusalem and Haifa) and often form uninterrupted urban contiguity. Under these conditions, undeveloped land available outside the metropolitan core should increase, at least theoretically, the attractiveness of more remote urban areas for private developers. In the absence of direct data, the distance from a metropolitan center can thus be used as a proxy for land availability.

Infrastructure. The availability of roads and engineering utilities in urban areas lowers construction costs, making areas of developed infrastructure more attractive for private investors. For example, there is a distinctive effect of high transport cost on the socio-economic development of urban areas (Clark 1982; Richardson 1977). Since the level of infrastructure development (e.g. transport network, engineering utilities) tends, in general, to diminish with advancing into peripheral areas (Newman 1993; Saini 1980; Portnov and Erell 1998), the attractiveness of these areas to private developers should also decline.

Government incentives. In a number of cases, the development of certain geographic areas (specifically, underdeveloped peripheral regions) is purposely encouraged by local governments on strategic, political or ideological grounds. Examples of such periphery-development policies are found in Northern Europe

(Sweden and Norway), Asia (Japan), and in the Middle East (Egypt and Israel). In Israel, for instance, economic incentives in so called "priority development areas" are offered to private developers in four basic ways: a) direct involvement of the state in infrastructure development and planning; b) financial incentives to private investors (tax exemptions, state loans and loan guarantees); c) allocation of public land, and d) location aid and buy-back guarantees (IMF 1996; Fialkoff 1992).

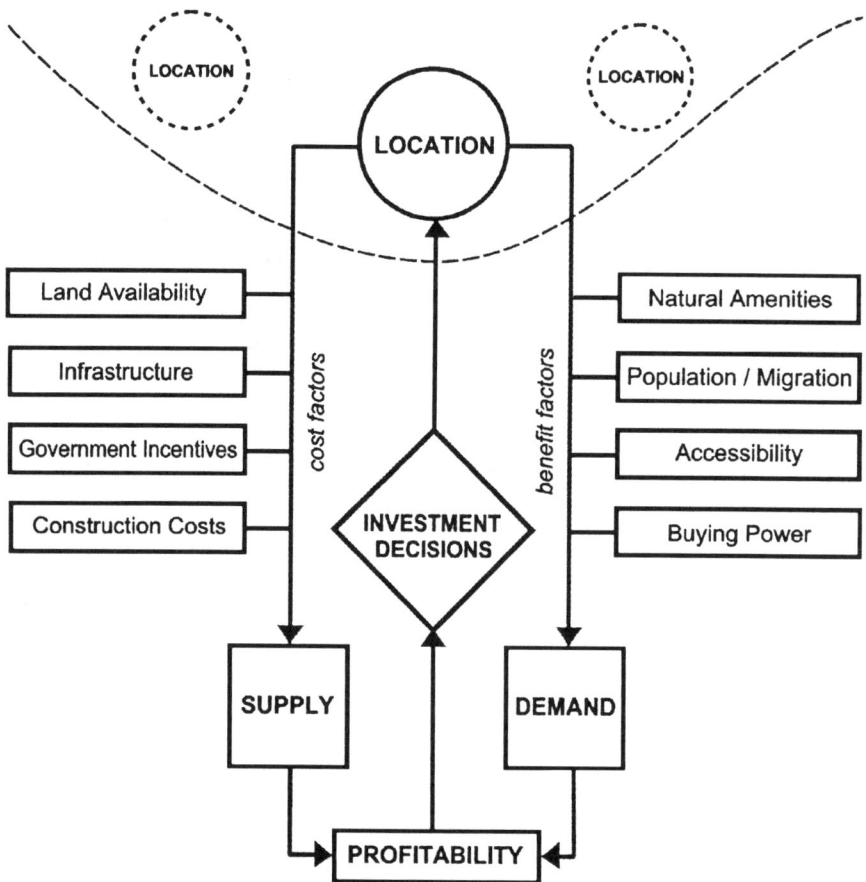

Fig. 4.4. A general model of factors determining the location of private construction across urban settlements

Construction costs. The direct expenses incurred in construction, primarily for building materials and labor, may vary by location according to the availability of physical and human resources. When combined with other location factors, these direct costs dictate the price of construction in a given urban settlement and influence the extent to which private developers are likely to invest in increasing the supply of building stock. Direct construction costs and their geographical variations are commonly documented by public or private building industry bodies for regions but are rarely available for individual urban settlements.

Natural amenities. The presence of natural amenities (climate, landscape, and vegetation) may significantly affect the population attractiveness of specific urban areas. In turn, this may enhance the attractiveness of these areas to private developers. On the other hand, the harsh climate of some geographic areas, which may be characterized by thermal stress, temperature extremes, blowing dust, or a lack of water resources and vegetation, may make areas lacking natural amenities somewhat less attractive for private developers.

Population/migration. The concentration of population and a positive migration balance in a settlement should attract, at least theoretically, private developers due to the fact that private construction naturally tends to the areas where demand is greater and thus the highest profit can be expected. This trend may, of course, be offset by other interfering factors, for instance, by land availability. The interrelation of the concentration of population and land availability may lead, for instance, to the concentration of private construction in *urban settlements of a certain size,* which are small enough to have considerable land resources available, and large enough to provide sufficient market demand for building. We may recall that this phenomenon is observed in Israel, as discussed in the section on the spatial patterns of private construction (see Fig. 4.3 and its discussion).

Accessibility. The rate of private construction should normally decline with increasing distance from major population centers, in which population and a great number of jobs and services are concentrated. In Israel, this assumption seems to be justified specifically for private residential construction (see Fig. 4.2). In the case of non-residential construction, this clear trend may, however, be offset by a lack of undeveloped land for the construction of territory-consuming industrial or business installations (industrial parks, shopping malls, etc.). On the other hand, business in remote hinterland areas may experience difficulties in the supply of skilled labor, which is far less available in remote areas than in large metropolitan centers (Mills 1972; Abe 1996). These "confronting" forces (land availability and availability of skilled labor) may thus lead to the absence of any "straight" distribution pattern (see the fit lines of private non-residential construction in Fig. 4.2).

The buying power of the local population is undoubtedly a key factor affecting the spatial distribution of private construction across urban areas. Since private construction is naturally aimed at the maximization of profit, its highest rates may be expected in the most affluent areas of a country. From the socio-economic standpoint, these areas are distinguished by a number of welfare indicators, including wages, unemployment, bank savings, the level of car ownership, etc.

According to the model suggested in Fig. 4.4, the aforementioned factors and forces form a dynamic "supply-demand" paradigm affecting potential profitability of investments and, therefore, investment decisions of private developers concerning alternative urban locations.

4.6
Case Study

To identify major factors and forces, which influence the rate of private construction across urban settlements in Israel, multiple regression analysis can be employed. For this analysis, the rates of private residential construction (PRHOUS) and non-residential construction (PRNHOUS) in urban settlements were considered as the dependent variables. At the same time, the following indicators were considered as explanatory variables (it is suggested that these quantitative variables can be considered as proxies for the largely qualitative factors discussed in the previous section):

- REMOTENESS is the aerial distance from a settlement to the closest metropolitan centre (either Jerusalem, Tel Aviv, or Haifa), km;
- POPULATION is the population size of a locality, thousands of residents;
- CLIMATE is an index of climatic harshness calculated for each urban locality as the mean annual number of days with summer heat stress (for definition see Bitan and Rubin 1991);
- SAVINGS is average personal bank savings, $US per capita, considered as a proxy for the economic affluence of a settlement's population;
- UNEMPLOYMENT is the average annual number of unemployed in a settlement, thousands of unemployed (together with per capita bank savings, this indicator is considered as a proxy for buying power of the local population);
- INFRASTRUCTURE is the annual rate of public road construction in a locality, km. (this index is considered as a proxy for governmental incentives to private developers), and
- MIGRATION is the annual net balance of migration in a settlement, thousands of migrants.

As a spatial basis for the analysis, 80 urban settlements with a population of over 10,000 residents were considered. Three points in time – 1992, 1993 and 1994 - were chosen to observe the phenomenon at hand using comparable and complete sets of data. The sample thus covered the most recent detailed statistical data published by the Israeli Bureau of Statistics in 1994 and 1996 respectively (ICBS 1994-96).

The linear regression model used for the analysis is as follows:

$$PC = B_o + B_1 \times F_1 + ... + B_n \times F_n + \varepsilon,$$

where PC is either the annual rate of residential or non-residential private construction in a particular urban locality; $B_o, B_1, ..., B_n$ are the respective regression coefficients, $F_1, ..., F_n$ - the above listed research variables, and ε is a random error term.

The analysis included to important procedures.

First, three functional forms of regression equation were tested: the linear form, semi-log form (only the explanatory variables are logarithmically transformed), and double-log form (both the left-hand and right-hand variables are transformed). As Table 4.2 shows, the linear models appear to provide, in most cases, the best fit. Reluctance to use this type of function form in the subsequent analysis was due to a heterogeneity-of-variance consideration, stemming from the fact that the *heteroscedasticity* of the raw data could theoretically cause instability of regression estimates. On the other hand, the logarithmic transformation helped to insure that variances are more homogenous.

Second, the issue of *multicollinearity* was analyzed (see collinearity statistics in Tables 4.3-4.5). Although partial collinearity of the explanatory variables could not be avoided, it was confirmed that the collinearity is, in general, within tolerable limits, and should not, therefore, cause significant instability of the regression estimates.

Table 4.2. Test of functional forms of the regression models

Functional form	Residential construction		Non-residential construction	
	R Square	F	R Square	F
1992				
Linear	0.671	21.247	0.795	27.051
Semi-log	0.641	18.652	0.508	10.784
Double-log	0.497	10.319	0.333	5.213
1993				
Linear	0.621	19.761	0.797	25.562
Semi-log	0.670	21.138	0.561	13.328
Double-log	0.490	10.007	0.346	5.515
1994				
Linear	0.694	23.652	0.613	16.524
Semi-log	0.602	15.797	0.395	6.815
Double-log	0.576	16.524	0.420	7.549

Note: All F values are significant at a 0.001 significance level

Table 4.3. Factors affecting the rate of private residential and non-residential construction in urban settlements of Israel in 1992

Factor	Residential construction				Non-residential construction			
	B	t	Collinearity statistics		B	t	Collinearity statistics	
			Tolerance	VIF			Tolerance	VIF
Remoteness	-1.638	-0.154	0.628	1.594	5.416	0.412	0.628	1.594
Climate	-25.956	-1.430	0.906	1.103	-13.497	-0.603	0.906	1.103
Population	72.513	5.788[a]	0.558	1.791	66.289	4.293[a]	0.558	1.791
Infrastructure	-6.702	-0.422	0.841	1.188	-23.765	-1.214	0.841	1.188
Savings	43.764	3.316[a]	0.557	1.795	51.766	3.182[a]	0.557	1.795
Unemployment	-32.144	-1.679[c]	0.734	1.362	-24.368	-1.033	0.734	1.362
Migration	18.692	1.203	0.864	1.158	20.364	1.064	0.864	1.158
Constant	-110.271	-2.076[b]			-184.752	-2.822[a]		
No of obs.	80				80			
R Square	0.641				0.508			
F	18.652[a]				10.784[a]			

Note: Right-hand variables are logarithmically transformed; [a] Indicates a two-tailed .01 significance level; [b] indicates a two-tailed .05 significance level; [c] indicates a two-tailed .10 significance level. VIF is the variance inflation factor, which expresses the variance of the regression coefficient. Large values of this indicator (VIF>4.0) indicate a significant degree of multicollinearity.

Table 4.4. Factors affecting the rate of private residential and non-residential construction in urban settlements of Israel in 1993

Factor	Residential construction				Non-residential construction			
	B	t	Collinearity statistics		B	t	Collinearity statistics	
			Tolerance	VIF			Tolerance	VIF
Remoteness	-10.676	-0.747	0.584	1.711	-5.820	-0.454	0.584	1.711
Climate	9.824	0.423	0.924	1.083	-9.793	-0.470	0.924	1.083
Population	130.860	5.403 [a]	0.255	3.917	92.450	4.258 [a]	0.255	3.917
Infrastructure	6.021	1.064	0.568	1.759	1.365	0.269	0.568	1.759
Savings	-2.930	-0.349	0.358	2.790	2.564	0.340	0.358	2.790
Unemployment	-59.517	-2.483 [a]	0.782	1.279	-39.002	-1.815 [c]	0.782	1.279
Migration	39.272	1.855 [c]	0.748	1.338	25.860	1.363	0.748	1.338
Constant	-68.607	-1.133			-45.377	-0.836		
No of obs.	80				80			
R Square	0.670				0.561			
F	21.138 [a]				13.328 [a]			

Note: See comment to Table 4.3

Table 4.5. Factors affecting the rate of private residential and non-residential construction in urban settlements of Israel in 1994

Factor	Residential construction				Non-residential construction			
	B	T	Collinearity statistics		B	T	Collinearity statistics	
			Tolerance	VIF			Tolerance	VIF
Remoteness	-10.689	-0.736	0.620	1.614	-0.692	-0.057	0.620	1.614
Climate	17.010	0.675	0.861	1.161	-4.867	-0.231	0.861	1.161
Population	74.633	4.157[a]	0.522	1.914	34.417	2.290[b]	0.522	1.914
Infrastructure	52.882	2.895[a]	0.795	1.257	10.914	0.714	0.795	1.257
Savings	39.023	2.488[a]	0.587	1.704	38.782	2.954[a]	0.587	1.704
Unemployment	-38.648	-1.411	0.875	1.142	-30.889	-1.347	0.875	1.142
Migration	14.216	1.066	0.858	1.166	10.645	0.954	0.343	1.166
Constant	-142.510	-2.139[b]			-83.641	-1.499		
No of obs.	80				80			
R Square	0.602				0.395			
F	15.797[a]				6.815[a]			

Note: See comment to Table 4.3

4.7
Research Results

The results of the regression analysis are presented in Tables 4.3-4.5. Nearly half of the regression coefficients in both models are significant at 0.05 and 0.01 levels and exhibit correct signs. The following discussion shall refer *only* to the factors, which are statistically significant at these confidence levels.

As seen, *population size, savings*, and *infrastructure* are the main factors that increase the rate of both private residential and non-residential construction in urban settlements, while private construction tends to be lower in settlements of high *unemployment*. Each of these factors will be discussed below.

Population size. Larger urban localities are generally attractive to private developers due to higher demand and, therefore, higher potential profitability of construction investment. This is in line with the previous discussion (see Fig. 4.4 and its discussion in the text).

Per capita bank savings and *unemployment* are, as suggested, two robust indicators of the affluence of the local population. It is quite natural, therefore, that private construction in Israel tends to the urban areas in which population is more affluent and less affected by unemployment. The general course of this trend was, for instance, argued in the framework of the above suggested development model, in which the aforementioned variables are included in the general group named "buying power of the local population" (see Fig. 4.4).

Infrastructure. In the models of residential construction, the statistical significance of infrastructure variable (see Tables 4.3-4.5), measured by the *t-statistic*, increased from $P>0.10$ in 1992 to $P>0.05$ in 1993, and to $P<0.01$. This trend can apparently be attributed to a more uneven distribution of road construction across particular urban settlements in 1994 than in previous years (see standard deviation of the infrastructure variable in Appendix 4).

Although the growing significance of infrastructure for private construction is not totally unexpected, the models suggested might be of significance for the country's future development planning. For instance, these models may help to justify the concentration of public resources on the development of intra-urban infrastructure as a precondition for attracting private developers to specific urban areas.

To emphasize the importance of the infrastructure variable in this context, a sensitivity test of the models to plausible changes of this and an other important policy variable – unemployment – was carried out. The results of the test are presented in Table 4.6. To perform this test, the respective models for 1994 (Table 4.5) were used. The actual values of the variables in question were plausibly adjusted, while the values of other explanatory variables remained unchanged. As seen, a 20% increase in the mean rate of road construction appears to lead to a 5.8% increase in the rate of private residential construction and to a 1.5% in that of non-residential building. Correspondingly, a 20% decrease in unemployment in a settlement may lead to a 3.8% increase in residential construction and a

5.3% increase in non-residential building. Together, these adjustments in unemployment and infrastructure may lead to almost 15% growth of private development in a settlement.

Table 4.6. Sensitivity test of the model to changes in critical policy variables (UNEMPLOYMENT and INFRASTRUCTURE), percent change

Indicator	Residential construction	Non-residential construction	Total
Infrastructure (+20%)	+4.3	+1.5	+5.8
Unemployment (−20%)	+3.8	+5.3	+9.1
Infrastructure (+20%) and Unemployment (−20%)	+8.1	+6.8	+14.9

4.8
Applications in Planning

Three outcomes of the present analysis of urban settlements may be useful for urban and regional planning. These are: a) the rate of private construction appears to be a meaningful indicator of a settlement's general socio-economic development, which reflects many more things that it directly measures; b) the tendency for maximum private construction rates to occur in urban settlements of a certain population size (70,000-80,000 residents), and c) the empirical models that provide a summary of the factors influencing the rate of private construction across urban areas.

With respect to private construction as a *general indicator of urban development*, three potential areas of use of this rate can be singled out:

1. *Improving the system of state statistics.* The key time-series monitored by the Israeli Central Bureau of Statistics for major urban settlements of the country includes, for example, the following six measures: population, natural growth, number of social workers per 1,000 residents, school enrolment, number of housing units constructed, and car ownership level (ICBS 1994-96). The present analysis indicates that some of these measures (for instance, car ownership and school enrolment) are merely indicative of the particular development aspects which they directly measure, while the factor of *private construction* appears to be a more general indicator of urban development, and could thus contribute to the set of indicators in the major development series.
2. *Gauging the progress of development in various settlements and regions.* The inclusion of private construction in the list of major development series may help to address a number of issues of regional development. Since the rate of private construction in a particular urban locality tends to correlate with a wide range of development factors (unemployment, settlement attractiveness for immigrants and in-country migrants, rates of annual growth, population

income), systematic comparison of these rates across different settlements and regions can serve as a useful tool for singling out "troubled" settlements and geographic areas, and thus better targeting governmental assistance such as public loans, loan guarantees or tax exemptions to private developers;
3. *Evaluating the efficiency of various planning policies aimed at developing particular settlements and regions.* Regional development policies are often aimed at encouraging socio-economic development in economically lagging peripheral areas. For instance, the policy of population dispersal in Israel encourages population growth and industrial development in the country' underpopulated periphery – the Negev and the Galilee (see the section on urban development in Israel). To gauge the effect of this and other similar development policies, a number of "conventional" indicators are used. As mentioned in the previous discussion, these include such factors as unemployment, migration balance, and capital movements. In light of the findings of the present study, the rate of private construction in a particular urban area may thus be considered as an effective supplement to the "conventional" list of policy evaluating data.

The tendency revealed during the analysis of urban settlements in Israel - that urban settlements in the country tend to become more attractive to private developers after reaching a particular size (70,000-80,000 residents) - seems to also be useful in determining an *alternative strategy for developing peripheral regions* which are currently sparsely populated. The basics of this strategy, which can be defined as the strategy of "redirecting priorities," are discussed by these authors in the previous chapter. This strategy assumes, in particular, that development resources should be primarily concentrated on a limited number of selected settlements in the frontier areas until they reach the above threshold and become considerably attractive to private developers. As soon as the above population threshold is achieved, the above support may be redirected on a stage by stage basis to other adjacent urban settlements. This process of the temporary and hierarchical concentration of resources can thereby be moved deeper and deeper into the frontier areas (for more particulars see Chap. 3 of this book).

The *empirical models* of the factors influencing the rates of private construction across urban settlements of Israel may also be of significance for the country's future development planning. Since these models emphasize the adverse effect of unemployment and the positive effect of infrastructure development on private construction in urban settlements, they may help to justify the concentration of public resources on the development of intra-urban infrastructure and reducing unemployment as a precondition for attracting private developers to specific urban areas.

References

Abe H (1996) New directions for regional development planning in Japan, In: Aden J, Boland P (eds), Regional Development Strategies: A European Perspective. Jessica Kingsley Publishers, London and Bristol, pp 273-295

Alonso W (1991) A theory of the urban land market. In: Cheshire PC, Evans AW (eds), Urban and Regional Economics. Cambridge University Press, Cambridge, pp 83-91

Andoh K, Ohta M (1997) A hedonic analysis of land prices in Yamanashi prefecture, Japan, Review of Urban & Regional Development Studies 9:146-158

Armstrong H, Taylor J (1993) Regional economics and policy. Harvester, New York

Benzaquen J (1998) Israeli success on the home front, Jerusalem Post, April 20

Bitan A, Rubin S (1991) Climatic atlas of Israel for physical and environmental planning and design. Ramot Publishing Co, Tel-Aviv

Bourne LS (1975) Urban systems: strategies for regulation. Claredon Press, Oxford

Clark C (1982) Regional and urban location. St.Martin Press, New York

Clawson M, Hall P (1973) Planning and urban growth: an Anglo-American comparison. The Johns Hopkins University Press, Baltimore

Diamond DR, Spence NA (1983) Regional policy evaluation: a methodological review and the Scottish example. Aldershot, Gover

Drabkin-Darin H (1957) Housing in Israel: economic & sociological aspects. Gadish Books, Tel-Aviv

Fialkoff C (1992) Israel's housing policy during a period of massive immigration. In: Golany Y, Eldor S, Garon M (eds) Planning and housing in Israel in the wake of rapid changes. Ministries of the Interior and of Construction and Housing, Jerusalem, pp 169-177

HABITAT (1997) Monitoring urban settlements with urban indicators. United Nations Center for Human Settlements, Nairobi

ICBS (1994-96) Local authorities in Israel: physical data. Israeli Central Bureau of Statistics, Jerusalem (in Hebrew)

ICBS (1997) Statistical abstract of Israel. Israeli Central Bureau of Statistics, Jerusalem

IMF (1996) Structure of investment incentives. Israeli Ministry of Finance, Jerusalem

Layton AP (1972) Some Australian experience with leading economic indicators. In: Lahiri K, Moore J (eds) Leading economic indicators: new approaches and forecasting records. Cambridge University Press, Cambridge, pp 211-230

Leeuw F de (1992) Toward a theory of leading indicators. In: Lahiri K, Moore J (eds) Leading economic indicators: new approaches and forecasting records. Cambridge University Press, Cambridge, pp 15-56

Levy JM (1985) Urban and metropolitan economics. McGraw-Hill, New York

Lipshitz G (1997) Immigrants from the former Soviet Union in the Israeli housing market: spatial aspects of supply and demand. Urban Studies 34(3):471-488

Markusen A (1996) Interaction between regional and industrial policies: evidence from four countries. International Regional Science Review 19:49-77

McGranahan D (1972) Development indicators and development models. In: Baster N (ed) Measuring development. Frank Cass, London, pp 91-102

MCH (1997) Construction and housing in Israel: monthly update. Ministry of Construction and Housing, Jerusalem (in Hebrew)

Mills ES (1972) Urban economics. Scott, Foresman and Company, London

Muth RF (1969) Cities and housing: a spatial pattern of urban residential land use. Chicago University Press, Chicago

Newman P (1993) The compact city: an Australian perspective. Built Environment 18(4): 285-300

Portnov BA (1998) The effect of housing on migrations in Israel: 1988-1994. J of Population Economics 11(3):379-394

Portnov BA, Erell E (1998) Development peculiarities of peripheral desert settlements: the case of Israel. International J of Urban and Regional Research 22(2): 216-232

Portnov BA, Pearlmutter D (1997) Sustainability of population growth: a case study of urban settlement in Israel. Review of Urban and Regional Development Studies 9(2): 129-145

Richardson HW (1977) Regional growth theory. MacMillan, London

Saini BS (1980) Building in hot dry climates. John Wiley & Sons, Chichester

Smith WF (1975) Urban development: the process and the problems. University of California Press, Berkeley

Wong C (1995) Developing quantitative indicators for urban and regional policy analysis. In: Hambleton R, Thomas H (eds) Urban policy evaluation: challenge and change. Paul Chapman Publishing Ltd, Cardiff, pp 111-122

Appendix 1
The effect of population size on per capita rates of private residential construction in different types of urban localities

Year	Population ≤ 80,000				Population > 80,000			
	B_0	t	B_1	t	B_0	t	B_1	t
1992	.315	2.91	.012	4.30[a]	1.645	6.75[a]	-.003	-3.11[b]
			R^2=0.305				R^2=0.708	
1993	-.156	.89	.032	3.21[b]	2.231	9.06[a]	-.004	-3.20[b]
			R^2=0.276				R^2=0.536	
1994	-.008	-.058	.018	5.33[b]	2.037	13.76[a]	-.003	-3.74[b]
			R^2=0.387				R^2=0.451	

[a] Indicates a two-tailed 0.01 significance level; [b] indicates a two-tailed 0.05 significance level

Appendix 2
Hierarchical cluster analysis (dendrogram using average linkage between groups)

```
Rescaled Distance
Variable   0         5         10        15        20        25
         +---------+---------+---------+---------+---------+---------+

POPUL    -+---+
UNEMPL   -+   +---------+
PRHOUS   -+---+         +---------+
PRNHOUS  -+        I         +-+
SAVINGS  ---------------+    I I
PUBHOUS  -------+------------------+ +-----+
PUBNHOUS -------+                  I       +---+
INFR     --------------------------+   I   +---+
IMMIGR   ------------------------------+   I   +-------+
CAROWN   ----------------------------------+   I       I
SCHOOLS  --------------------------------------+       I
MIGR     -------------------------------+--------------+
NGROW    -------------------------------+
```

Appendix 3
Mean proximity and correlation among the research variables

Variable	Mean distance	Mean correlation with the rest of the set[a]
PRHOUS	107.04	0.386
PRNHOUS	98.85	0.422
POPUL	98.23	0.439
SCHOOLS	151.81	0.117
IMMIGR	129.69	0.190
MIGR	164.37	0.172
NGROW	170.72	0.124
CAROWN	139.88	0.137
UNEMPL	101.00	0.429
SAVINGS	121.43	0.252
PUBHOUS	108.04	0.273
PUBNHOUS	119.66	0.328
INFR	123.07	0.242

[a]Based on absolute values

Appendix 4
Statistical parameters of the research variables

Variable	1992		1993		1994	
	Mean	Std.Dev.	Mean	Std.Dev.	Mean	Std.Dev.
HOUSING	47.56	56.77	51.35	73.48	52.37	75.56
NON-HOUSING	28.33	60.17	30.31	57.14	31.62	51.29
REMOTENESS	31.76	36.53	31.76	36.53	31.76	36.53
CLIMATE	51.66	18.54	51.66	18.54	51.66	18.54
POPULATION	51.41	79.75	52.56	81.00	53.78	81.79
INFRASTRUCTURE	0.56	1.25	1.22	2.19	0.94	2.04
SAVINGS	1383.85	3079.36	1043.19	2842.50	1314.55	3045.22
UNEMPLOYMENT	24.53	11.37	22.48	11.52	20.20	12.81
MIGRATION	1.87	3.51	.82	1.92	2.77	4.73

5 The Effect of Remoteness and Isolation on Development of Peripheral Settlements[1]

Boris A. Portnov and Evyatar Erell
J. Blaustein Institute for Desert Research, Ben-Gurion University of the Negev, Sede-Boker Campus, 84990, Israel

5.1 Introduction

As differentials in development potential have increased in recent years (Wong 1995), overcoming inequalities in the level of socio-economic development has become a key issue for urban and regional planners worldwide. These inequalities are often the result of the spatial characteristics of the settlement pattern, reflecting differences between core and peripheral areas.

In many peripheral areas, specifically in peripheral desert regions, the local population is denied access to social amenities that are concomitant with a larger settlement size. While the inhabitants of some small, peripheral settlements may prefer to preserve the rural character of their communities, and therefore resist further population growth, many welcome the advantages that a larger community allows: as the population of a community increases, it crosses the threshold for higher-level services and offers more varied opportunities for employment, social services and leisure.

The development of peripheral areas may also be driven by *push* factors, rather than by pull. Peripheral areas may provide an alternative for the residents of core areas, if urban infrastructure in the periphery is developed and if the level of social amenities in peripheral districts is acceptable. When this is not the case, over-population of core areas may result in severe social and environmental problems.

Whether or not the development of peripheral areas is desirable in a given context, the factors affecting the growth prospects of a particular settlement or region must be understood. To assess the degree of sustainability exhibited by peripheral settlements in their population growth and economic development, effective monitoring indicators are needed. Given, however, the complex and multi-faceted nature of urban growth, direct assessments alone are unlikely to be sufficient. As Portnov and Pearlmutter (Chap. 3 of this volume) show, the fact that the overall rate of annual population growth in a particular urban community reaches a certain level does not characterize either the present socio-economic state of the settlement or its future growth potential. Indeed, a particular rate of overall growth may result from various combinations of natural growth rates (the excess of births over deaths) and migration balance.

---

[1] The chapter is based on a shorter article published by these authors in Review of Urban and Regional Development Studies (1998) 10(2): 123-141.

In previous chapters of this book (Chaps. 2 and 3), it was argued that population growth which is based primarily on the attractiveness of a locality to newcomers (i.e. the locality has a positive migration balance that surpasses the rate of its natural growth) signals the sustainability of population growth in the future. The present analysis applies this concept to the analysis of the growth patterns of small urban settlements in peripheral regions in Israel, in order to investigate the *effects of spatial isolation and distance from major urban centers on the long-term sustainability of population growth*. Two other indicators of socio-economic development − the rate of private construction and the rate of unemployment − are also used to study the combined effect of the above spatial parameters of the urban field (*isolation* and *remoteness*) on socio-economic development of peripheral desert and non-desert localities. In particular, the following questions will be addressed by the analysis:

1. To what extent do the spatial characteristics of the urban field in peripheral areas affect the degree of sustainability exhibited by small urban centers in their population growth and economic development?
2. Which planning policies and strategies are conducive to sustaining urban growth in underdeveloped peripheral areas?

5.2
Sustainable Population Growth of Urban Settlements: Components and Research Paradigms

Additional urban growth, particularly in densely inhabited core areas, is often undesirable. Peripheral regions, however, are often underdeveloped and underpopulated, and may require sustainable urban growth as a precondition for achieving other aspects of development: diversity of employment, services, and leisure opportunities. Urban and regional planners must therefore be able to gauge the degree of sustainability of urban growth, particularly in peripheral urban settlements. The following sections will clarify the definition of the term *sustainability* as used in this study, provide a brief overview of accepted indicators for measuring urban growth, and propose indicators for the subsequent analysis.

5.2.1
Population Growth

In Chap. 3 of this volume, the ration between migration balance and natural growth (the MB/NG index) was introduced as an integrated indicator of sustainable population growth. In the following analysis, this indicator will be used as a major quantitative tool for gauging the degree of sustainability exhibited by specific urban settlements in their population growth.

As Portnov and Pearlmutter (Chap. 3 of this book) point out, the importance of the MB/NG indicator clearly surpasses its direct use as a simple measure of population increase: The ongoing migration attractiveness of a particular settlement implies its sound economic development, favorable physical environment and other preconditions which are essential for sustainable, i.e. continuing, urban growth in general. It is also important that the indicator in question allows normalization for natural growth rates, which may vary substantially from one settlement or region to another.

5.2.2
Measuring Economic Development

The economic performance of a region or a particular urban settlement may be estimated by several types of indicators, including labor, capital, economic output and construction.

Labor criteria. In general, employment may be measured by two aggregated groups of data: unemployment rates and employment change.

- The *unemployment rate* is easily available and is widely recognized, but it suffers from a number of drawbacks. First, there are several definitions of the term, primarily because of the difficulty in establishing the appropriate definition of the working population. Second, the overall (headline) rate of unemployment gives no information about its composition. For instance, a high unemployment rate in a particular settlement may result from a mismatch between the demand for particular skills and the actual qualification of the local labor force. This phenomenon is known as structural unemployment. If a town suffers from *structural* unemployment, the population of the settlement may still grow steadily due to an influx of skilled migrants from elsewhere, whereas if unemployment is high in all the sectors of the local economy, there is no incentive for job seekers to migrate to the locality. A high overall rate of unemployment in a locality may thus be often misleading.
- *Changes in employment,* unlike measures of unemployment, do not require complex definitions, but they too, in general do not differentiate between employment gains and losses in different economic sectors (for instance, in manufacturing and service industries).

In spite of these inadequacies, employment change and unemployment rate remain the most inclusive indicators of urban and regional development.

Capital criteria. These criteria are used to trace the overall investment flow to a settlement or region due to expansions or movements of commercial firms. The major disadvantage of this indicator is that it does not take into account the possibility of subsequent closures of the firms, i.e. their survival rates (Henderson 1980). The indicator in question also says nothing about actual changes in employment/unemployment associated with relocation of firms (Diamond and Spence 1983). In addition, the data on capital transfers between settlements and geographic areas are not always available. In Israel, for instance, these data are not monitored by official statistics (ICBS 1996).

Economic output criteria. These criteria include three quantitative indicators: 1) overall growth of output, 2) growth of output per worker, and 3) growth of output per capita. The overall (absolute) growth of output can be used as an indicator of the growth of productive capacity, which depends (at least, in part) on the extent to which a settlement or region attracts capital and labor from elsewhere. At the same time, the growth of output per worker is often used as a direct measurement of industrial productivity, whereas output per capita is frequently used as a general indicator of socio-economic welfare (Armstrong and Taylor 1993). The usefulness of the above indicators for economic output in the study of urban and regional processes is nevertheless limited, since they often provide contradictory readings. As Armstrong and Taylor (1993) suggest, a region may, for instance, exhibit low output growth and rapid growth of output per capita simultaneously if there is significant out-migration.

Construction criteria. The overall rate of construction is widely accepted as a key indicator of socio-economic development (Levy 1985; Smith 1975). This is due to two main reasons: First, the effect of construction on socio-economic development in urban localities is typically time-lagged, and this datum may thus be used as an indicator of a locality's future micro-economic performance and growth potential. Second, the rate of housing construction in a settlement tends to have a direct effect on migration. For example, a large number of housing units started in the locality may result in the influx of migrants from other settlements and regions (Portnov 1998). The differences between public and private construction are also important:

- The *rate of private construction* is often given separately from the rate of public construction. As Portnov and Pearlmutter (Chap. 4) justly argue, the particular importance of this indicator as a general development datum is primarily due to the fact that, from the micro-economic point of view, the annual rate of private construction in a given urban area is a good indicator of its overall "investment climate."
- The *rate of public construction*, on the other hand, may reflect social or political considerations rather than attractiveness for investment. In the case of Israel, for instance, private construction is "naturally" concentrated in the Central, Tel Aviv, and Northern districts, which are in fact the most populated areas of the country. At the same time, public construction is greatly "skewed" toward the Jerusalem and Southern districts, and to Judea, Samaria and Gaza. This trend is due to the government policy of population dispersal from the country's overpopulated core to the underpopulated periphery and other "strategically sensitive" areas (for more details see Chap. 4 of this book).

In conclusion:

In view of the aforementioned considerations, the rate of private construction may be used in the present analysis of peripheral urban settlements as a general indicator of economic development. This indicator will be supplemented with the unemployment rate commonly considered one of the most inclusive and easily available development data. Taken together, these indicators -- *private construction* and *unemployment* – can provide a useful base for the analysis of the overall socio-economic development of peripheral urban settlements.

5.3
Spatial Characteristics of Urban Development in Peripheral Areas

Quantitative spatial analysis of urban and regional development deals with three main characteristics, each requiring suitable indicators: a) the distribution of population within a given geographical framework; b) the remoteness of a specific settlement in relation to an established center; and c) the degree of isolation of a particular town in relation to other settlements in the region.

5.3.1
Distribution of Population and Settlement Location

The distribution of population within a given geographic area is commonly measured using two major indicators: population density and percentage of the population living in localities of certain size.

Population density is an important regional development datum whose significance is traditionally advocated by location economics (Levy 1985; Smith 1975). In Israel, this indicator of urban growth is included in the index of major time-series monitored by the Israel Central Bureau of Statistics (ICBS 1996). However, *mean* density figures for a given region often hide significant variations in local population density. The usefulness of such data for planning are thus limited, unless the scale of the statistical areas is sufficiently detailed to allow a spatial analysis of the region in question.

Percentage of the population living in localities of certain size is a robust indicator of regional development used in a number of urban and regional studies (Bourne 1975, Clawson and Hall 1973). It should be noted, however, that the applicability of this indicator to entire geographic regions appears to be restricted. Knowing, for instance, that 50% of a district's population live in urban settlements of 50,000 residents or more tells us little about actual patterns of urbanization in the area. Indeed, all the population in question may be concentrated either in one large urban center of 1,000,000 residents or in ten urban localities of 100,000 residents each. From the standpoint of the present analysis, these two cases represent completely different issues, while the above indicator of urban development treats them alike.

In Chap. 2 of this volume, two indicators of settlement location – *remoteness* and *isolation* – were introduced. As suggested, the effects of remoteness may be related to several measures of distance from the established urban centers: aerial distance, distance by road, time required to travel to the center, etc. Aerial distance, since it is "objective" and simple to measure, is probably the most useful of these measures, and therefore will be used in the following analysis. If the extent of the infrastructure and quality of service are more or less uniform throughout the area under study, it does not introduce an undesirable bias in the results.

The choice of the established urban center is somewhat arbitrary, and depends on the type of interaction in question. The services provided by cities of a particular population size may vary from country to country, depending on local economic and social patterns as well as on transportation infrastructure. A functional definition is, however, the most appropriate one in this context.

In the case of Israel, the notion of remoteness is somewhat relative, given the size of the country as a whole. The actual aerial distances between peripheral settlements and the major urban centers – some 100-150 km – are indeed relatively small. However, there are two factors which affect the importance of this spatial indicator even in a small country: First, even these distances exceed those normally considered practicable for daily commuting; and second, the perception of remoteness may affect investment decisions or movements of population no less than the real distances involved. Thus, it is the "relative remoteness" of the country's peripheral areas that may be the influencing factor, rather than absolute distance.

As Portnov and Erell (1998) argue, the considerable distances often found between the established urban localities in peripheral areas are likely to cause a shortage of joint intra-regional educational and recreational structures, and limit the choice of job opportunities. The effects of isolation may, however, be related to the distance to other communities of similar size in the region, as well as to the number of such communities. Both measures are an indication of the potential for intra-regional economic and social interaction.

5.4
Research Method

In Chap. 2 of this book, the "sample-control" method was used to study the long-term development of peripheral vs. core settlements of different population size. In the present case study, a similar approach was used to study the effect of remoteness and spatial isolation on the patterns of population growth and economic development of urban communities. In contrast to the former analysis (see Chap. 2), the sample of urban settlements in the present case study was formed as to include only small urban localities (10,000-40,000 residents) representing three geographic regions of the country: the center, the north (the Galilee) and the south (the Negev desert).

Five small urban settlements located in the central, densely populated, part of the country formed Set 1 (Population figures in brackets are for 1970/1994, respectively) These are: *Yavne* (10,000/26,000), *Qiryat-Ono* (15,000/24,000),

Or-Yehuda (12,000/24,000), *Nes-Ziyyona* (12,000/22,000), and *Rosh-H'Ayin* (12,000/23,000).

Ten other towns of similar size located in outlying districts of the country - either in the *Galilee* or in the *Negev* – formed respective Sets 2 & 3 (Fig. 5.1). Specifically, *Nazerat-Illit* (15,000/38,000 residents), *Afula* (17,000/34,000*), Migdal-H'Emeq* (9,000/21,000), *Qiryat-Sh'mona* (15,000/19,000), and *Bet-She'an* (12,000/15,000) were included in Set 2, while Set 3 was formed by the following urban localities of the Negev: *Dimona* (23,000/30,000 residents), *Ofaqim* (9,000/19,000), *Arad* (5,000/19,000), *Sederot* (8,000/17,000) *and Netivot* (5,000/14,000).

All these towns were established within a short period in the 1950's - early 1960's, and, all else being equal, are presumed to be at similar stages in their development.

The overall size of the sample - 15 settlements - is relatively small. However, the absence of other settlements of comparable size, specifically in the Negev region, prevented the investigation of a larger number of settlements. The need to establish sufficient socio-economic similarity of the settlements investigated, particularly concerning their ethnic makeup, further restricted the sample size. While the ethnic makeup of the towns in this study is not absolutely identical, an attempt was made to select communities with a similar proportion of the population of oriental descent (Asian-African born) as opposed to those born in either Europe or America. The statistical differences between the samples with respect to the above indicator are minor, and should not affect the result of the analysis.

In view of the limited size of the sample, a comparison of its statistical characteristics with those of the population (i.e. all 45 urban localities of similar size in Israel) was carried out (see Appendix). The test indicated that the differences between the sample and the population with respect to major development characteristics – population size, housing, unemployment, and location – are minor.

The relatively low Z-scores (P>0.10) do not indicate any statistically significant bias of the sample with respect to the above indicators of settlement development.

In contrast to the approach used in Chap. 2, twelve biannual points in time - 1970 through 1994 - were selected for the present case study using fully comparable and complete sets of data. The time-series for the analysis were drawn from the respective annual publications of Israeli Central Bureau of Statistics, "Local Authorities in Israel: Physical Data."

Three quantitative parameters of the settlements' development discussed in the previous sections were included in the analysis as dependent variables:

1. The rate of private construction, ['000 m^2 per annum];
2. The rate of unemployment, measured as percentage of unemployed in the overall population of a settlement;
3. The structure of population growth, as indicated by the MB/NG index.

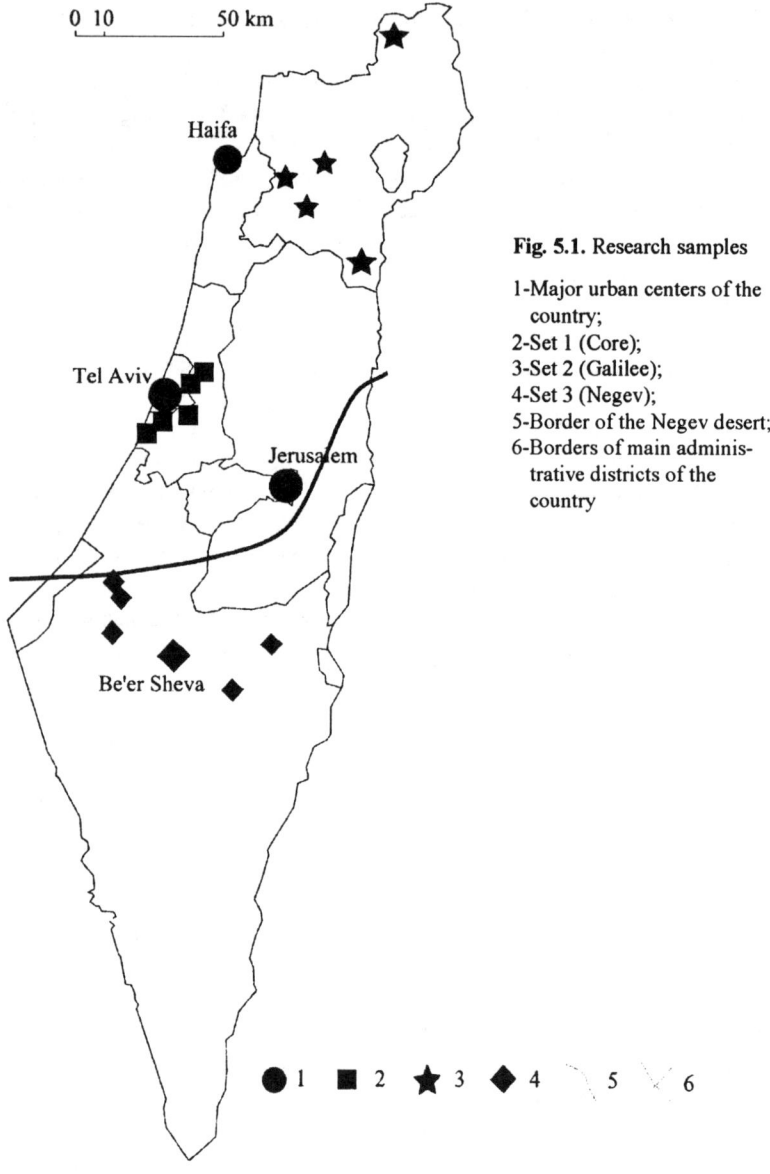

Fig. 5.1. Research samples

1-Major urban centers of the country;
2-Set 1 (Core);
3-Set 2 (Galilee);
4-Set 3 (Negev);
5-Border of the Negev desert;
6-Borders of main administrative districts of the country

The explanatory variables were represented by two spatial factors whose importance was hypothesized in the previous sections. Following is a short description of quantitative indices used to measure these variables:

- *Remoteness.* This indicator was defined as the aerial distance from a settlement to the closest major urban center (either Tel Aviv, Haifa or Be'er Sheva), in km.
- *Spatial isolation* was computed for each settlement as the number of other urban settlements located within a practical range for daily commuting (assumed to be 20 km).

Investigating the effect of location on the socio-economic development of urban settlements required, however, controlling for other factors that may influence urban growth in a specific geographic area. A number of such controls are suggested, and other than *climate*, they are characterized as being independent of a specific location or environment: *housing, employment, employment growth, services, regional center,* and *population size*. The following subsection deals with each of these parameters briefly, and indicates why these factors may be of significance in influencing development patterns of peripheral settlements.

5.5
Controls

To control for other factors that may influence population growth in small urban localities, the following "control" factors were included in the analysis:

Housing. Availability of housing is traditionally considered a key factor affecting the rates of inward and outward migrations (Kirschenbaum and Comay, 1974; Newman, Gradus and Levinson, 1995; Portnov and Pearlmutter, 1997). Since it is suggested that sustainable population growth requires a positive migration balance, the availability of housing is seen as a key factor promoting growth. A high rate of housing construction also affects unemployment in a settlement directly and indirectly due to new employment opportunities created in construction, and in related industries (infrastructure, maintenance, etc.). The availability of housing is commonly approximated by the overall size of housing stock in a locality, by the vacancy rate, and by the overall rate of new housing construction. The proxy chosen to reflect the condition of the housing market in a given locality was the *annual rate of new housing construction* ['000 m^2 of floor area per annum]. To avoid the effect of endogeneity between dependent and explanatory variables (specifically, between private construction and housing), the rate of housing construction was used for modeling population growth (the MB/NG indicator) and unemployment only.

Unemployment. Availability of employment is considered one of the strongest motives for internal migration, as well as affecting the choice of immigrants concerning their first place of residence upon arriving in the country. High unemployment rates, on the other hand, tend to discourage new immigrants from settling in a given city, while encouraging local residents to migrate to other locations. The key role of this factor in influencing migration patterns is advocated by a number of scholars (Fischer 1976; De Jong and Fawcett 1981; Moore and Rosenberg 1995). A high rate of unemployment may also have an adverse effect on the rate of private construction in a settlement, since private construction normally tends to occur in the most affluent areas, where a higher profit can be expected. The index chosen to reflect the condition of the employment market was the *mean annual unemployment rate* [%] in each town. This variable was used in the analysis of population growth (the MB/NG indicator) only.

Employment growth. This index reflects annual changes in the unemployment rate in a settlement [%] and is considered as a proxy for growth/decline of employment opportunities. As in the previous development datum, this indicator was used only in the analysis of population growth.

Population size of urban settlements. The role of this factor in influencing internal migration and new immigrants' initial distribution is widely acknowledged (More and Rosenberg 1995; Fischer 1976; De Jong and Fawcett 1981). The importance of this factor is attributed to the fact that a certain population size is an essential prerequisite for maintaining a sufficient level of diversity in employment and living conditions of urban settlements (Doxiadis 1977). A number of studies (Bourne, 1975; Portnov and Pearlmutter 1997), suggest that reaching a certain population threshold may trigger a "built-in mechanism" which ensures sustainable population growth of an urban settlement in the future. This factor is therefore expected to have a significant influence on two other indicators included in the study area: unemployment and private construction. This indicator was measured in thousands of residents living in the locality.

Regional center. The functional characteristics of a major city with respect to the surrounding region may have a major influence on their respective growth patterns. However, there may also be a qualitative difference in development patterns of the corresponding hinterland areas, which is caused by disparities in population size between the major urban centers of a country. It is thus suggested that the effect of population size of the central cities should be included in the analysis as a separate explanatory variable. The size of the closest urban center was measured in thousands of residents.

Climate. The harsh environment of some geographic areas (heat stress, temperature fluctuations, blowing dust, a lack of water resources and vegetation, etc.) places considerable limitations on urban amenities and human comfort. Urban settlements located in unfavorable climatic conditions might, therefore, be less desirable and attractive (at least theoretically) to new immigrants and to internal migrants. Harsh climatic conditions may also require the installation of expensive climate-control systems. This, in turn, may substantially lower the attractiveness of climatically unfavorable areas to private developers. The climatic differences between geographic areas may be described by a number of quantitative indicators: the number of rainy days, mean summer and winter temperatures, amount of precipitation, etc. In Israel, over-heated conditions are generally perceived as an important source of thermal discomfort. Therefore, it was suggested to use the *mean annual number of days with heat stress* (Bitan and Rubin 1991) as an index of climatic harshness.

Services. The quality of social services and facilities (schools, shopping facilities, health services) is often mentioned as a major factor influencing internal migration and other aspects of urban development (Kirschenbaum and Comay 1974; Newman et al. 1995). A number of qualitative measures can be used as a proxy for the quality of services and facilities in a locality. These include the overall number of facilities of each kind in the settlement, the average annual number of customers per establishment, the number of pupils per class, etc. Since other data are not available, the latter indicator, although relatively weak, was used as a proxy for the average level of services and facilities in a locality.

5.6
Analysis Procedure

Statistical significance of the above factors in influencing the sustainability of population growth (as indicated by the MB/NG ratio) and the economic development of settlements (measured by rates of private construction and unemployment) may be identified and measured by analysis-of-variance. With respect to the analysis procedure, it is to be noted that in order to ensure the homoscedasticity of errors, a logarithmic transformation was applied to original explanatory variables.[2] Hartley's F-Max test for violations of the homoscedasticity assumption confirmed that the variances are indeed homoscedastic.

Three other considerations also deserve mention.

The development of Israel has been characterized by periods of rapid economic growth, usually associated with a great influx of new immigrants from abroad, followed by years of much slower growth. The effect of the economic cycle at the national level, as well as that of changes in the rate of immigration, have a direct bearing on the changes in population growth in various parts of the country. An economic recession, inflation, rising housing prices and other economic processes in the country as a whole may have an adverse impact on the overall rates of internal migration, as well as on patterns of population growth in particular urban settlements. The role of external economic factors in influencing the patterns of growth of peripheral localities was pointed out by Kneese (1978) and Portnov and Erell (1997). However, the relationship between macro-economic factors, which are (by definition) not spatial, and the growth patterns of peripheral communities, is beyond the scope of this study.

The overall rate of foreign immigration to the country has been subject to substantial annual changes. For instance, during three recent decades (1967-96), yearly rates of foreign immigration to Israel varied from 9,500 immigrants in 1986

---

[2] Nearly all variables included in the research scope (housing, employment, services, remoteness, population size, climate, and spatial isolation) appear to have only positive values. This fact is of importance, since logarithmic transformations were required. As for employment growth, whose original values have different signs, an adjustment was needed. The values of this variable were readjusted using the minimal value as the conditional baseline.

to some 199,500 immigrants in 1990 (ICBS 1996). Since the rate of immigration to a particular settlement is a key component of its migration balance, it may be expected that the above fluctuations of the overall number of new immigrants directly affect the values of the suggested indicator of sustainable population growth(the MB/NG index). In order to reduce the impact of the annual changes in the rate of foreign immigration on the outcome of the analysis, initial immigration data for each settlement were normalized. For this purpose, the average annual immigration to the country for the entire time-span in question (42,000 new immigrants) was assumed as the conditional baseline.

Lastly, the effect of some of the factors described above is normally time-lagged, since it is the perception of reality, rather the actual conditions, which affects the decision making process of individuals. This is particularly true if the relevant information is not easily available, e.g. by first hand knowledge. Potential migrants, for instance, often become aware of changes in conditions at their prospective destination long after they have actually occurred. Therefore, the values of some of the explanatory variables (housing, population size, employment, employment growth, and services) were lagged by two years in order to reflect this process.

5.7
Research Results

The following subsection introduces the models suggested to explain the degree of sustainability exhibited by urban settlements selected for the present case study in their population growth (the MB/NG indicator) and economic development (private construction and unemployment)

5.7.1
Population Growth

The results of the analysis of variance are shown in Table 5.1. The table shows that the dominant factors affecting the structure of population growth ($F>3.5$, $P<0.05$) are, as expected, housing construction, and the distance from main urban centers, i.e. the remoteness of the community in question. The effect of the second spatial factor, spatial isolation, appears to be relatively insignificant. This indicates that while there may be opportunity for interaction between neighboring urban centers, few social, educational or economic exchanges actually occur. There is thus insufficient evidence to support the hypothesis that grouping a number of small urban centers may, by itself, improve their chances of achieving sustainable population growth.

Table 5.1. Factors influencing sustainability of population growth of selected urban settlements (analysis-of-variance)

Factor (see text for explanations)[a]	F-Ratio	F-Probability
Climate	3.272	0.073
Employment	0.834	0.478
Employment growth	0.023	0.879
Housing	17.904	0.000
Index of isolation	0.501	0.480
Remoteness	3.876	0.023
Population size	0.367	0.694
Regional center	0.084	0.733
Services	2.911	0.037
Combined effect	5.006	0.000
Mean Square (explained)	13.797	
Mean Square (residual)	2.756	
No of obs.	180	

[a] Factors' initial values were logarithmically transformed to comply with the homogeneity-of-variance assumption.

Contrary to our initial theoretical assumption (see Subsection 5.5), *population size* did not appear to have a statistically significant effect on population growth of the urban settlements investigated in our study. This unexpected result may have two possible explanations. First, one should keep in mind that the population size of urban localities selected for the present case study was restricted to a range of 5,000-40,000 residents. Within this restricted range, the effect of population size may indeed be less profound than within a wider range, say up to a size of 250,000 residents. In addition, as mentioned above, diversity in employment and living conditions is an important precondition for sustaining the population growth of urban localities. It is possible, thus, to suggest that this condition can hardly be met by individual settlements of the above population range, particularly in peripheral areas in which overall patterns of urbanization are scattered.

In spite of the apparent lack of evidence in this study for a relationship between the spatial isolation of communities and their population growth, the theoretical importance of this factor led us to investigate the combined effect of this factor with that of remoteness. The justification for developing such an indicator may be explained as follows: In centrally located settlements, the presence of neighboring communities of similar size does not increase significantly the potential for intra-regional contact, because social and economic life is dominated by the metropolitan core. However, in peripheral communities, the lack of a dominant urban center should result in more links between the smaller communities. The combined effect of remoteness and isolation may thus be represented in a single index accounting for their inter-relationship.

The index in question can conditionally be named the *index of clustering* (IC). It represents a derived indicator measuring the two separate spatial parameters of urban development described above, and may be expressed as a simple ratio (Fig. 5.2):

$$IC = IS / IR,$$

where *IS* and *IR* are respectively *spatial isolation* (the number of other urban settlements located within a practical range for daily commuting, assumed here to be 20 km) and *index of remoteness* (aerial distance from a settlement to the closest major urban center, in kilometers). The derived indicator thus tends to have a high value in central, densely populated areas, where distances from metropolitan centers are small and the urban field is dense, while its values tend to be lower in remote peripheral areas in which urban centers are more scattered.

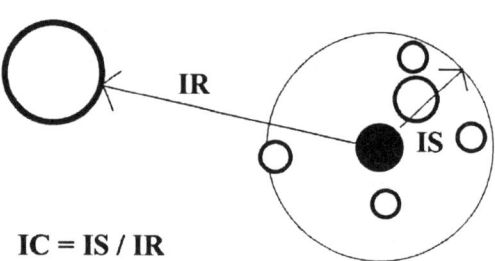

Fig. 5.2. Spatial components of the index of clustering

The impact of this index on the patterns of settlements' growth is expected to be twofold:

- In *sparsely populated peripheral areas*, the presence of small, neighboring urban communities may increase their chances to sustain their population growth and economic development due to socio-economic interaction.
- In *core areas*, where social and economic life is dominated by a major metropolitan center, dense clusters of small urban localities may reduce the rate of migration to a given settlement and its relative attractiveness to investors due to "inter-town competition."

One reservation is required: The index in question does not differentiate between, for example, the two following situations: A town with five other urban settlements within commuting distance, located 25 kilometers from a metropolitan center; and another town, located 50 kilometers from a major urban center, but which has ten other towns within commuting distance. In both cases, the proposed index of clustering would have a value of 0.2, despite the fact that these two cases are not identical with respect to their development patterns. However, the identical values of this index do suggest that with respect to the sustainability of population growth and economic development these settlements may exhibit some similarities.

The integral effect of remoteness and isolation, measured by the index of clustering (IC) on the sustainability of population growth in small urban communities was investigated by repeating the analysis of variance presented in the previous section, changing only the representation of the spatial parameters so that the combined index replaced the separate measures. Table 5.2 shows the results of the modified analysis.

Table 5.2. Modified analysis of variance showing the effect of the index of clustering on the sustainability of population growth of selected urban settlements

Factor (see text for explanations)[a]	F-Ratio	F-Probability
Climate	6.329	0.013
Employment	1.268	0.288
Employment growth	0.106	0.745
Housing	18.411	0.000
Index of clustering	6.552	0.012
Population size	0.180	0.835
Regional center	0.002	0.967
Services	2.762	0.045
Combined effect	5.640	0.000
Mean Square (explained)	15.478	
Mean Square (residual)	2.744	
No of obs.	180	

[a] Factors' initial values were logarithmically transformed to comply with the homogeneity-of-variance assumption.

Three factors appear to be highly significant statistically ($F>6.0$, $P\leq0.01$) in influencing the ratio of MB/NG, which is taken as the indicator of sustainable population growth. These are *housing, climatic harshness* and the *index of clustering*. While the effect of housing on inter-urban migration is not surprising (see *inter alia* Lipshitz 1997; Portnov and Pearlmutter 1997; Portnov 1998), the two other factors - *climatic harshness* and the *index of clustering* - deserve some analysis, and will be discussed below, in some detail.

5.7.2
Index of Clustering

As we may recall, this indicator combines two intrinsic spatial parameters of urban development: *remoteness* from major urban centers of the country, and density of the urban field in the area (*isolation*). The relationship between the suggested measure of sustainable population growth, the ratio MB/NG, and that of the settlement location, as described by the *index of clustering*, is shown in Fig. 5.3. (In order to emphasize the trends occurring in peripheral areas of the country, the diagram displays only the settlements included in the appropriate data sets, Sets 2 and 3).

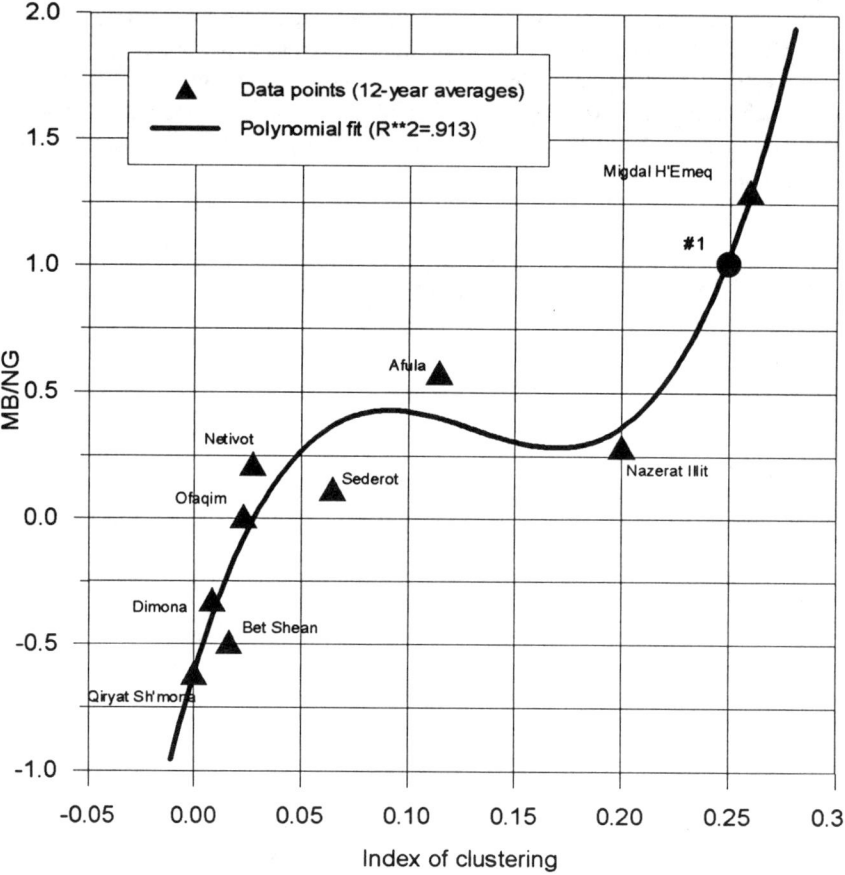

Fig. 5.3. Sustainability of population growth (the MB/NG indicator) as function of peripheral settlements' location (Index of Clustering)

Each symbol in the diagram represents twelve-year averages computed for the selected peripheral localities (Sets 2-3). The filled dot (#1) corresponds to the point beyond which migration balance (MB) tends to become the predominant source of a settlement's population growth.

The graph indicates that the sustainability of the population growth of urban settlements in peripheral areas of the country tends to increase with increasing values of the index of clustering. When the value of this index exceeds 0.25, migration balance in a community becomes the major component of its population growth. In other words, if the distance from an urban settlement to the closest urban center of the country equals, for example, 50 km, approximately 12 other urban localities within a commuting range appear to be needed to make the population growth of the settlement sustainable (50 x 0.25 = 12.5). Keeping in mind that only settlements whose population exceeds 10,000 residents were included in the analysis, we can tentatively identify the *minimal* population size of the above settlement cluster: 12 x 10,000 = 120,000 residents, or 130,000 residents including the settlement at hand. Analogously, the growth of a similar cluster of settlements 100 km away from the closest urban center may become sustainable once the total population exceeds 240,000-260,000 residents.

The relatively complex (step-like) shape of the above trend line (see Fig. 5.3) also deserves a comment. To understand why two inflection points (those corresponding to 0.10 and 0.20 on the index of clustering axis) occurred, some attendant socio-economic changes associated with growing density of the urban "field" should be taken into consideration. As the population size of a settlement cluster grows, it may eventually cross the threshold of a higher level of socio-economic development. In particular, this cluster might include a relatively large retail market, transportation center, developed medical and sport facilities, and a complex job market that a single town or a smaller settlement cluster can not in most cases sustain. Upon reaching this level of socio-economic development, no qualitative changes may occur until the cluster surpasses a higher threshold which allows it to sustain more complex urban functions such as, for instance, an airport, a large recreation center, theater, etc. This stepped form of urban and regional development was, in fact, described in the early 1960's by Doxiadis (1977) in his *Ekistic* concept. The trend in question is also in agreement with Christaller's hierarchy of urban places, which assumes "natural breaks" between settlements of different size (Clark 1982).

Alternatively, the "step-like" form of the curve may be a result of the spatial relationships between the towns in the specific sample, and not a general phenomenon. The two settlements which lie near the above infection points - *Afula* and *Nazerat Illit* – (Fig. 5.3), are both part of Group 2 in the sample; Both are close to Haifa, which is the largest urban center of the region. And, they are also close to another fairly big town – Nazareth (54,000 residents, as of 1994). This particular spatial arrangement may thus exert a great influence on the population growth of the localities in question due to the above-mentioned "inter-city competition" for potential migrants. Further studies in other countries may thus be required to establish the typical form of the curve, and to determine the range of values the index of clustering may have in order to produce sustainable population growth in a region.

5.7.3
Climatic Harshness

The statistical significance of *climatic harshness* as a factor affecting sustainable population growth is quite unexpected. None of the urban communities studied suffer from extremely harsh climatic conditions, compared with the climates of hot, dry deserts in other countries. Our previous analyses of urban growth in Israel (see Chaps. 2 and 3 of this book, as well as Portnov and Pearlmutter 1997; Portnov and Erell 1997; Portnov 1998) did not indicate that this factor had a major influence on rates of development or population growth in various geographic areas of the country. While this finding should be treated with caution, pending a more extensive study focused on this issue, the following explanations are suggested.

Previous studies by the authors were based on statistical data for the most recent period of urban growth (1988-1994) while the present analysis deals with a 25-year time span (1970-1994). In the early 1970's - the 1980's, development towns included in the analysis were in the early phases of their urbanization. The lack of vegetation coupled with poorly developed public urban space at the time, contributed to an unappealing image which deterred possible migrants. While the actual meteorological conditions in some of the settlements are in fact overheated for some of the day during the summer months, the perception of these communities as being hot, sometimes dusty places is perhaps more important. The effect of adverse climate may therefore have been very real.

The situation has been slowly changing in recent years: Rising income levels in Israel, accompanied by an improving standard of living, led to the widespread installation of air-conditioning in offices as well as in residential buildings, beginning in the late 1980's. A similar process heralded the development of large urban centers in the hot areas of the south and southwestern United States. The ability to provide thermal comfort, at least in building interiors, is thus a great leveling force, allowing the development of regions previously considered too inhospitable for human habitation. At about the same time, increasing affluence also led to more attention being paid to the condition of outdoor space, both urban and private. More resources were devoted to landscaping, with the result that previously dusty and unappealing "leftover space" was transformed into gardens, parks, and landscaped pedestrian zones. It is therefore possible that climatic differences between geographic areas have in fact had an influence on the development and population growth rates of small urban communities in Israel over the past two or three decades. However, the effect of climate on future urban development in Israel is expected to decline in importance.

5.8
Economic Development

The results of the analysis of the factors affecting the rate of private construction in urban settlement are represented in Table 5.3.

Only one factor - *employment growth* - appears to have a statistically significant effect on the rate of private construction in the settlements ($F=6.0$; $P<0.05$). This is not surprising, since the growth of employment in a settlement implies increasing affluence of the local population. On the other hand, the relatively low statistical significance of the index of clustering ($F=3.3$; $P<0.1$) clearly contradicts our initial assumption. Nevertheless, since the absence of any statistically significant relationships between the location of urban settlements and their attractiveness to private developers seemed to be unlikely, the relatively low F-value was attributed to the non-linear relationship between the factors.

Table 5.3. Factors influencing the rate of private construction in selected urban settlements (analysis-of-variance)[a]

Factor (see text for explanations)	F-Ratio	F-Probability
Climate	2.614	0.130
Employment	0.055	0.814
Employment growth	6.019	0.016
Index of clustering	3.310	0.073
Population size	2.243	0.138
Regional center	0.493	0.485
Services	1.364	0.260
Combined effect	2.614	0.011
No of obs.	180	

[a] Initial values of all explanatory variables and those of dependent variable were logarithmically transformed to comply with the homogeneity-of-variance assumption and improve fit.

As Fig. 5.4 shows, this assumption appears to be justified. The rate of private construction in urban settlements initially grows as clustering of the urban field increases. Then, upon reaching a certain threshold (IC = 1.0-1.5), the rate of private construction starts to decline. We may recall that the ∩-shaped functional relationship between the index of clustering (IC) and the socio-economic development of urban communities was, in fact, predicted in the previous discussion. It was assumed, in particular, that an increase in urban clustering may have a positive effect on the socio-economic development of peripheral localities, while in densely populated core areas, the dense patterns of urbanization may have an adverse effect on the socio- economic development of individual urban communities due to "inter-town competition" for potential migrants and private developers.

As Table 5.4 shows, the effect of the index of clustering on the *unemployment* rate in small urban settlements appears to be significant. The nature of this relationship is further clarified by Fig. 5.5 which shows that the rate of unemployment clearly tends to decline with increasing values of the index of clustering.

This and the previously discussed trend (clustering vs. private construction) naturally lead to the following question: *Why does the index of clustering appear to have such a pronounced effect on economic development of the localities?*

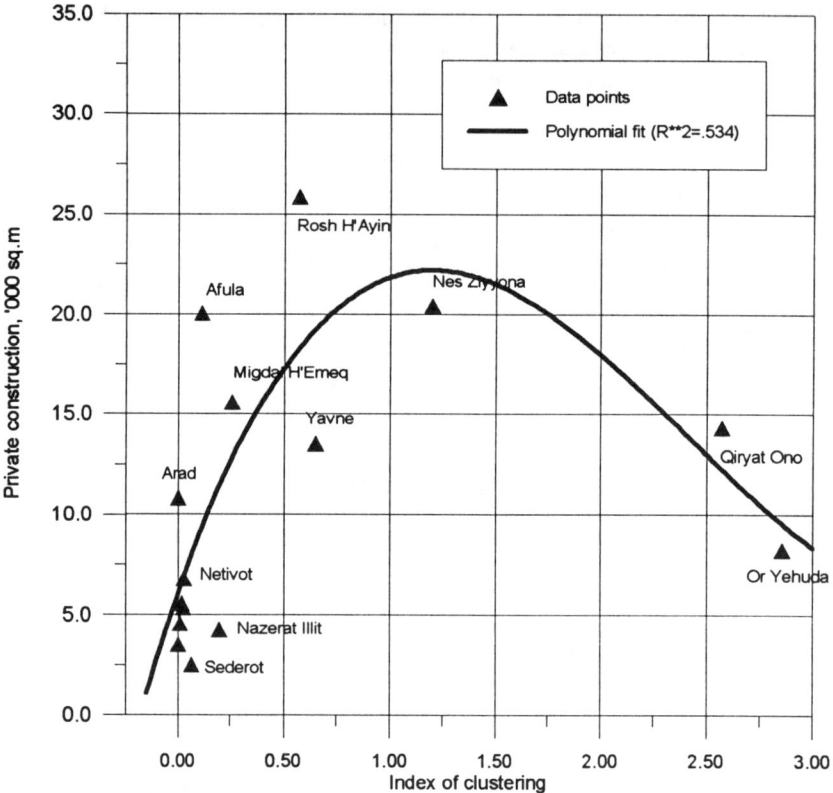

Fig. 5.4. The effect of the clustering of the urban field on the rate of private construction in selected urban localities

Note: Each symbol represents twelve-year running averages computed for respective urban settlements

This question was, in fact, also addressed in the previous discussion. The considerable distances often found between the established urban localities in peripheral areas are likely to cause a shortage of joint intra-regional educational and recreational structures, and to limit the choice of job opportunities. Thus, for instance, employees of an industrial enterprise shut down in a central, densely populated, district often find similar jobs in the same city or in adjacent urban settlements without changing their current place of residence. Concurrently, employees of a similar company in a small, peripheral community may have to leave the district for another area where similar employment is available. The combined effect of remoteness and isolation, reflected in the index of clustering appears, therefore, to be a valuable indication of the intra-regional economic and social interaction in peripheral areas.

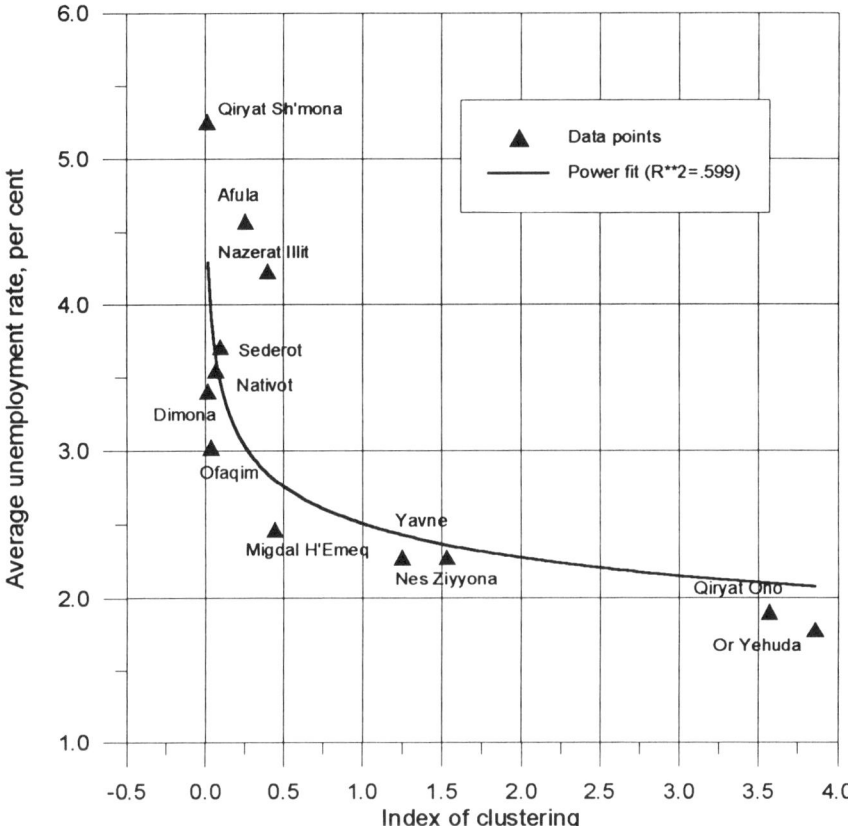

Fig. 5.5. The clustering of the urban field vs. unemployment rates in selected localities

Table 5.4. Factors influencing the rate of unemployment in selected urban localities

Factor (see text for explanations)[a]	F-Ratio	F-Probability
Climate	0.001	0.974
Housing	3.332	0.022
Index of clustering	4.181	0.043
Population size	7.399	0.001
Regional center	9.388	0.003
Services	0.305	0.822
Combined effect	5.959	0.000
No of obs.	180	

[a] Factors' initial values are logarithmically transformed to comply with the homogeneity of variance assumption

5.9
Conclusions and Policy Implications

There are a broad variety of regional policies aimed at redirecting population growth from overpopulated core regions to underdeveloped peripheral areas. Such development policies are advocated in Northern Europe (Sweden and Norway), Asia (Japan), and in the Middle East (Israel and Egypt), as well as in other countries.

Following such a policy, urban & regional planning in Israel has been for many years aimed at population dispersion from the densely populated urban center of the country along the Mediterranean Sea to its underpopulated hinterland areas. As mentioned in Chaps. 2 and 3 of this book, this goal was pursued mainly by means of the development of a broad network of new towns in the peripheral districts of the country.

The present analysis suggests an alternative strategy for the development of sparsely populated peripheral regions, wherever this objective is desirable. According to this view, policy should be directed at achieving a certain density of the urban field, as determined by the value of the proposed index of clustering. Possible approaches to implementing this policy were already mentioned in Chap. 2

These include:

1. *Development clusters with a clearly expressed urban core,* which seems to be relevant to peripheral areas which already have existing regional centers represented by relatively big urban localities;
2. *Development clusters of small urban settlements having no dominant urban core* (this pattern of urbanization seems to be relevant to hinterland areas whose current settlement patterns are less intensive, and where existing small settlements are widely scattered across the area).

The present analysis (see Fig. 5.3) leads to the conclusion that the population size of these development units may vary (with allowance for the distance from the closest metropolitan center) from 120,000 to 250,000 residents (Fig. 5.6).

An unexpected finding of this study was the effect of unfavorable climatic conditions on the sustainability of population growth in relatively recently established urban settlements. The lessons for urban and regional planners elsewhere should be that while the means exist to overcome extreme climatic conditions, an attempt should be made to select appropriate locations for new cities. Urban planning should also be concerned with creating an urban fabric that is responsive to the surrounding environment from the very first stages in the development of the new settlement, in order to reduce the adverse effects of the surroundings.

Although the concrete findings of this study - the minimal size of a settlement cluster and the intensity of the factors influencing sustainable population growth of urban localities - are specific to the conditions in Israel at this time, the mode of analysis and its applications for planning policy may be applicable to regional and urban planning elsewhere.

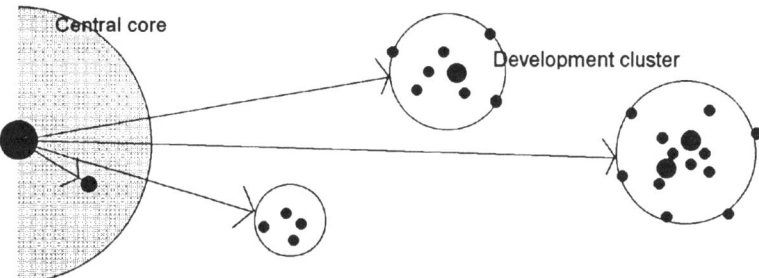

Fig. 5.6. The diagram illustrating the increase in the size of settlement cluster as a function of the area's distance from the major population centers of the country

In particular, the strong interrelationship found between the *clustering of the urban field* and the *sustainability of population growth and economic development in peripheral areas* is probably characteristic of development processes in many countries. The application of this insight may be of value to planners and decision-makers in any country that experiences acute problems of inter-regional inequalities in socio-economic and population growth.

References

Armstrong H, Taylor J (1993) Regional economics and policy. Harvester, New York, 2nd edition
Bitan A, Rubin S (1991) Climatic atlas of Israel for physical and environmental planning and design. Ramot Publishing Co, Tel-Aviv
Bourne LS (1975) Urban systems: strategies for regulation. A comparison of policies in Britain, Sweden, Australia and Canada. Clarendon Press, Oxford
Clark C (1982) Regional and urban location. St.Martin's Press, New York
Clawson M, Hall P (1973) Planning and urban growth: an Anglo-American comparison. The Jonhs Hopkins University Press, Baltimore and London
De Jong GF, Fawcett JT (1981) Motivation for migration: an assessment and a value-expectancy research model. In: De Jong GF, Gardner RW (eds) Migration decision making. multidisciplinary approaches to microlevel studies in developed and developing countries. Pergamon Press, London, pp 13-53
Diamond DR, Spence NA (1983) Regional policy evaluation. A methodological review and the Scottish example. Aldershot, Gover
Doxiadis K (1977) Ecology and ekistics. Westview Press, Boulder
Fischer CS (1976) The urban experience. Harcourt Brace Jovanovich, Inc, New York
Henderson RA (1980) The location of immigrant industry within a U.K. assisted area: the Scottish experience, Progress in Planning 14(2).
ICBS (1996) Statistical abstract of Israel. Israeli Central Bureau of Statistics, Jerusalem

Kirschenbaum A, Comay Y (1974) Dynamics of population attraction to new towns - the case of Israel. In: Pruschansky J (ed) Dialogue in development - natural and human resources. Proceedings of the 3rd World Congress of Engineers and Architects, Jerusalem

Kneese AV (1978) The economic and economically related aspects of new towns in arid areas. In: Golany G (ed) Urban planning for arid zones. John Wiley & Sons, New York, pp 123-138

Levy JM (1985) Urban and metropolitan economics. Mcgraw-Hill Book Company, New York

Lipshitz G (1997) Immigrants from the Former Soviet Union in the Israeli housing market: spatial aspects of supply and demand. Urban Studies 34(3): 471-488

Moore EG, Rosenberg MW (1995) Modeling migration flows of immigrants groups in Canada. Environment and Planning A 27: 699-714

Newman D, Gradus Y, Levinson E (1995) The impact of mass immigration on urban settlements in the Negev 1989-1991. Working Paper Series No 3. The Negev Center for Regional Development, Be'er Sheva

Portnov BA (1998) The effect of housing on migration in Israel. J of Population Economic, 11(3):379-394

Portnov BA, Erell E (1998) Long-term development peculiarities of peripheral desert settlements: the case of Israel. Int J of Urban and Regional Research 22(2):216-232

Portnov BA, Erell E, Pearlmutter D (1997) Development peculiarities of urban settlements in the Negev: cross-regional analysis. Working Paper Series No 9. The Negev Center for Regional Development, Be'er Sheva

Portnov BA, Pearlmutter D (1997) Sustainability of population growth: a case study of urban settlement in Israel. Review of Urban and Regional Development Studies 9(2): 129-145

Smith WF (1975) Urban development: the process and the problems. University of California Press, Berkeley

Wong C (1995) Developing quantitative indicators for urban and regional policy analysis. In: Hambleton R, Thomas H (eds) Urban policy evaluation: challenge and change. Paul Chapman Publishing Ltd, Cardiff, pp 111-122

Appendix
A test of the relation between the settlement sample and the population

Variable	Population		Sample		Z-Score
	Mean	Std Dev	Mean	Std Dev	
Employment	0.46	0.38	0.57	0.39	1.10
Index of clustering	0.67	0.83	0.74	1.09	0.25
Index of isolation	9.55	6.98	9.73	8.88	0.08
Housing	15.16	11.43	13.20	9.46	0.80
Population	23.26	7.57	23.15	6.81	0.06
Remoteness	27.82	21.04	26.80	17.77	1.02

6 Modeling the Migration Attractiveness of a Region

Boris A. Portnov
J.Blaustein Institute for Desert Research, Ben-Gurion University of the Negev, Sede-Boker Campus, 84990, Israel

6.1
Introduction

If a region exhibits neither a significant influx of migrants nor an outflow of its current residents, this state of migration can be defined as *neutral*. Knowing the country-specific preconditions for an area's *"migration neutrality"* can help planners and decision-makers to determine regional development policies aimed at a more balanced distribution of a country's population. This may be achieved by both encouraging an influx of migrants into peripheral desert areas, and preventing further concentration of migrants in overpopulated core regions.

In this chapter, interregional migration is considered as a function of the *interrelationship between employment growth and housing availability in a region*. It is argued that as long as employment and housing are in sync, there is little change in net migration. Migration occurs when, because of a scarcity of land, a large influx of immigrants, economic conditions, or a governmental policy, employment and housing are not in balance.

This thesis will be examined by studying the patterns of cross-area migration in two relatively small and densely populated countries – Israel and Japan. The analysis of interregional migration in these countries will attempt to answers the following questions:

1. Is there a general "mechanism" through which physical disparities in regional development affect the patterns of cross-district migration?
2. Which aspects of regional inequalities (climate, employment, housing availability, etc.) have the most profound effect on the rates and direction of inter-area migration?
3. Which planning policies and strategies are conducive to increasing the migration attractiveness of peripheral development regions?

The chapter begins with a brief review of factors and forces affecting cross-district migration, followed by an overview of the general patterns of regional development and migration in Israel and Japan. A general model of the factors affecting cross-district migration is suggested, and regression analysis is used to explain the factors influencing the rate of interregional migration in the two countries. In the concluding sections, the empirical models of *"neutral mi-*

gration" are proposed to determine a specific balance of employment and housing which is conducive to migration attractiveness of a region.

6.2
Employment and Housing Factors of Interregional Migration

In the general literature on migration, employment and housing are considered as essential components of a *"push-pull continuum"* (George 1970) affecting migration decision-making. This continuum distinguishes between two types of migrations: 1) migrations caused by necessity or obligations, and 2) moves stemmed from socio-economic and cultural needs. While the first type of migration incentives largely relates to area-of-origin political and/or religion (*push*) factors, the second group of motives is assumed to be determined by economic hardships of the area of origin (*push factors*) and economic opportunities of the area of destination (*pull factors*).

De Jong and Fawcett (1981) single out six distinctive motives for migration. These include: a) economic motive (employment and earnings), b) social mobility/social status motive, c) residential satisfaction (housing) motive, d) motive to maintain community-based social and economic ties, e) family and friend affiliation motive, and f) motive of attaining life-style preferences. In this classification, employment-related factors (availability of employment in the area, employment growth, inter-area wage differentials) are considered to be the strongest motives for interregional mobility, as well as affecting the choice of immigrants concerning their first place of residence in the country of their destination.

Friedmann (1973) suggests an alternative classification of migration variables based on his popular core-periphery paradigm: 1) employment opportunities; 2) accessibility (i.e. distance between the sources and spatial targets of migration); 3) the number and intensity of prior contacts with the migrant's place of destination, and 4) the educational level of the population at points of migrant origin. This classification also considers employment opportunities as a major factor of long-distance migration, and suggests that other "urban attributes" (social services, opportunities for education and housing) are simply concomitant with the availability of employment in a region.

The predominant role of employment-related factors in interregional migration is also emphasized by a neo-classical theory of regional development. According to this theory (Perroux 1983; Richardson 1977), inter-area migration is a "readjustment process" that tends to reduce disequilibrium within local labor markets. As Richardson (1977:90) notes, "migrants tend to move from low-wage to high-wage areas and from areas of labor surplus to those with labor shortages." This point of view is also expressed in some more recent migration studies (Greenwood and McDowell 1991; Lipshitz 1992; Michel et al. 1996).

Armstrong and Taylor (1993) however, strongly criticize this thesis by arguing that labor is not perfectly mobile and does not respond readily to regional differences in either wage rates or unemployment, which are only two of the many influences acting upon the geographic movements of workers. Armstrong and Taylor (*ibid.*) also suggest that the migration of labor does, in fact, increase regional disparities, rather than decrease them, since it causes the outflow of skilled and educated migrants from economically depressed areas.

In empirical studies of interregional migration (LaLonde and Topel 1997; Moore and Rosenberg 1995; Lipshitz 1997), the effect of employment on interregional migration is, nevertheless, widely acknowledged. It is argued, in particular, that high unemployment rates and low wages tend to discourage migrants from settling in a given area or settlement, while encouraging local residents to migrate to other locations.

In a number of empirical studies on interregional migration, the *availability of housing* is also considered an important factor affecting the rates and direction of migration (Lipshitz, 1997; Burnley et al. 1997; Portnov and Pearlmutter 1997; Portnov 1998). A recent survey of in-country migrants in Australia (Burnley et al. 1997) demonstrates that the desire to own an affordable home is the main reason for migration from Sidney to its periphery.

In his study of spatial distribution of foreign immigrants in Israel, Lipshitz (1997) concluded that government housing construction in the peripheral districts of the country and fairly low housing prices prompted some categories of new immigrants to move to the country's peripheral regions despite a lack of employment in these areas.

Another study of population migration in Israel (Portnov and Pearlmutter 1997) also argues that in the wake of mass immigration of 1990-91, the population of the major metropolitan centers of the country tended to move to urban localities of smaller size in which housing is more available and affordable.

The effect of housing availability and affordability on migration is also found in Japan. As Abe (1996) points out, high land and housing prices in the core encourage people to settle in local towns outside the major metropolitan areas and to commute daily for work in the central cities.

In their analysis of long-distance migration in Canada, Moore and Rosenberg (1995) report that the percentage change in the size of housing stock in the area of destination is one of the most statistically significant factors influencing the migration choices of both foreign immigrants and Canadian-born migrants.

6.3
Modeling the Migration Behavior

The neo-classical *human capital model* (Borjas 1989; Poot 1996) represents a common approach to modeling long-distance migration. This model suggests that individuals and families compare the present value of earnings at the present location with that at alternative locations and the costs of move. Richardson (1977) justly argues, however, that this cost-benefit approach appears to be oversimplified. In supporting this conclusion, he points out that this model rests upon several critical assumptions (homogenous labor, constant returns to scale, perfectly competitive labor markets, and absence of other, non-economic migration motives), which appear to be in many cases clearly unjustified.

De Jong and Fawcett's (1981:47-48) *value-expectancy model* attempts to explain inter-area migration as a function of a broad array of both economic and non-economic (environmental, cultural and social) factors. This model, however, requires a precise specification of the *personally valued goals* that "might be met by moving...and an assessment of the perceived linkage, in terms of expectancy, between migration behavior and the attainment of goals in alternative locations." Although this model may provide a perfect explanation of the behavior of individual migrants, its ability to explain more general migration patterns appears to be considerably restricted.

Numerous empirical models of inter-urban and cross-district migration can also be found in the recent literature on migration decision-making (see *inter alia* Michel et al. 1996; Portnov 1998; Portnov and Pearlmutter 1997; Moore and Rosenberg 1995; Greenwood and Stock 1990). These empirical models cover a wide range of migration variables (*viz.* employment, housing, the level of urbanization, and climatic differences between geographic areas), and commonly employ multiple regression analysis as an analytical tool. While this technique often provides a good fit for specific statistical data at hand, it remains unclear, however, whether these models can hold beyond their original "time-area" framework.

6.4
Housing-employment Paradigm of Interregional Migration

A simplified model of the way in which the combination of employment and housing factors may affect the patterns of interregional (in-country) migration is represented in Fig. 6.1.

Suppose that labor is required in the area. This may be obtained in three ways: a) by employing commuters from other localities; b) by employing current residents of the area (either currently unemployed, or workers from other industries in a settlement), or c) by attracting migrants from elsewhere (either residents of other districts or foreign immigrants).

The arrival of newcomers from other areas boosts the demand for housing. The response of the local market to this demand may, however, be time-lagged or inadequate if the local resources are not sufficient. This is often the case if land for development is not readily available in, for instance, overpopulated core areas. The shortage of land for development may, in turn, increase housing prices and cause an outflow of current residents who cannot afford decent housing in the area to areas where housing is more available and affordable (see *inter alia* Portnov and Pearlmutter 1997). High housing prices may also hamper the influx of migrants to the region despite the availability of employment.

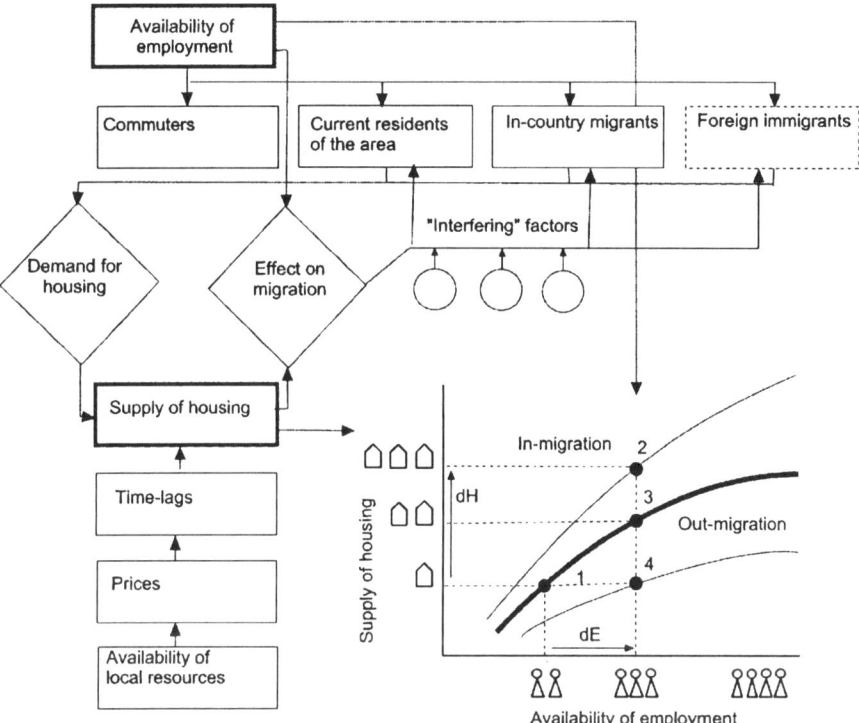

Fig. 6.1. Employment-housing paradigm of interregional migration

The effect of this housing-employment imbalance on migration may be either weakened or strengthened by various "interfering" factors. These may include various local amenities (climate, natural and built environment), macroeconomic performance of the country as a whole, and mass immigration (Milne 1993; Walker et al. 1992; Frey 1995; Fulton et al. 1997; Portnov and Erell 1998).

The chart incorporated in the model (see Fig. 6.1) provides a simplified illustration of this *"employment-housing" link* and its effect on migration. Suppose that a region has an established balance of employment and housing availability and relatively low net migration (Point #1). Now, assume that additional employment opportunities became available in the area (dE), and this employment growth is accompanied by an increase in housing availability (dH). Under these conditions, an influx of migrants to the area can be expected (Point #2).

Although the response of housing market may be lagging or insufficient (Point #3), a decrease in housing availability *may not* affect the existing level of a district's migration attractiveness: Potential migrants and current residents of the area may still tolerate high housing prices in the region as a "necessary charge" for better employment opportunities, compared to other regions.

Now assume that due to, for instance, land shortage, supply lags or a high rate of foreign immigration, the availability of housing in the area dropped beyond this "tolerability" level (Point #4). In this case, "housing-driven" migration from the area may become increasingly likely. This hypothetical threshold of "tolerability," on which Points #1 and #3 are located, can thus be termed the *"neutral migration curve"* (see the bold curved line in the chart): The area beneath this curve corresponds to negative net migration (*out-migration*), while the area, which lies above it, is likely to match the state of predominant *in-migration* to the region (see Fig. 6.1).

6.5
Israel and Japan: General Patterns of Regional Development

The theoretical model discussed in the previous section (Fig. 6.1) can be tested using migration statistics available for two small and densely populated countries – Israel and Japan. Although these country differ substantially in respect to a number of general development indicators – land area, population size, the overall rate of population growth, and demographic make up (see Table 6.1), – they share a number of similarities which are essential for the present analysis. In both countries, the overall density of population is considerably high (250 persons per km^2 in Israel and 340 persons per km^2 in Japan). In addition, the core areas of these countries are among the most densely populated areas of the world. In the Tokyo prefecture of Japan, for instance, the population density approaches 5,400 people per km^2, while in the Tel Aviv district of Israel, the overall density of population exceeds 6,700 residents per km^2 (ICBS 1997; JSB 1997). Since it is expected that the aforementioned effect of housing-employment imbalance on migration should presumably occur if the local *land resources are not sufficient to accommodate an additional demand for new affordable housing*, the two countries can be considered good examples for the present case studies.

The restrictions on data availability should also be taken into consideration. Although high population densities and regional inequalities in population distribution can be found elsewhere, in both developed and less developed countries, the list of major development series monitored by the national statistics in Israel and Japan (ICBS 1997; JSB 1997) exhibits many similarities that are extremely important for the present comparative analysis.

Table 6.1. Selected socio-economic indicators for Israel and Japan

Indicator	Israel	Japan
Population size [1,000]	5,500	125,500
Land area [km^2]	21,946	374,744
Overall population density [per km^2]	246	335
Population density in the core[a] [per km^2]	6717	5384
Overall population growth rate [%]	2.11	0.21
GDP [$US per capita]	15,500	21,300
Percent of population of 65 years and over	10.0	15.0
Telephones [per 1,000 residents]	470	510
Maximum and minimum rates of interregional migration in 1985-95 [per cent]	-0.91/5.87	-0.38/0.46
Interregional inequalities:[b]		
– population density	1.545	1.002
– availability of employment	0.360	0.914
– housing construction	0.645	0.901

Sources: CIA 1997; ICBS 1997 and JSB 1997. [a] The Tokyo prefecture in Japan and the Tel Aviv district in Israel; [b] Williamson's unweighted inequality index (for more details see Williamson 1975)

6.5.1
Patterns of Urbanization in Israel and Recent Development Policies

As mentioned in Chapter 3 of this book, the majority of urban settlements in Israel are concentrated along the Mediterranean coast, or in close proximity to its major metropolitan centers - *Tel Aviv, Jerusalem,* and *Haifa*. The overall population of these urban centers along with their immediate hinterland (the Tel Aviv, Central, Jerusalem, and Haifa districts) amounts to nearly 70% of the country's population (ICBS, 1997). Realizing the extent of the problems, posed by over-concentration of the country's population in few densely populated urban regions, government and planning officials in Israel have favored at various times the implementation of a policy promoting dispersal of the population to the peripheral areas of the country. From the early 1950's to late 1960's, this policy was primarily sustained by directing new immigrants to sparsely populated areas of the country, while since the early 1970s this policy was gradually replaced by geographically selective "aid programs" (for more details on this policy and its implications for the country's regional development see Chapter 7 of this volume).

6.5.2
Current Issues of Regional and Urban Development in Japan

The core area of Japan's industrial development is concentrated in three highly urbanized clusters: Tokyo and Yokohama; Kobe, Osaka, Kyoto and Nagoya; Hiroshima, Kitakyushu and Fukuoka (see Fig. 6.3). Population density within these urban areas often exceeds 10,000-12,000 people per km^2 (JSB 1997). As Vining (1982) argues, one reason the population density is so high is that almost all the flat land for building in Japan is on its eastern coast. The only substantial areas of available flat land are on the northern island of Hokkaido, which has a severe climate with harsh winters.

Until the early 1970s, urban development in Japan was characterized by large-scale "job-driven" migration from local regions to three metropolitan areas – the Tokyo, Osaka and Nagoya regions (Abe 1996). Since the first oil crisis of 1973 to the late 1980s, the return migration from metropolitan areas to the local regions reduced significantly. From the end of the 1980s and into the 1990s, net migration into the Tokyo metropolitan region again increased, and the overcrowding of population and economic activity in Tokyo remains one of the most serious regional problems in Japan (Abe 1996; Markusen 1996).

In light of ongoing concentration of population and economic activity in few metropolitan centers, three major regional development policies were adopted by the government: 1) the construction of a national network of transportation and communication to form "national development axes" along the eastern and western coasts; 2) the creation of growth poles in local regions; and 3) the relocation of industries from the metropolitan regions to the peripheral areas (Abe 1996).

6.6
General Patterns of Interregional Migration

Although Israel and Japan differ substantially in their population size and development patterns, the general patterns of inter-area migration in these two countries exhibit, as we shall see in this section, considerable similarities.

6.6.1
Israel

Surprisingly, the "skewed" concentration of population and economic activity towards the core districts of Israel (see Section 6.5.1) does not appear to be directly reflected in the patterns of in-country migration. As Fig. 6.2 shows, since the late 1980s, the attractiveness of the country's core areas (the Tel Aviv, Jerusalem, and Haifa districts) to internal migrants tended to decline, while that of the peripheral Southern district appeared to grow.

To explain this phenomenon, a number of factors should be taken into consideration. First, it should be kept in mind that a great portion of the overall population growth in Israel is due to *foreign immigration*. During the recent wave of mass immigration from the former Soviet Union in 1990-91, immigration accounted for as much as *80 percent* of population increase (ICBS 1997).

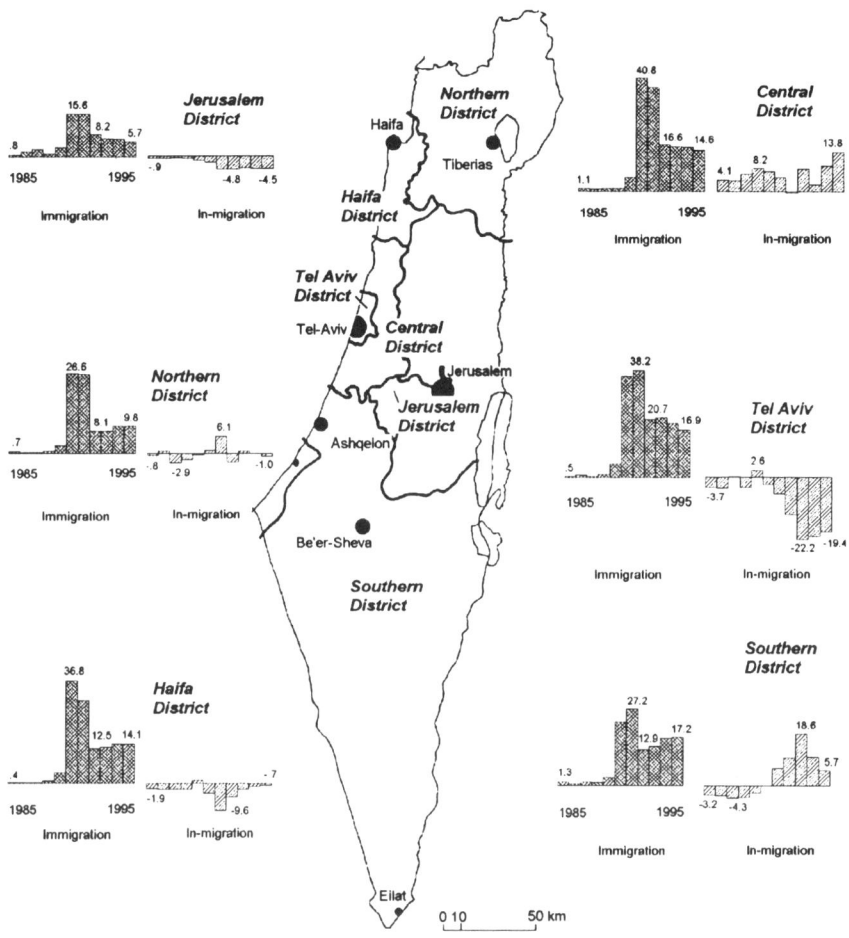

Fig. 6.2. Foreign immigration and in-country migration by district of Israel in 1985-1995, thousands of migrants

As Fig. 6.2 shows, during these years, the overwhelming majority of the new immigrants (NI) settled in the country's central areas -- the *Haifa*, *Tel-Aviv*, and *Central* districts. Concurrently, two out of three of the above central districts (the *Tel-Aviv* and *Haifa*) experienced in these years a substantial loss of their established population due to a negative annual balance of in-country migration (see Fig. 6.2). The influx of *new immigrants* (NI) to a particular district and the outflow of *in-country migrants* (IM) from this district thus appear to be interrelated. For example, the inception of mass migration from the Tel Aviv district, which occurred in 1990-91, clearly coincides with the beginning of the above immigration wave (Fig. 6.2). Even in the Central district, which can be considered as the suburban "buffer" of the Greater Tel Aviv, the balance of internal migration dropped below zero level in 1991 when the district accommodated some 37,000 NI (see Fig. 6.2). Though indirectly, these parallel processes of internal migration and NI's initial distribution may correlate *via* housing prices: The mass influx of new immigrants to a particular district boosts housing demand and increases housing prices. This, through a spiral process, may eventually cause the outflow of current residents who cannot afford decent housing in the metropolitan centers to areas of the country where housing is more widely available and more affordable (Portnov and Pearlmutter 1997).

In the area of *employment,* a similar "push" effect of immigration on internal migrants was noted by a number of North American studies (see among others Fulton et al. 1997; White and Liang 1993; Walker et al. 1992). The aforementioned *"housing competition"* between immigrants and in-country migrants seems, however, to be a specific Israeli phenomenon caused by extremely high immigration rates, government housing subsidies to newcomers, and a shortage of land for new development.

In explaining inter-regional patterns of population growth in Israel, the *relatively small land area* of the country (some 21,500 km^2) should also be taken into account. Relatively short aerial distances between its various regions make it possible to commute on a daily basis across regional borders. Such commuting is, for instance, practicable from the Central district and the Ashqelon sub-district (which is a part of the Southern district) to the core areas of the country – the Tel Aviv and Jerusalem districts – in which the most jobs are available. The relatively high level of migration attractiveness, exhibited by the Central and Southern districts in recent years (see Fig. 6.2), appears to support this assumption.

6.6.2
Japan

General patterns of in-country migration in Japan during the past decade (1985-95) are graphically represented in Fig. 6.3.[1]

This figure shows two general trends. First, similarly to that in Israel, *relative migration attractiveness* of the core districts of Japan (the Kanto, Tokai, and Kinki) in the past decade appeared to decline, while that of peripheral areas (the Hokkaido, Tohoku, and Kyushu) tended to grow. Second, the *absolute overall rate of net migration* in *both core and peripheral districts* of the country tended to decrease.

A number of possible explanations of these trends can be suggested. First of all, as Abe (1996) argues, the relative level *of housing rent* in the core districts of Japan grew in recent years at a considerably high rate due to continuously grow-

---

[1] The data for the analysis were drawn from the annual publications of Japanese Statistics Bureau (JSB), *"Japan. Statistical Yearbook."* Since the disaggregated data on the distribution of foreign immigrants across geographic areas of the country are not available, the diagram represents overall in-country migration rates

ing land prices. At the same time, the relative level of housing rent in virtually all the peripheral areas of the country appeared to drop (JSB, 1997).

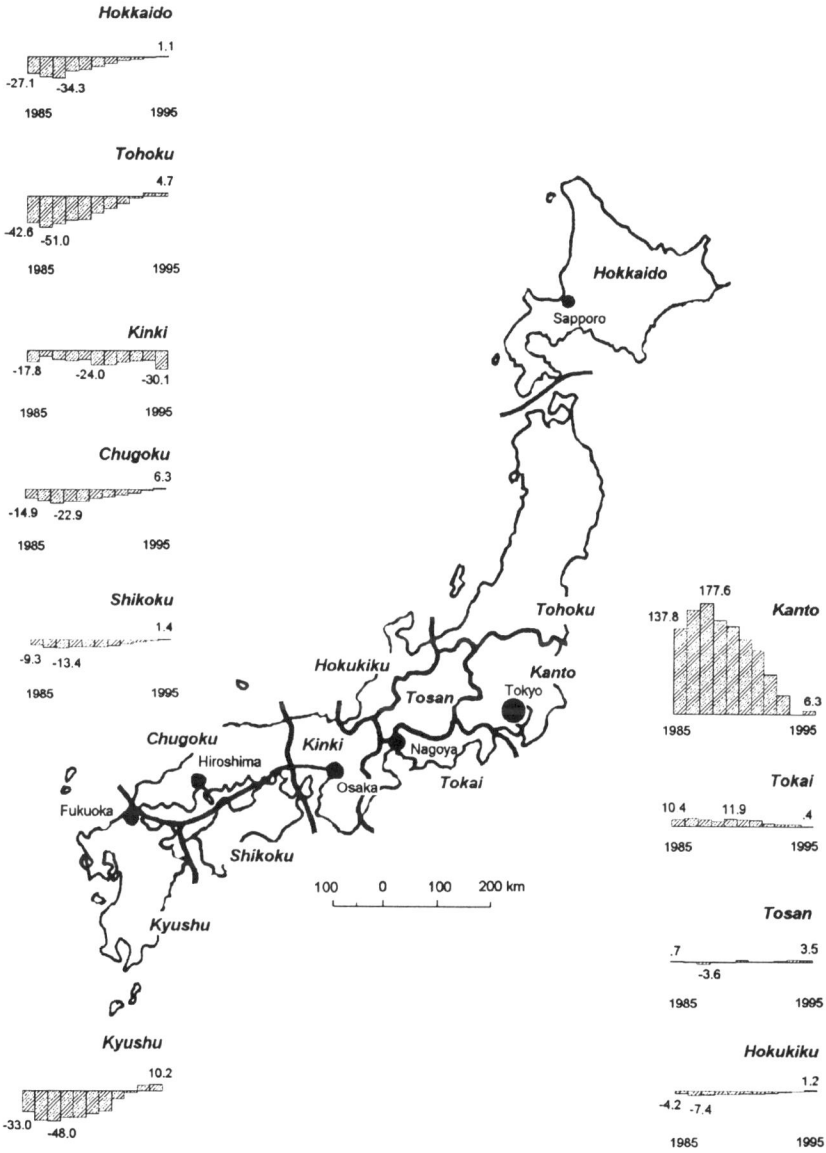

Fig. 6.3. Migration balance by district of Japan in 1985-1995, thousands of migrants

Other influencing factors can also be mentioned. These are the overall slowdown of the Japanese economy (which became apparent after 1991), regional policy of Japanese government aimed at encouraging population growth and industrial development in the periphery, and infrastructure development. As Milne (1993) argues, the business cycle has a significant influence on interregional migration flows and people's propensity to migrate. In particular, the overall rate of in-country migration tends to decrease in "bad" economic years when people prefer to refrain from costly long-distance moves (Portnov and Erell 1997). This explains why the overall rates of net migration by districts of Japan tended to drop. On the other hand, a decline in migration attractiveness of Japan's metropolitan regions can be attributed to the policy of Japanese government aimed at preventing excessive accumulation of population and economic activities in the country's core areas (Abe 1996; Markusen 1996).

Advanced transportation system of Japan may also help to explain the above drop in migration attractiveness of the country's core. Since the area practicable for daily commuting covers considerable distances from the country's core, the population of this area can commute to jobs in the overpopulated core maintaining relatively cheap housing in the periphery.

6.7
Research Method

To identify and measure the significance of the factors affecting the patterns of cross-district migration in the two countries in question, the present analysis was carried out in two main phases.

The first phase included the collection and descriptive analysis of statistical data for ten administrative districts of Japan (Fig. 6.3), and six administrative districts of Israel (Fig. 6.2).[2] The data for the analysis covered both a) physical parameters of the districts' development (employment, housing construction, housing prices, and climate), and b) the districts' migration rates.

The data for the analysis were drawn from the following sources:

- Israeli Central Bureau of Statistics: *Statistical Abstract of Israel* (1985-1996), and *Construction in Israel* (1985-1996).
- Japanese Statistics Bureau: *Japan. Statistical Yearbook* (1985-1997).

In addition, climatic indices for Israel were drawn from Climatic Atlas of Israel (Bitan and Rubin, 1991).

In *the second phase* of the research, net migration rates in a district in a current year (percent) were regressed on the following explanatory variables:

---

[2] Three statistical districts of Israel -- Judea, Samaria, and the Gaza area -- are deliberately excluded from the analysis due to an assumption that migrations to and from these areas are affected by political and ideological considerations rather than by "routine" factors of in-country migrations

- CLIMATE: As a proxy for climatic harshness of the area, two different indicators were used. Since in Israel, overheated conditions are commonly perceived as the main source of thermal discomfort, the mean annual number of days with heat stress (Bitan and Rubin, 1991) was used as an index of a district's climatic harshness. At the same time, in Japan, the mean annual number of snowy and rainy days (JSB 1997) was used as a proxy for climatic harshness of an area;
- ECONOMY: The annual change of per capita GDP was included in the analysis as an explanatory variable. As suggested in the previous discussion, the rate of cross-district migration can be affected by the overall economic performance of a country as a whole;
- EMPLOYMENT CHANGE: the annual change in the number of employees in a district, per cent;
- HOUSING CONSTRUCTION: the area of gross building of housing in a given year, '000 m^2;
- HOUSING CHANGE: the annual growth in the rate of housing construction in a district compared to the previous year;
- HOUSING PRICES: Since fully comparable data for Israel and Japan were unavailable, two proxies for housing prices were used. In Israel, housing prices in a district were approximated by the average market price of a standard housing unit, while in the case of Japan, the average housing rent in private sector was used. To narrow the differences between these two different approaches, the absolute values for a district were normalized using the national average in a respective year as the conditional baseline;
- IMMIGRATION: Since in Israel, foreign immigration is a subject to substantial annual changes (see Subsection 6.6.1), the overall rate of immigration in a given year was included in the analysis as an additional explanatory variable. (Due to the relatively low rates of foreign immigration to Japan, this variable is included only in the model for Israel).

The list of variables for the analysis thus covered all the main factors which a hypothetical immigrant or an in-country migrant could expectedly employ in the process of migration decision-making: *housing, employment, the overall economic situation, and climatic harshness of the area*. A "standard" list of possible motivations (see *inter alia* De Jong and Fawcett 1981; Moore and Rosenberg, 1995) was in the case supplemented with indicators of *climatic harshness*, and with *data on housing construction* and *housing prices* in a district. At the same time, some "traditional" components of the motivation list (level of social and health services, and inter-area wage differentials) were dropped due to restrictions on the data availability and comparability.

The analysis also required to test three important statistical assumptions: independence of error terms, independence of explanatory variables, and homoscedasticity of errors.

First, the data employed for the analysis include both *cross-district* statistics (six administrative districts in Israel and ten administrative districts in Japan), and *time-series* measurements: 10 points in time – 1985 through 1995 – for Israel and three time points in time – 1985, 1990 and 1995 – for Japan.[3] This provided us with 60 and 30 valid observations, respectively. The use of time-series required, however, to test the assumption that error terms are statistically independent. To this end, the *Box-Ljung autocorrelation* test was carried out.[4] This test showed that serial correlation for all the research variables is within standard error limits. The *Durbin-Watson test* of the independence of regression residuals was also performed.[5] The results of this test are reported in Tables 6.2-6.4. This test also indicated that autocorrelation is not statistically significant.

Second, it was assumed that the actual effect of some of the above explanatory variables is *time-lagged*, since it is the perception of reality, rather than the actual conditions, which affects the decision making process of individuals (De Jong and Fawcett 1982; Portnov and Pearlmutter 1997). The values of some of the explanatory variables (housing construction, housing prices, and immigration) were, therefore, lagged by one-year in order to reflect this process.

Third, *multicollinearity* of the explanatory variables was tested. For Israel, this test also did not indicate any significant collinearity that could cause instability of regression coefficients. In the case of Japan, significant collinearity between some explanatory variables (employment change and housing construction, housing price and housing construction, etc.) was, however, detected. This did not allow us to introduce simultaneously all the explanatory variables into the regression equation. While various combinations of explanatory variables were tested, only the best performing model is reported (see Table 6.4). The collinearity diagnostics reported in Tables 6.2-6.4 confirms that the variance inflation factors for all the variables introduced in the equations is within tolerable limits and, therefore, should not cause instability of regression estimates.

Lastly, in order to ensure the *homoscedasticity* of errors, a logarithmic transformation was applied to original explanatory variables.[6] A subsequent test confirmed that variances after the transformation are indeed similar.

---

[3] Employment statistics for Japan are available only on a five-year basis. The most recent employment data are available for 1985, 1990, and 1995. This restricted the analysis to these three time periods.

[4] The statistic tests the hypothesis that a set of sample autocorrelations is associated with a random series.

[5] The statistic tests the null hypothesis that the residuals from regression are independent, against the alternative that the residuals follow a first-order autoregressive process. The values of Durbin-Watson statistic range from 0 to 4. A value near 2 indicates very little autocorrelation, while a value toward 0 or 4 indicates strong positive and negative autocorrelation, respectively.

[6] Nearly all the variables included in the analysis (housing construction, employment, climate, density, and housing prices) appear to have only positive values. This fact is of importance, since logarithmic transformations were required. As for employment growth, and GDP change, whose original values had different signs, an adjustment was needed. The values of these variables were readjusted using their minimal values as the conditional baseline.

6.8
Influencing Factors

Tables 6.2-6.4 show the results of the regression analysis. First, it should be noted the regression model explaining the factors that affect the overall rate of net migration across districts of Israel (Table 6.2) appears to be a relatively good fit ($R^2=0.706$). For Japan (Table 6.4), the fit is somewhat lower but also reasonable ($R^2=0.438$). The tables also show that, in both countries, nearly half of explanatory variables are statistically significant (T>2.0, P<0.05). In the following discussion, we shall consider *only* variables that are significant at this significance level.

Table 6.2. Factors affecting the overall rate of interregional migration in Israel (multiple regression: linear-log form)[a]

Variable	B	Beta	Collinearity Diagnostics		T	Sig. T
			Tolerance	VIF[b]		
Employment change	-0.0342	-0.0105	0.6807	1.469	-0.122	0.903
Climate	-0.6281	-0.1670	0.8722	1.147	-2.191	0.032
Housing construction	0.4782	0.1800	0.7229	1.383	2.150	0.036
Economy	-0.5650	-0.1574	0.6316	1.583	-1.756	0.084
Immigration	2.3269	0.6574	0.5271	1.897	6.703	0.000
Housing growth	0.3165	0.1546	0.7584	1.319	1.891	0.064
Housing prices	-1.2376	-0.2092	0.9133	1.095	-2.808	0.007
(Constant)	2.7842				0.988	0.324
No of obs	60					
R^2	0.7059					
F	19.8906					
Durbin-Watson	1.9660					

[a] The initial values of the explanatory variables are logarithmically transformed to comply with the homoscedasticity of variance assumption; [b] variance inflation factor

As Table 6.2 shows, net migration balance in a district of Israel increases with increasing the overall *immigration* to the country and the overall rate of *housing construction* in the area. On the other hand, net migration in a region appears to be aversively affected by high *housing prices* and harsh *climate* of the area. None of these relationships is totally unexpected. For instance, the effect of *immigration* on the migration balance of a region seems to be clear since foreign immigration is an essential component of the country's overall population growth (see Subsection 6.6.1): When foreign immigration increases, a district has, on the average, a better chance to improve its migration balance.

Table 6.3. Factors affecting the rate of in-country migration in Israel (regression model: linear-log form)[a]

Variable	B	Beta	Collinearity Diagnostics		T	Sig. T
			Tolerance	VIF		
Employment change	0.2880	0.1849	0.6807	1.469	1.524	0.133
Climate	-0.4040	-0.2245	0.8722	1.147	-2.093	0.041
Housing construction	0.6869	0.5403	0.7223	1.383	4.588	0.000
Economy	-0.2345	-0.1364	0.6316	1.583	-1.083	0.283
Immigration	-0.2903	-0.1714	0.5271	1.897	-1.243	0.219
Housing growth	0.2234	0.2279	0.7584	1.319	1.982	0.052
Housing prices	-0.8520	-0.3009	0.9133	1.095	-2.872	0.006
(Constant)	0.7211				0.380	0.705
No of obs	60					
R^2	0.4184					
F	5.9616					
Durbin-Watson	2.5479					

[a]See comments to Table 6.2

Table 6.4. The significance of the factors affecting interregional migration in Japan (regression model: linear-log form)[a]

Variable	B	Beta	Collinearity Diagnostics		T	Sig. T
			Tolerance	VIF		
Employment change	0.0702	0.4042	0.8881	1.126	4.544	0.000
Climate	-0.1447	-0.1497	0.9180	1.089	-1.712	0.091
Housing construction	0.2541	0.4676	0.9601	1.042	5.466	0.000
Economy	-0.0580	-0.0892	0.9789	1.022	-1.053	0.295
(Constant)	-0.1010				-0.167	0.868
No of obs	30					
R^2	0.4379					
F	15.5789					
Durbin-Watson	1.9213					

[a] See comments to Table 6.2

The effect of *housing* (the rate of construction, housing prices) on migration rates is also anticipated and is in line with our initial research assumptions (see Section 4). It is surprising, however, that the effect of *employment change* on migration balance of the region does not appear to be statistically significant (Table 6.2). This phenomenon has two possible explanations. First, it can be attributed to the heterogeneity of migrant population, specifically to the presence of foreign immigrants that are less familiar with the local employment situation than in-country migrants. To test this assumption, the regression model was recalculated in order to include *only* internal migrants (Table 6.3). In this model, the factor of employment change is indeed more significant compared to the initial model (Table 6.2). Second, relatively short aerial distances between various parts of the country should also be taken into consideration. As mentioned in Subsection 6.6.1, these distances make it possible to commute on a daily basis across regional borders. This, in turn, may reduce the effect of employment change in a specific district on the general patterns of inter-area migration.

Indirectly, the latter assumption appears to be justified in the case of Japan (Table 6.4). Since commuting distances in this country are far greater compared to those in Israel, it is thus not surprising that the effect of both *housing construction* and *employment change* on the districts' attractiveness to in-country migrants in Japan appears to be highly statistically significant (T>4.5, P<0.001).

6.9
Employment-housing Balance

In this section, we shall attempt to construct the curves of *"neutral migration"* (see Fig. 6.1) using statistical data available for the two countries in question. This task requires a number of preparatory steps.

At the first step, only the cases with *a relatively low level of net migration* (±0.25 percent for Israel and ±0.10 percent for Japan) were selected. There were found *eight* such cases among 60 valid observations available for Israel and 12 cases among 30 valid observations available for Japan.

Considering the actual rates of inter-area migration in the countries under consideration (see Table 6.1), the net balance of migration in these cases can conditionally be assumed as a *neutral migration level.*

At the second step of the analysis, the selected subset of cases was used to compare the overall annual increase in housing construction with the actual growth in the number of jobs available in the area.

The results of the analysis are graphically represented in Fig. 6.4, which portrays only the cases that match the above condition of *"migration neutrality."*

The models appear to be a reasonably good fit and are relatively easy to interpret. In the case of Japan, for instance, a *1.2 per cent* increase in employment should be accompanied by a *four per cent* increase in new housing construction in order to maintain the "migration neutrality" of the area (see Fig. 6.4b). Concurrently, lower rates of housing construction in the region are likely to cause out-migration. At the same time, for Israel, preconditions for the above *"migration neutrality"* are as follows: If the number of jobs in a district increases by some *six percent*, this employment growth should be accompanied by a *10 percent* increase in new housing construction in order to maintain the existing migration balance of a region (see Fig. 6.4a). The rates of construction above this level may boost in-migration to the area, while lower rates are likely to increase out-migration.

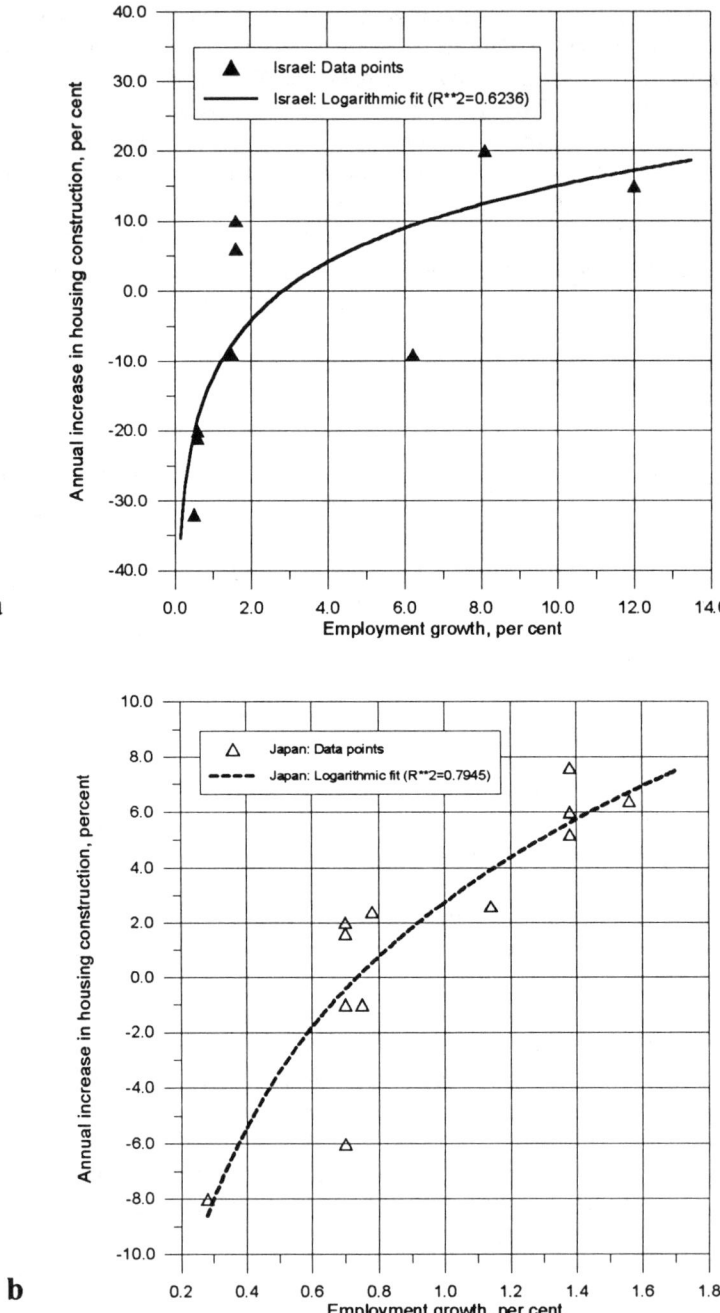

Fig. 6.4. Neutral migration curves for Israel (a) and Japan (b)

6.10
Conclusion and Policy Implications

According to a general migration theory, the predominant motive of interregional migration is job-related: long-distance migrants move to areas where more jobs are available, and wages are higher (LaLonde and Topel 1997). The present comparative analysis of interregional migration in Israel and Japan makes it possible to partially reconsider this notion: While the core regions of these countries concentrate the bulk of available jobs, the attractiveness of these areas to in-country migrants appeared to decline substantially in recent years.

To explain this phenomenon, a *"housing-employment"* paradigm is suggested: The availability of employment in the core areas attracts in-migrants, and this, in turn, boosts housing demand. The response of the local market to this demand is, however, often inadequate due to primarily a lack of land for development. This increases housing prices and causes the outflow of current residents (who cannot afford decent housing in the area) to areas where housing is more available and affordable. Expensive housing also hampers a further influx of migrants to the core regions despite the availability of employment. It seems to be likely that as a result of these interrelated trends, net migration balance of the core areas in the two countries tended in past years to decline, while that of peripheral region appeared to grow.

Based on this "housing-employment" paradigm, the notion of *"neutral migration"* is suggested. This state of migration corresponds to such a balance of housing and employment availability, which does not cause *either a substantial influx of newcomers or an outflow of migrants from the area.*

The analysis of empirical data made it possible to develop *neutral migration curves* (NMC) for the two countries included in the study area. Although further studies in other countries are required to establish the typical form of these curves, and to determine a country's settings that influence their parameters, these models may represent a useful tool for urban and regional development planning. The areas of the potential use of these curves can be identified as follows:

1. *Allocation of public housing construction.* NMC developed for a specific country and continuously updated may help the local decision-makers to determine the amount of public construction to be allocated in priority development areas in order to encourage inward migration.
2. *Restricting further population growth in overpopulated, specifically core areas.* NMC may assist in formulating regional development policies aimed at stimulating out-migration from over-populated core regions. Knowing, for instance, country-specific preconditions for out-migration may lead to developing a set of restrictive regional policies aimed at preventing further concentration of migrants in overpopulated core areas. These specific policies may in-

clude higher rates of taxation, stricter land-use regulation, restrictions on industrial expansion and other regional policy mechanisms.
3. *Targeting economic activity towards economically distressed areas.* NMC may help to identify "better-off" regions and regions with a high risk of out-migration. These data may inform public policies on the more effective distribution of geographically specific development priorities favoring regions that are currently in distress and other potentially "problematic" geographic areas.

References

Abe H (1996) New directions for regional development planning in Japan. In: Aden J, Boland P (eds), Regional development strategies: a European perspective. Jessica Kingsley Publishers, London and Bristol, pp 273-295

Armstrong H, Taylor J (1993). Regional economics and policy. Harvester, New York

Bitan A, Rubin S (1991) Climatic atlas of Israel for physical and environmental planning and design. Ramot Publishing Co, Tel-Aviv

Borjas GJ (1989) Economic theory and international migration. International Migration Review 23(3): 457-485

Burnley IH, Murphy PA, Jenner A (1997) Selecting suburbia: residential relocation to outer Sidney. Urban Studies 34(7): 1109-1127

CIA (1997) The 1996 World Fact Book. Internet Edition

De Jong GF, Fawcett JT (1981) Motivation for migration: an assessment and a value-expectancy research model. In: De-Jong GF, Gardner RW (eds), Migration decision-making. Multidisciplinary approaches to microlevel studies in developed and developing countries. Pergamon Press, New York, pp 13-53

Frey WH (1995) Immigration and internal migration 'flight' from US metropolitan areas; towards a new demographic Balkanization. Urban Studies 32(4-5): 733-757

Friedmann J (1973) Urbanization, planning and national development. SAGE Publications, Beverly Hills and London

Fulton JA, Fuguitt GV, Gibson RM (1997) Recent changes in metropolitan-non-metropolitan migration streams. Rural Sociology 62(3): 363-384

George P (1970) Types of migration of the population according to the professional and social composition of migrants. In Clifford A, Jansen J (eds) Reading in the sociology of migration. Pergamon Press, Oxford, pp 39-47

Greenwood M, Stock R (1990) Patterns of change in the international location of population, jobs and housing: 1950 to 1980. J of Urban Economics 28(2): 243-276

Greenwood MJ, McDowell JM (1991) Differential economic opportunities, transferability of skills, and immigration to the United States. Review of Economics and Statistics 73(4): 612-623

ICBS (1997) Statistical abstract of Israel 1996. Israeli Central Bureau of Statistic, Jerusalem

JSB (1997) Japan. Statistical Yearbook. Japanese Statistics Bureau, Tokyo

LaLonde RJ, Topel RH (1997) Economic impact of international migration and the economic performance of migrants. In: Rosenberg MR, Stark O (eds) Handbook of population and family economics. Elsevier, Amsterdam, pp 800-850

Lipshitz G (1992) Divergence versus convergence in regional development. J of Planning Literature 7(2): 123-138

Lipshitz G. (1997) Immigrants from the former Soviet Union in the Israeli housing market: spatial aspects of supply and demand. Urban Studies 34(3): 471-488

Markusen A (1996) Interaction between regional and industrial policies: evidence from four countries. International Regional Science Review 19(1): 49-77

Michel F, Perrot A, Thisse JF (1996) Interregional equilibrium with heterogeneous labour. J of Population Economics 9(1): 95-113

Milne W (1993) Macroeconomic influences on migration. Regional Studies 24(7): 365-373

Moore EG, Rosenberg MW (1995) Modeling migration flows of immigrant group in Canada. Environment and Planning A 27: 699-714

Perroux F. (1983) A new concept of development: basic tenets. Croom Helm, London and Canberra

Poot J. (1996) Information, communication and networks in international migration systems. The Annals of Regional Science 30:55-73

Portnov B. (1998) 'The effect of housing construction on population migration in Israel. J of Ethnic and Migration Studies 24(3) (in press)

Portnov B, Erell E (1998) Long-term development peculiarities of peripheral desert settlements: the case of Israel. Intern J of Urban and Regional Research 22(2): 216-232

Portnov B, Pearlmutter D (1997) Sustainability of population growth: a case study of urban settlement in Israel. Review of Urban and Regional Development Studies 9(2): 129-145

Richardson HW (1977). Regional growth theory. McMillan, London

Vining DR Jr (1982) Migration between the core and the periphery. Scientific American 247(6): 37-45

Walker R, Ellis M, Barff R (1992) Linked migration systems: immigration and internal labour flows in the United States. Economic Geography 68(3): 234-248

White MJ, Liang Z (1993) The labor market competition of immigrants and the native-born: insights from internal migration. Papers of American Sociological Association (ASA)

Williamson JG (1975) Regional inequalities and the process of national development: a description of the patterns. In: Friedmann J, Alonso W (eds) Regional policy. The MIT Press, Cambridge, pp 158-200

7 Investigating the Effect of Public Policy on Population Growth in Peripheral Areas[1]

Boris A. Portnov
J. Blaustein Institute for Desert Research, Ben-Gurion University of the Negev, Sede-Boker Campus, 84990, Israel

7.1 Introduction

The *policy of population dispersal* (PPD) in Israel is an example of a broad variety of regional policies aimed at redirecting population growth from overpopulated core regions to underdeveloped peripheral areas. Similar development policies are found in Northern Europe (Sweden and Norway), Asia (Japan), and elsewhere in the Middle East (Egypt).

The main objective of this policy – *settling the underpopulated areas of Israel through population dispersal* – was announced in 1949 in response to the predominant concentration of the country's population in a few metropolitan areas. To achieve this goal, population growth of the country's periphery in the 1950s-1960s was primarily sustained by directing new immigrants to so called "priority development zones" (PDZ). Geographically, the spatial frontiers of these zones loosely coincide with two peripheral districts of the country -- the Northern and Southern district (see Fig. 6.2 in Chapter 6 of this book).

Since the early 1970s, the aforementioned approach of immigrant location was gradually replaced by various economic incentives. These incentives are of four basic types (Fig. 7.1):

- planning and development;
- financial incentives to private investors;
- allocation of public land, and
- housing and location aid (IMF 1996; Lerman and Lerman 1992).

Although 50 years have passed since the policy in question was announced for the first time, its actual effect on the inter-regional distribution of the country's population has not been yet sufficiently studied and understood.

While numerous studies (Drabkin-Darin 1957; Shefer 1990; Gradus and Krakover 1977; Lipshitz 1996) were carried out to trace the changes in the population balance between Israel's core and peripheral districts, the main question as to *whether these changes can be attributed to the policy itself* (rather than to other exogenous and endogenous factors) remains unanswered. This question

---

[1] The chapter is based on a shorter manuscript written by this author in cooperation with Y. Etzion and D. Pearlmutter

is, however, crucial for the overall evaluation of the policy's actual performance in the past and for formulating policy adjustments to be made in the future.

The present chapter attempts to address the issues of population change and the policy's actual effect in their interrelation, and, specifically, answer the following questions:

1. What changes have occurred in the population balance between the core and peripheral regions of Israel over the past decades?
2. To what extent can these changes be attributed to the government policy of population dispersal?

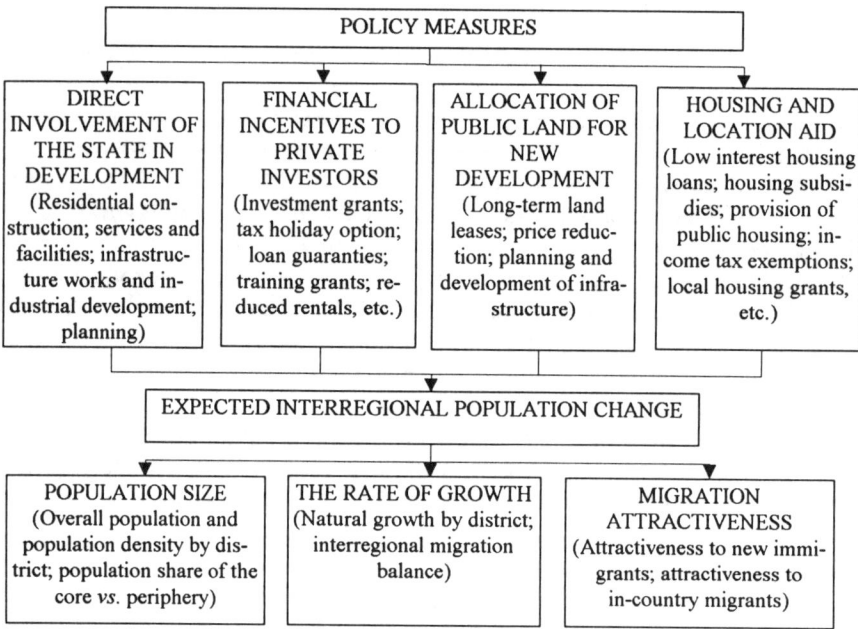

Fig. 7.1. Major components of the policy of population dispersal in Israel and its expected effects on the inter-regional population change

7.2
Regional Policy Evaluation: Contemporary Trends

Approaches to regional policy evaluation can conditionally be divided into three categories:

1. *Indirect methodologies* that use simple statistical techniques to identify relationships between the intensity of regional policy and movements in particular indicators;

2. *Partial methodologies* which attempt to isolate the specific effects of policies on changes in selected evaluation indicators, and
3. *Cost-benefit methodologies* that measure and compare the whole range of benefits and costs of policy intervention on welfare or national efficiency grounds (Diamond and Spence 1983).

A simple and efficient approach to measuring the impact of regional policy was developed by Moore and Rhodes (1973). This approach is based on the assumption that the effect of a regional policy can be expressed as the difference between the *actual values* of selected development indicators and the *values that would have been achieved in the absence of the policy*. The *policy effect* over time thus represents the cumulative difference between two sets of values.

In contrast, the *quasi-experimental control group method* (Isserman and Merrifield 1987) compares selected development indicators between a given geographic area and its controls, i.e. *a set of places whose economic development enables measurement of what would have happened in the place under study without policy intervention*. The authors of this method acknowledge, however, that the use of this technique in planning may be impeded by a lack of suitable controls, specifically if the number of comparable geographic units is considerably limited.

Balchin (1990) suggests an alternative approach to investigating the success or failure of a regional policy. In his book on regional policy in Britain, he examined the changes in selected development indicators (the growth of employment, unemployment rates, production, income, expenditure, labor supply, housing, health, and the standards of living) in the northern and southern regions of the U.K. during *"policy-on"* and *"policy-off"* periods. This analysis allowed him to conclude that during the phasing-out of regional policy in 1979-87, unemployment rates in the North increased more rapidly than in the rest of the country, while manufacturing output in this region appeared to decline sharply.

It should be noted, however, that these evaluation techniques were primarily designed for assessing the impact of regional policy on *economic development*, as expressed in terms of unemployment, employment, and capital movement. The applicability of this approach to gauging the effect of regional policies on changes in *population-related indicators* (inter-regional population balance, migration, and overall population growth) has not hitherto been examined.

In addition, an apparent weakness of the above discussed and similar evaluation techniques is that these methodologies consider only the criteria related to the policy itself *(policy measures)* and to "outcome" indicators *(policy effects)*. At the same time, other, *non-policy related factors* appear to be underestimated. These exogenous factors such as overall rates of foreign immigration, existing patterns of urban development, and inter-area rates of natural growth may, however, either substantially weaken or enhance the overall effect of a regional policy. In contrast, the methodology suggested in this chapter attempts to identify and incorporate these "controls" in the general evaluation framework.

7.3
Research Methodology

The model suggested for evaluating the effects of regional development policy on the patterns of population change is graphically represented in Fig. 7.2. The model includes four major components: *policy targets and measures, objects of intended impact, controls,* and *policy effects.*

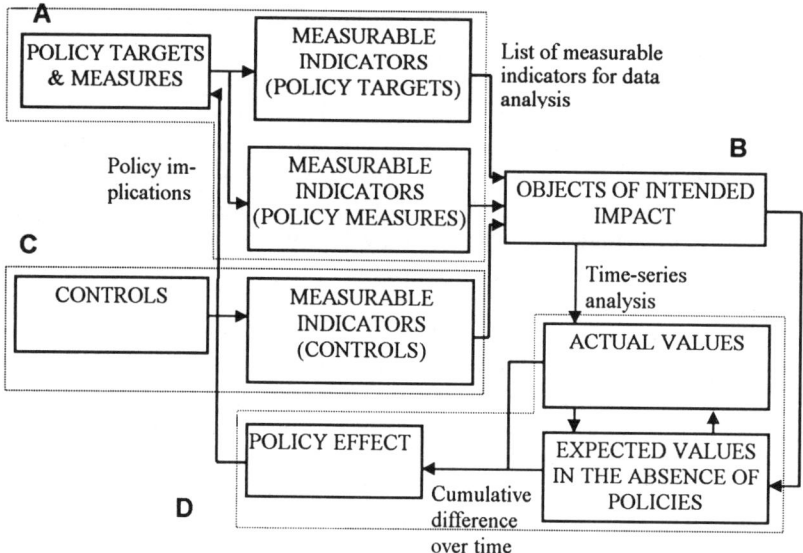

Fig. 7.2. The model suggested for regional policy evaluation

Policy targets and measures. This category contains two aggregated groups of data: a) indicators of population balance *(policy targets),* and b) factors related to *policy measures*. The former group *(policy targets)* encompasses quantitative measures of population change: overall inter-regional balance of population, and rates of annual population growth in various geographic areas. The latter group of data *(policy measures)* may include the following policy-related indices: funds allocated for training grants, annual public investments in infrastructure development, and annual rates of public construction.

Objects of intended impact. In some cases, geographic targets of regional policies are clearly specified. These may include specific settlements or geographic regions. As for PPD in Israel, the spatial targets of this policy are somewhat less clearly defined. Since this policy explicitly refers to overpopulated and underpopulated areas, the spatial targets of this policy can be expressed in terms of a *"core-periphery"* paradigm. To "demarcate" these areas, the *population density* criterion can be used.

Controls. A theoretical model is suggested to describe major exogenous factors affecting population growth in various geographic areas (Fig. 7.3). According to this model, the factors in question fall into two functional groups: population and economy.

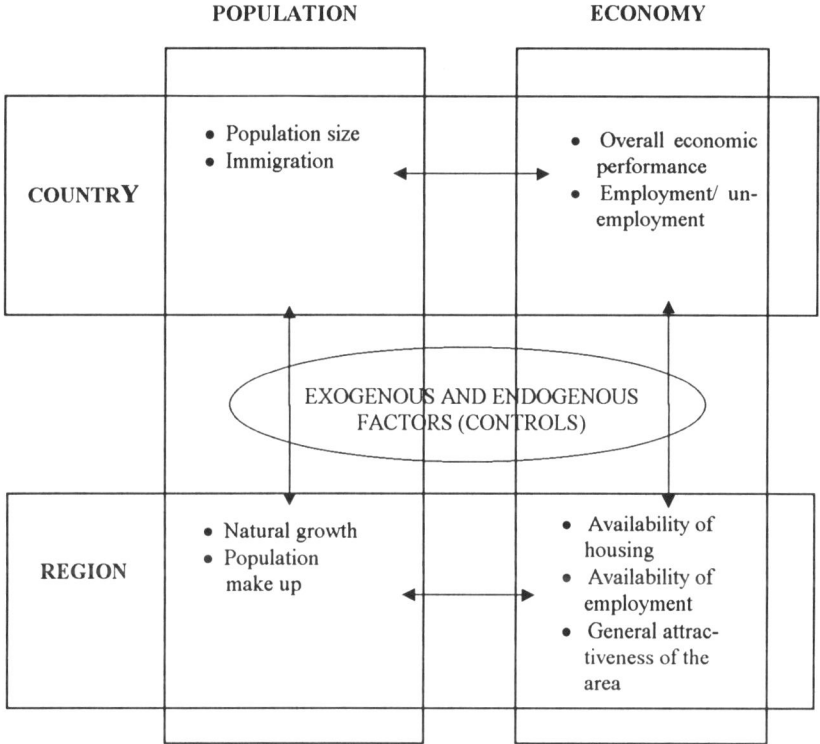

Fig. 7.3. Factors affecting inter-regional population growth (controls)

Variations in the *population composition* across various geographic areas have a substantial effect on the patterns of inter-regional population growth. Differences in fertility and mortality rates between regions (natural growth) may significantly influence the overall patterns of areas' population growth. These disparities are often due to the differences in the ethnic composition of the local populations (Gradus 1984; Portnov and Erell 1998). In a number of regional studies (Lipshitz 1997; Portnov and Pearlmutter 1997; Portnov 1998), it is also argued that a mass influx of immigrants to a country may trigger a chain of long-term population exchanges between the areas.

The overall *economic performance* of the country as a whole and specifically that of individual regions (as expressed by employment, unemployment, housing, and overall attractiveness to private developers) is a key regulator of inter-regional population growth.

- *Availability of employment* is commonly considered one of the strongest motives for inward migration to a certain settlement or region (see Chap. 6 of this book). High unemployment rates, on the other hand, tend to discourage migrants from settling in a given city or region, while encouraging local residents to migrate to other locations.
- *Availability of housing* is also traditionally considered a key factor affecting the rates of inward and outward migrations, and of population growth in general (see inter alia Lipshitz 1997; Portnov and Pearlmutter 1997).
- *Population attractiveness* of a region represents an integral measure which is a function of the area's socio-economic and physical characteristics, including climate, natural landscape, population size, and remoteness. It is suggested that as a proxy for this attractiveness, the rate of private construction in the area can be considered. The latter datum appears to be a good indicator of the "investment climate" in the area, as well as its attractiveness to investors and migrants (for more details on this indicator see Portnov and Erell 1998 and Chapter 4 of this volume).

Policy effects. The effect of a regional policy can be expressed as the cumulative difference between the actual values of selected measurable indicators *(policy targets)* and the expected values of these indicators in the absence of the policy (Fig. 7.2). While changes in the *actual values* of policy indicators can be traced using available time-series data, the *expected values* of the indicators in question can be obtained by using statistical modeling. According to this approach, population growth in a particular geographic area can be estimated as a function of the area's socio-economic and physical characteristics. In addition to policy measures, such characteristics may include groups of controls as diagrammed in Fig. 7.3: macro-economic performance, population, immigration, and local economy. The relative importance of each of these groups of factors may be identified and measured using a multiple regression model.

7.4
Research Approach

Following the approach suggested in the previous section, the analysis of the *policy of population dispersal* in Israel (PPD) was carried out in four phases.

Phase 1. Definition of the samples. Six geographic regions of the country (the Jerusalem, Northern, Haifa, Central, Tel Aviv and Southern administrative districts) were divided into two contrast groups: the "core" and "periphery". The core regions were defined as those with the greatest density of population, while the remainder of the country was termed as the periphery (see Appendix 1).[2]

Phase 2. Analytical structuring. The annual rates of population growth in various geographic areas (GROWTH) were included in the analysis as the dependent variable. In so doing, in order to control for the differences between fertility and mortality rates in various geographic areas, the actual rates of natural growth in an area were normalized using the rate of natural growth for the country as a whole in a given year as the conditional baseline.

To measure the intensity of the policy under consideration, two quantitative policy measures were used:

- PUBLIC CONSTRUCTION: the annual rate of public construction in the area including both residential and non-residential construction initiated by the government, local authorities and companies entirely controlled by these institutions (thousands of m^2), and
- INFRASTRUCTURE: the average annual rate of road construction in a district (km).

Phase 3. Selection of controls. To control for other factors presumably influencing the annual rates of population growth in the core and periphery of the country, the analysis included the following indicators whose influence was hypothesized in the previous section:

- ECONOMY: the annual change in the gross domestic product (percent) considered as a proxy for a macro-economic performance of the country as a whole;
- EMPLOYMENT CHANGE: the annual change in the number of persons employed in a district (thousands of employees);
- IMMIGRATION: the annual number of foreign immigrants to the country (thousands of new immigrants).
- PRIVATE CONSTRUCTION: the annual rate of private construction in a district (thousands of m^2).

Statistical parameters of the research variables are represented in Appendix 2.

Phase 4. Evaluation of the policy's effect. Biannual data for 1948 through 1996 were used to trace the *general trend* of population change between the core

---

[2] The "core-periphery" dichotomy is an important concept in social science developed by Friedmann (1966; 1973). According to this model, development primarily originates in a relatively small number of "foci" located at the points of highest potential interaction within a communication field. Innovations diffuse from these centers to areas of lower potential interaction. In this framework, core regions are perceived as major centers of innovative change, which derive their power from their ability to centralize economic activity, decision-making, and other functions related to the development process. Concurrently, the periphery represents a subsystem which is territorially and socially of low accessibility to the core, and is characterized by limited access to markets, means of production, services, cultural facilities, and sources of economic and political power (Gradus 1984; Ewers and Nijkamp 1990). Among various criteria to identifying spatial limits of core areas, the population density criterion is one of the most commonly used (see inter alia Vining 1982; Lipshitz 1992; Champion 1988).

and periphery of the country (see Subsection 7.5). The respective time-series were drawn from the annual publications of the Israel Central Bureau of Statistics (ICBS 1970-97). At the same time, for the estimation of PPD's effect on inter-regional population growth (Subsections 7.6 and 7.7), biannual data for 1970-96 were employed. The reduction of the research sample at the second stage of the analysis was determined by restrictions on data availability and comparability.

7.5
Preliminary Results and Discussion

The rates of population growth in various geographic areas of the country are represented in Fig. 7.4, while the changes in the population size of the core and periphery over the past five decades (1948-95) are illustrated in Fig. 7.5.

As Fig. 7.4 shows, since the founding of the state in 1948, the highest rates of population growth have predominantly occurred in the country's periphery. In particular, the southern subdistricts of the country (Be'er Sheva and Ashqelon) continuously exhibited some of the highest rates of annual population increase. Thus, in the wake of the 1990-91 mass immigration from the former Soviet Union, the population of the Ashqelon subdistrict (see Fig. 7.4) grew each year by 8-10 per cent, i.e. three times as fast as that of the country as a whole. High rates of population increase were also observed in other peripheral subdistricts, including Zfat, Golan, and Yizre'el. At the same time, annual growth in the most of the areas of the country's core was, during the period in question, somewhat less substantial. In the Tel Aviv district, for instance, the population grew by less than *two per cent* in 1970-75, and less than *one per cent* in 1990-95. *Does this indicate that the spatial distribution of the country's population gradually became more even?*

As Fig. 7.5 shows, the answer to this question is rather negative. Until 1990-95, the gap in population size between core and peripheral areas of the country tended to increase. This gap was equal to 170,000 residents in 1948, then increased to 400,000 in 1960; to 500,000 in 1980 and to 700,000 in 1990. Between 1990 and 1995, this gap, however, decreased to 600,000 residents. This decrease was primarily attributed to the recent patterns of outward migration from the country's overpopulated core to its periphery where housing is more available and affordable (Lipshitz 1996; Portnov and Pearlmutter 1997).

In general, between 1948 and 1995, the population of the core grew by some 2.5 million residents, while that of the periphery increased by only 2.0 million people (see Fig. 7.5).

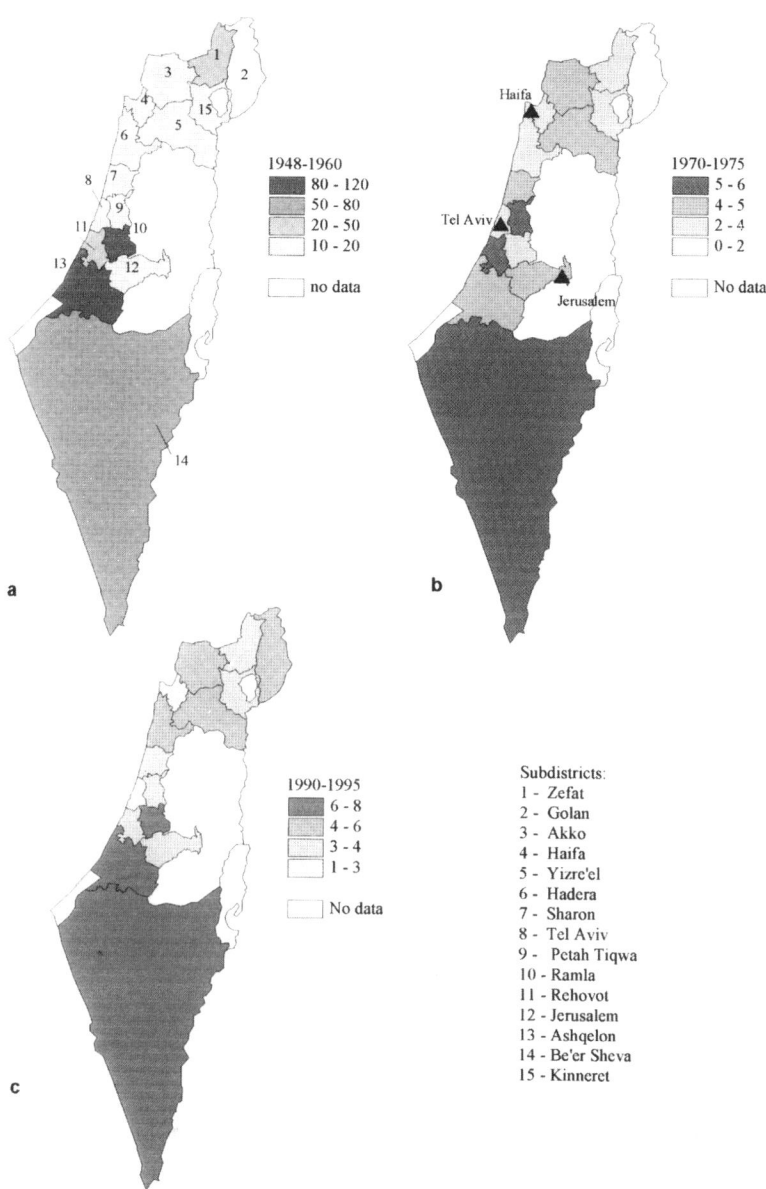

Fig. 7.4. Average annual rates of population growth by subdistrict of Israel in 1948-1995

This trend thus clearly contradicts the main objectives of PPD which was intended from its outset to *restrict overpopulation of the core, by redirecting future growth to the periphery* (see Section 7.1).

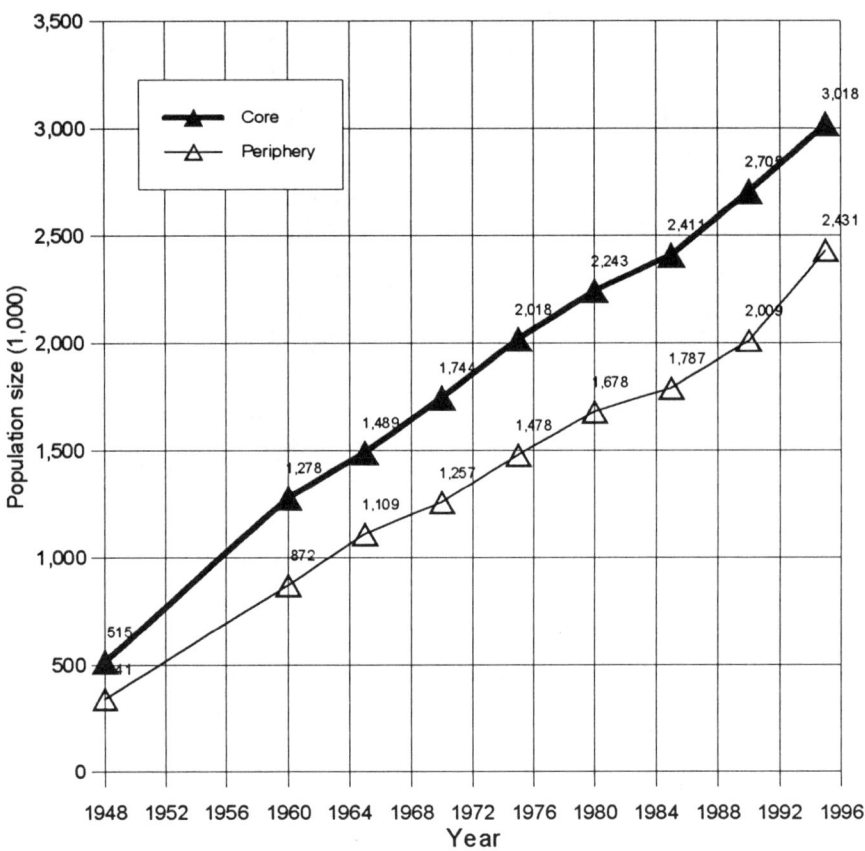

Fig. 7.5. Changes in the population size of core and peripheral areas of Israel in 1948-95

Should the fact that the population gap between regions increased lead us to the conclusion that the policy of population dispersal was generally ineffective? The answer to this question is not straightforward. The growing gap in *absolute population size* between the core and periphery was accompanied by a steady increase in the *proportional share* of the periphery in the overall balance of the national total. In 1948, population of the periphery amounted to only 40 per cent of the country's population. By 1970, the share of the periphery increased to 42 per cent, and at the beginning of 1996 it accounted for 45 per cent of the national total. However, the proportional population share of the periphery cannot grow indefinitely while the gap in absolute population size between the respective regions also tends to increase.[3]

While it is clear, then, that the absolute population gap has increased and the proportional share has been only slightly moderated, at this point of the analysis we cannot confidently say *whether or not the policy in question prevented the gap in population size between the central core and periphery from becoming even wider.* A multiple regression analysis may help to answer this important question.

7.6
Modeling Procedure

In Section 7.4, quantitative policy variables (PUBLIC CONSTRUCTION and INFRASTRUCTURE) and their controls (IMMIGRATION, ECONOMY, PRIVATE CONSTRUCTION, and EMPLOYMENT CHANGE) were introduced. For the sake of the present analysis, these indicators are considered as explanatory variables. Using these variables, regression models were separately computed for the core and peripheral districts of the country (Table 7.1) using the respective overall rates of population growth in the core and periphery (GROWTH$_1$ and GROWTH$_2$, respectively) as the dependent variables. The linear regression model employed for the analysis is as follows:

$$GROWTH_i = B_o + B_1 \times F_1 + ... + B_n \times F_n + \varepsilon,$$

$B_o, B_1,..., B_n$ are the respective regression coefficients, $F_1,..., F_n$ - the above listed research variables, and ε is a random error term.

It is assumed that the actual effect of some of the above explanatory variables is time-lagged, since it is the perception of reality, rather than the actual conditions, which affects the decision making process of individuals (De Jong and Fawcett 1982). The values of some of the aforementioned explanatory variables (private and public construction, immigration, and employment change) were, therefore, lagged in order to reflect this process.

As mentioned in Section 7.4, data employed for the analysis include both *cross-district* data (six administrative districts), and *time-series* measurements (14 points in time). This required testing the assumption that error terms are statistically independent. To this end, the *Box-Ljung* autocorrelation test was carried out.[4] This test showed that serial correlation for all the research variables is within standard error limits. The test of *spatial correlation* and the *Dur-*

---

[3] As our analysis shows, the growth of the periphery's share in the overall balance of the country's population (PS) with the passage of time (T) can be best expressed in terms of an inverse function: PS=A+B/T. This function has a certain limit beyond which no substantial increase in the proportional share of the periphery can be expected. Thus, as long as the current trends of "core-periphery" growth are preserved into the future, the share of the periphery may not exceed 45-46 per cent.

[4] This statistic tests the hypothesis that a set of sample autocorrelations is associated with a random series.

bin-Watson test of the independence of regression residual were also performed.[5] The results of these tests are reported in Appendix 3 and Table 7.1, respectively. As seen, this analysis also does not indicate significant autocorrelation. Lastly, *multicollinearity* of the independent variables was tested (see Appendix 4). This test also did not indicate significant collinearity of the explanatory variables that could cause an instability of the regression coefficients.

Table 7.1. Factors affecting the overall rates of population growth in core and peripheral districts of Israel (multiple regression)

Factor	Core		Periphery	
	B	t	B	t
Economy	0.8378	2.077[b]	0.5368	1.756
Employment change	0.2200	2.326[b]	0.2907	2.715[a]
Immigration	0.1083	5.213[a]	0.1505	9.423[a]
Infrastructure	0.0415	0.675	-0.0166	-0.767
Private construction	0.0210	7.488[a]	0.0068	2.101[b]
Public construction	-0.0057	-0.583	0.0114	3.561[a]
Constant	-2.5436	-0.952	0.6830	0.284
No of obs.	42		42	
R^2	0.8126		0.8465	
Durbin-Watson	1.6914		2.0065	

[a] indicates a two-tailed 0.01 significance level; [b] indicates a two-tailed 0.05 significance level

7.7
Influencing Factors

As Table 7.1 shows, the model fit measured by R^2 is relatively high: $R^2 = 0.813$ for the core districts, and $R^2 = 0.847$ for the peripheral areas. This implies that the rates of population growth in respective geographic areas reflect the qualities of these areas relatively well. Nearly all the estimated coefficients fall within expected signs and are significant at 0.05 and 0.01 levels. In the following discussion, we shall refer only to the factors that are statistically significant at these levels of probability.

Immigration, private construction, the growth of employment, and *economy* are the main factors that increase the rates of *population growth* in the *core districts* of the country. None of these relationships is totally unexpected. The effect of *employment-related factors* on migration influx to a particular area was emphasized by numerous migration studies carried out elsewhere (Lipshitz 1997; LaLonde and Topel 1997).

---

[5] This statistic tests the null hypothesis that the residuals from regression are independent, against the alternative that the residuals follow a first-order autoregressive process.

The difference between the factors influencing the rates of population growth in the country's core and those affecting population growth in the periphery is less expected. The influence of *private construction* does not appear to be highly statistically significant in influencing the patterns of population growth in the country's periphery (t=2.1, P=0.05). However, the effect of *public construction* on population growth in the periphery is significant at a 0.01 level (t=3.6). This positive effect implies that the involvement of the state in development of the country's periphery indeed stimulates population growth of the peripheral areas and thus *prevents the gap in population size between the central core and periphery of the country from becoming even wider.*

This "straight" conclusion may, however, raise a number of legitimate questions. It might be argued that public construction simply "*follows*" population growth rather than causes this growth in the periphery. In the case in question, this assumption does not seem to be justified due to the following considerations. First, as mentioned in Section 7.6, nearly all the explanatory variables, including *public construction*, were time-lagged to reduce the possible effect of the dependent variable (population growth) on its independent determinants. Second, unlike that in other developed countries, the distribution of public construction in Israel is heavily affected by political considerations, rather than by actual demand. Data for 1985 and 1995 represented in Appendix 5 help to illustrate this spatial "bias." While private construction was "naturally" concentrated in the Central, Tel Aviv, and Northern districts, which are in fact the most populated areas of the country, public construction was greatly "skewed" toward the Jerusalem and Southern districts, Judea, Samaria and the Gaza Area (see Appendix 5). This trend is clearly due to the governmental policy of settling the country's periphery and other "strategically sensitive" areas. In view of these considerations, our conclusion concerning the *causal effect* of public construction on population growth in specific geographic areas of Israel seems to be fully justified.

7.8
Alternative Scenarios

The results of the analysis discussed in the previous section naturally lead to the question of *what would happen to the population gap between "core" and "periphery" if the latter's population growth were not "artificially" stimulated?* To answer this question, the approach suggested in Section 7.3 (see Fig. 7.2) can be used.

Since the effect of the *infrastructure* variable is statistically significant (see Table 7.1), we shall focus our attention on another policy variable included in the analysis -- *public construction*, -- and on another control variable *(private construction)* whose rates in particular geographic areas could be *indirectly* affected by governmental policy.

Since the average *per capita* rates of *public construction* in peripheral districts of the country have exceeded those in the core (see Appendix 6), the above general question can be reformulated in a more specific form: *What would happen to the population size of the periphery if the per capita rate of public construction in this area were reduced to the level of that in the core?*

Assuming the population size of the periphery in 1970 as a conditional baseline, the trend in question can be predicted using the regression model computed for the periphery (see Table 7.1) and respective annual rates of construction in the core (Appendix 6). The results of the test are presented in Table 7.2 (*Scenario A*).

Table 7.2. Overall population growth of the periphery (expected from regression)[a]

Year	Expected growth, '000 (regression estimates)	Expected growth, '000 (Scenario A)	Percentage difference	Expected growth, '000 (Scenario B)	Percentage difference
1972	47.03	44.96	-4.39	44.68	-4.98
1974	31.93	27.57	-13.65	25.62	-19.76
1976	31.82	25.65	-19.38	22.16	-30.33
1978	37.73	32.76	-13.15	25.68	-31.93
1980	31.46	26.44	-15.98	21.40	-32.00
1982	21.82	18.73	-14.16	8.08	-63.00
1984	26.60	24.06	-9.57	19.29	-27.50
1986	25.06	24.27	-3.14	17.68	-29.45
1988	22.11	23.16	4.74	24.04	8.72
1990	115.38	114.49	-0.77	115.24	-0.13
1992	64.28	63.96	-0.50	61.02	-5.07
1994	67.38	43.90	-34.84	41.25	-38.78
1996	62.80	60.98	-2.91	58.10	-7.50
Total:	(585.40)	(530.93)	(-9.30)	(483.22)	(-17.45)
Overall growth[b]	(1170.80)	(1061.86)		(966.44)	

[b] Includes "missing" odd years

Although the effect of infrastructure development on the pattern of population growth was not found to be statistically significant (see Table 7.1), it is suggested that public investments in infrastructure may have had an indirect effect on the above patterns through private development. As Portnov and Pearlmutter (1998) argue, each km of new roads constructed in Israel causes, on the average, an additional increase in the rate of private construction in a region by some 10,000 m^2. In the absence of more specific data, this estimate was used in the analysis for evaluating another development scenario (*Scenario B*, Table 7.2) which assumes that the country's periphery did not receive in the past years any preferential "treatment" in either *public construction* or *infrastructure* development.

Although the reduction of the overall population growth in the periphery could be expected due to the positive sign of the respective variables in the regression model (see Table 7.1: periphery), the above tests make it possible to "quantify" this expected effect. As Table 7.2 shows, the expected decrease over the entire time-span in question would reach 9.3 per cent according to Scenario A, and 17.5 per cent according to Scenario B. As a result of the latter decline, for instance, the gap in population size between the core and periphery would reach 1,000,000 residents by 1996, while in reality it was equal to about 600,000 residents (see Figure 7.5). This result is thus in line with the above conclusion that public involvement in construction and infrastructure development has indeed stimulated population growth in the periphery, albeit to a relatively small extent.

7.9
Conclusion

There are many policy approaches aimed at encouraging socio-economic development and population growth in underpopulated peripheral areas. The impact of these policies, however, is not clear. First, there is a deficit of theoretical impact models. Beyond mere descriptions of the official objectives of such policies, there is little knowledge of the long-term and indirect impacts, the side effects and the unintended and eventually unwanted consequences. Secondly and consequently, there is a deficit of identified indicators that express the impacts of regional development policies. If we are not able to measure the benefits of these policies, we do not have the means to judge their effectiveness.

While the actual changes in regional development over time can easily be estimated using available statistical data, the main question is *whether these changes can be attributed to a policy itself* (rather than to other exogenous and endogenous factors). To answer this question, an evaluation model is suggested. This model is based on the approach according to which the effect of a regional policy can be evaluated as the *difference between the actual values of selected development indicators* and the *values that would have been achieved in the absence of the policy*.

To test this methodology, the *policy of population dispersal* (PPD) in Israel was considered. This policy represents an example of regional policies aimed at redirecting population growth from overpopulated core regions to underdeveloped peripheral areas. To gauge the actual impact of this policy on inter-regional population change, a set of quantitative indicators was used. These include: a) criteria indicating population change (population size of core and peripheral areas, and the annual rate of population growth), b) policy measurements (the rate of public construction and infrastructure development), and c) development controls (overall immigration rate, employment change, and the rate of private construction in the area).

While the actual values of these indicators of population change were traced using available time-series data, the expected values of these indicators were obtained using hedonic-type modeling. According to this modeling approach, population growth in various geographic areas is estimated as a function of these areas' development characteristics.

The analysis indicated that although the gap in the population size between the center and periphery of the country tends to increase, the policy of population dispersal (PPD), and specifically, the involvement of the government in construction in development areas, appears to prevent a further increase of this gap. According to an alternative scenario analyzed, this gap would reach 1.0 million people by 1996, while in reality it was 600,000 residents. This shows that spatial public policy can in fact aid in redistributing population, thus easing the severity of the core-periphery imbalance, and thus result in more sustainable regional development.

Although the present analysis was restricted to PPD in Israel, the evaluation model suggested and its applications for planning policy may be applicable to urban and regional physical planning elsewhere.

References

Andoh K, Ohta M (1997) A hedonic analysis of land prices in Yamanashi prefecture, Japan. Review of Urban & Regional Development Studies 9(2): 146-158

Balchin PN (1990) Regional policy in Britain: the north-south divide. Paul Chapman Publishing, London

Champion AG (1988) The reversal of the migration turnaround: resumption of traditional trends? Int Regional Science Review 11(3): 253-260

De Jong GF, Fawcett JT (1981) Motivation for migration: an assessment and a value-expectancy research model. In: De Jong GF, Gardner RW (eds) Migration decision-making. multidisciplinary approaches to microlevel studies in developed and developing countries. Pergamon Press, New York, pp 13-53

Diamond DR, Spence NA (1983) Regional policy evaluation: a methodological review and the Scottish example. Gover, Aldershot

Drabkin-Darin H (1957) Housing in Israel: economic and sociological aspects. Gadish Books Co, Tel Aviv

Ewers HJ, Nijkamp P (1990) Sustainability as a key force for urban dynamic. In: Nijkamp P (ed) Sustainability of urban systems. Avenbury, Aldershot, pp 3-16

Friedmann J (1966) Regional development policy: a case study of Venezuela. MIT Press, Cambridge

Friedmann J (1973) Urbanization, planning, and national deveiiopment. SAGE Publications, Beverly Hills/London

Gradus Y (1984) The emergence of regionalization in a centralized system: the case of Israel. Environment and Planning D: Society and Space 2: 87-100

Gradus Y, Krakover S (1977) The effect of government policy on the spatial structure of manufacturing in Israel. J of Developing Areas 11: 393-409

ICBS (1970-1997) Statistical abstract of Israel, Annual. Israeli Central Bureau of Statistics, Jerusalem

IMF (1996) Structure of investment incentives. Israeli Ministry of Finance, Jerusalem (Internet addition)

Isserman AM, Merrifield JD (1987) Quasi-experimental control group methods for regional analysis: an application to an energy boomtown and growth pole theory. Economic Geography 63(1): 3-19

LaLonde RJ, Topel RH (1997) Economic impact of international migration and the economic performance of migrants. In: Rosenberg MR, Stark O (eds) Handbook of population and family economics. Elsevier, Amsterdam, pp 800-850

Lerman R, Lerman E (1992) A comprehensive national outline plan for construction, development and absorption of immigrants - N.O.S # 31. In: Golani Y, Eldor S, Garon M (eds), Planning and housing in Israel in the wake of rapid changes. Ministries of the Interior and of Construction and Housing, Jerusalem, pp 29-47

Lipshitz G (1992) Divergence versus convergence in regional development. J of Planning Literature 7(2): 123-138

Lipshitz G (1996) Spatial concentration, and deconcentration of population: Israel as a case study. Geoforum 27(1):87-96

Lipshitz G (1997) Immigrants from the former Soviet Union in the Israeli housing market: spatial aspects of supply and demand. Urban Studies 34(3): 471-488

Moore B, Rhodes J (1973) Evaluating the effects of British regional economic policy. Economic Journal 83: 87-110

Portnov BA (1998) The effect of housing construction on population migrations in Israel, J of Ethnic and Migration Studies 24(3): 541-558

Portnov BA, Erell E (1998) Development peculiarities of peripheral desert settlements: the case of Israel. Int J of Urban and Regional Research 22(2): 216-232

Portnov BA, Pearlmutter D (1997) Sustainability of population growth: a case study of urban settlements in Israel. Review of Urban & Regional Development Studies 9(2): 129-145

Portnov BA, Pearlmutter D (1998) The rate of private construction as a comprehensive indicator of urban growth. Paper presented at the 2nd International Conference on Urban development: A Challenge for Frontier Regions, April 4-7. Be'er Sheva, Israel

Shefer D (1990) Innovation, technical change and metropolitan development: an Israeli example. In: Nijkamp P (ed) Sustainability of urban systems: a cross-national evolutionary analysis of urban innovation. Aldershot, Avebury, pp 167-182

Vining DR Jr (1982) Migration between the core and the periphery. Scientific American 247(6): 37-45

Appendix 1
Core and peripheral districts of Israel (1996)

District	Population (1,000)	Land area, km^2	Density of population per sq.km
A. Core districts			
Tel Aviv	1,139.7	170.0	6,704.4
Central	1,257.5	1,242.0	1,012.5
Jerusalem	677.2	627.0	1,080.1
B. Peripheral districts			
Haifa	758.2	854.0	887.8
Northern	977.9	4,501.0	217.3
Southern	798.7	14,107.0	56.6

Source: ICBS, 1997

Appendix 2
Statistical parameters of the research variables

Variable	Core districts		Peripheral districts	
	Mean	Std Dev	Mean	Std Dev
Overall growth	16.26	11.36	15.16	10.69
Employment change	1.51	12.67	1.47	9.04
Infrastructure	23.74	18.98	37.77	36.10
Private construction	710.49	375.11	482.44	233.31
Public construction	195.40	125.34	267.32	292.84

Appendix 3
Spatial collinearity of the research data[a]

A. Core

District	Jerusalem	Central	Tel Aviv
Jerusalem	1.0000	-0.1993	0.1965
Central	-0.1993	1.0000	-0.2257
Tel Aviv	0.1965	-0.2257	1.0000

B. Periphery

District	Haifa	Northern	Southern
Haifa	1.0000	.4131	.4641
Northern	.4131	1.0000	.3461
Southern	.4641	.3461	1.0000

[a] Only spatial variables (public construction, private construction, employment change, infrastructure, and overall growth) are included

Appendix 4
Collinearity of the explanatory variables (Pearson correlation coefficients)

A. Core

	V3	V4	V5	V6	V7
V2	-0.0548	0.1811	0.0321	-0.1303	0.3686
V3		-0.0588	0.4170	0.4370	0.1023
V4			-0.1354	-0.1795	-0.1133
V5				0.3574	-0.0924
V6					-0.3877

B. Periphery

	V2	V3	V4	V5	V6
V1	-0.0830	0.1811	-0.1079	0.0182	0.3289
V2		-0.0312	-0.2185	0.2735	0.0335
V3			-0.3041	-0.0035	-0.1290
V4				0.1035	-0.0894
V5					-0.5417

V1: Immigration; V2: Infrastructure; V3: Economy; V4: Private construction; V5: Public construction; V6: Employment growth

Appendix 5
Distribution of population and residential construction by administrative district of Israel in 1985 and 1995, per cent

District	1985			1995		
	Population	Public construction	Private construction	Population	Public construction	Private construction
Jerusalem	12.0	28.1	8.6	11.8	22.1	5.6
Northern	16.6	7.4	30.0	16.9	3.1	20.7
Haifa	13.7	4.3	11.1	13.2	9.9	9.3
Central	21.0	17.9	28.2	21.6	22.9	37.6
Tel Aviv	23.5	4.5	15.9	20.3	2.7	13.1
Southern	12.0	6.7	5.9	13.7	31.9	10.7
Judea, Samaria and Gaza Area[a]	1.2	31.1	0.3	2.5	7.4	3.1
Total:	100.0	100.0	100.0	100.0	100.0	100.0

Source: ICBS, 1987 and 1997; [a] Jewish localities.

Appendix 6
Average annual rates of public construction in core and peripheral districts of Israel in 1970-96

Year	Core		Periphery	
	overall rate, 1,000 m^2	per capita, m^2	overall rate, 1,000 m^2	per capita, m^2
1970	516.0	0.27	295.0	0.40
1972	894.0	0.45	711.0	0.90
1974	1231.0	0.53	929.0	1.09
1976	1480.0	0.61	1026.0	1.12
1978	737.0	0.29	748.0	0.76
1980	636.0	0.27	518.0	0.50
1982	709.0	0.31	540.0	0.50
1984	450.0	0.18	309.0	0.28
1986	381.0	0.14	124.0	0.10
1988	257.0	0.09	195.0	0.16
1990	300.0	0.10	168.0	0.13
1992	1307.0	0.41	2519.0	1.86
1994	700.7	0.21	365.3	0.24
1996	1214.2	0.32	796.9	0.50

8 Ecological-oriented Options for the Sustainable Development of Drylands[1]

Uriel N. Safriel

J. Blaustein Institute for Desert Research, Ben-Gurion University of the Negev, Sede-Boker Campus, 84990, Israel

8.1
Desert and Development

"Desert" is a range of terrestrial ecosystems. An ecosystem is a production enterprise: its biota appropriates from the environment raw materials such as minerals and water, and energy such as solar energy, for the production PROCESS. From an ecological standpoint "development" is an induced increase of ecosystem production. From the human standpoint ecosystems also provide "services," such as the sequestration of carbon. "Sustainable development" is a process that increases ecosystem production without impairing the generation of ecosystem services.

In terrestrial ecosystems, water originates as precipitation, stored in the soil and moves from there to the biota. Only a small portion becomes incorporated in the product. Most flows through the biota and transpires to the atmosphere. In addition, solar energy removes water from the soil by evaporation. In dryland terrestrial ecosystems, both transpiration and evaporation are so high relative to precipitation that they become the controlling factors of production. Production ceases when in the production machine, the plant, water is lost to the atmosphere, during the production process, at a rate that is faster than the rate of replacement. Development can either (a) increase water inputs, e.g. by storage or transportation; (b) reduce evaporation, e.g. by run-off management; or (c) reduce transpiration, by crop selection, provided that the cost of each of these measures or their combination, including the cost in reduced ecosystem services does not surpass the benefit of the increased production. Whether development is profitable and sustainable depends on the type of dryland, within the range of dryland ecosystems.

Drylands are classified by their Aridity Index (UNEP 1992) - the mean annual (1951-1980) ratio of actual water input (precipitation) to potential losses by evapo-transpiration. Most extreme drylands are the hyperarid ones, where precipitation is <5% of potential evapo-transpiration. Seven per cent of global land is hyperarid, mainly in the Saharo-Arabian region. The next categories are arid and the semiarid drylands, mainly in Australia, Asia and North America. The

---

[1] The chapter is based on a paper on this topic published by this author in Journal of Arid Land Studies (1995) 5: 351-354. Reprinted with permission of the Japanese Association for Arid Land Studies

dry-subhumid drylands have an aridity index of 0.51-0.65 and occur mainly in sub-tropic regions. Altogether 47% of global land is dryland.

8.2
Development of Hyperarid Drylands

With their extremely low precipitation and high potential evapo-transpiration, production of hyperarid drylands can increase only with reliably increased water inputs: from locally available groundwater, by storage of flood water, or by transporting water from elsewhere. Thus, development totally depends on irrigation. Since all developed water resources are more saline than those of direct precipitation, and given the high evaporation rates, there is a salinization risk, which eventually reduces production irreversibly. Water resource development requires high technology and capital investments that can be provided only by an appropriate socio-economic and political structure. These are rarely available in many developing countries with hyperarid drylands. Even when available, and the cost of water resource development and prevention of salinization is lower than that of the increased production, the cost of food production in the hyperarid dryland remains higher than in other drylands and non-dryland ecosystems. It is cheaper to increase food production of already high-productivity ecosystems, than of ecosystems with inherently low productivity. Therefore, rather than develop hyperarid drylands for food production, alternative uses should be considered, that can generate more income if carried out in hyperarid drylands than in other ecosystems. This income will enable the inhabitants of the hyperarid drylands to import food from ecosystems where its production is more profitable.

Three uses are proposed. First, solar energy is more abundant, its supply is more stable, and there is more space to intercept and concentrate it in hyperarid drylands than elsewhere. Investments in solar energy research and technology can make the inhabitants of hyperarid drylands exporters of solar energy. Second, many crops thrive on the desert's abundance of solar energy and warm temperatures and produce more in hyperarid drylands than in other ecosystems, provided water losses are reduced and/or replenished. High salinity water often abundant in hyperarid drylands but useless for most crops, is ideal for certain algae and fish when combined with sun and warmth. Third, hyperarid drylands are of a unique scenic value that together with their vast open expanses, makes them attractive to leisure activities of affluent populations, which reside in the non-drylands. Conventional or eco-tourism may generate more income to the inhabitants of the drylands than any other type of development.

The three development options of the hyperarid drylands depend on ecosystem services rather than on its productivity properties. The role of ecology is in regional planning based on knowledge of ecosystem properties and behavior under the three types of development. The planning process should result in an optimal land allocation for each of the three development types, that guarantees the sustainability of each. This development requires research, high technology, extension and training for the people of the hyperarid drylands. These factors, as well as water resource development, require capital investments, which in turn depend on an appropriate socio-economic-political structure. Until these are available, development of hyperarid drylands is not recommended.

8.3
Development of Arid Drylands

Cash crops, solar energy and tourism are appropriate development options for the arid drylands too. But the higher productivity of arid ecosystems relative to hyperarid ones provides additional options, which utilize not just the services but also the productivity of these ecosystems that occupy 12.1% of global land. Arid drylands are traditionally used as rangelands, by nomad pastoralists. In most regions they are not overpopulated and hence not overgrazed. Rather than increasing their productivity, measures should be taken to avoid overexploitation resulting from nomadism. This too is a socio-political rather than an ecological issue.

A development option for arid drylands is the tapping of an often unutilized resource, biodiversity: wild plants with a potential economic value, not as forage or food species, but as sources for potentially useful natural compounds. Plant species richness is higher in the arid than in the hyperarid drylands, although it is far lower than in many other non-dryland ecosystems. However, arid plants have evolved adaptations to cope with a diversity of extreme conditions - radiation, heat, drought, salinity, floods, and their spatio-temporal fluctuations and unpredictability.

Due to these extreme conditions, arid dryland plants produce seeds that are able to last a long time in the soil. Also, due to their relative scarcity, arid dryland plants are attractive to herbivores. To withstand the physical conditions, to make seeds durable, and to resist predation, arid dryland plants have evolved interesting compounds, which are there to be discovered, characterized and commercialized. The role of ecology is to weigh the benefit of developing arid rangelands against the cost in the reduction of their peculiar biodiversity, and its potential economic significance.

8.4
Development of Semiarid Drylands

As compared to other drylands, semiarid ones occupy the largest area - 17.7% of global land. Pastoralism and subsistence agriculture are the traditional uses of these drylands. With mounting population pressures these uses take the form of overexploitation, bringing about land degradation. It is expressed in an apparently irreversible reduction of ecosystem productivity down to levels of the arid drylands - desertification. Development of semiarid drylands should not only (a) restore desertified areas and (b) arrest further desertification, but also (c) increase productivity to cater for the growing population, yet prevent the desertification associated with attempts to increase productivity of semiarid drylands. Having more precipitation and less potential evapo-transpiration than hyperarid drylands, the cost of measures to increase water inputs and reduce losses is not as high. In Israel, for example, landscape management practice for achieving effective runoff harvesting, coupled with afforestation ("savannization"), is currently researched and developed as a measure to arrest desertification and promote range quality, biodiversity and leisure uses.

The research and development of savannization and other modes of semiarid dryland uses may be slower than population growth. Furthermore, food production of the semiarid drylands may still be more expensive and risky to the environment than in the less arid ecosystems. Therefore, the three high-tech development options for the hyperarid drylands may be viable options for the semiarid drylands as well, provided the appropriate socio-political infrastructures exist.

However, the significance of semiarid biodiversity may surpass its significance in other ecosystems. Semiarid drylands, especially in the Middle and Near East, are rich in progenitors, relatives, and primitive races of modern cultivated crops. The significance of these "biogenetic resources" is widely appreciated, although the fact that these natural assets for guaranteeing future global food security are inhabitants of the drylands, particularly the semiarid ones, most prone to desertification, is often overlooked. The role of ecology is to explore methods of dynamic conservation and utilization of these dryland resources - a demanding charge, because unlike the conservation of species, the theory and practice of conserving a naturally evolving genetic diversity are still in their infancy.

An additional significance of semiarid drylands' biodiversity has emerged only recently, when the threat of global change due to the "greenhouse effect" has become a serious concern. Semiarid regions often serve as transitions between arid and non-arid biogeographical provinces. Species whose centers of distribution is in regions which are more xeric or more mesic than the semiarid one, are represented in the semiarid region by their peripheral populations. It is conceivable that a semiarid peripheral population comprises more genotypes than a core population of the same species (Safriel et al. 1994). This is because a semiarid region "sandwitched" between a more arid and a less arid region, fluctuates in climatic conditions more than other areas. It is the "desert edge" that pulsates spatio-temporally across the semiarid region, as is the case in Israel, for example (Joseph and Ganor 1986). Therefore, an annual plant of an arid species that resides in the periphery of its species distribution, in the semiarid dryland, and that its genotype is adapted to arid conditions but not to mesic ones, produces more seeds in dry years, whereas another plant of the same species but with a genotype adapted to more mesic conditions produces more seeds in the wetter years. Neither genotypes become extinct due to the fluctuating conditions. In the core, though, only genotypes adapted to arid conditions survive. Peripheral populations may therefore be more likely to withstand global warming whereas core populations may become extinct. The extinction of core species will impair the functioning of their ecosystems, and the generation of their services. Semiarid ecosystems should be therefore preserved to serve as repositories of the resistant peripheral populations. These could be used for restoration and rehabilitation of ecosystems elsewhere, which are expected to be affected detrimentally by global change. Because semiarid regions are most prone to desertification, their development impairs the capacity of mitigating future climate change. It is the role of ecology to explore optimal land allocation of the semiarid dryland for sustainable development that allows sufficient space to be set aside for the conservation of peripheral populations.

8.5
Development of the Dry-Subhumid Drylands

These comprise 10% of global land and are the least xeric of all drylands. Their traditional use is the transformation of scrubland to cropland in the valleys and plateaus, and pastoralism on sloping scrub or park plant formations. Current development is engaged in transforming mountainous rangelands to croplands. Tourism and solar energy development may be less and food production more profitable in the dry-subhumid drylands than in other drylands. But much of dry-subhumid agriculture requires irrigation and thus investment in water resource development and infrastructure coupled with risks of salinization. The role of ecology is to address:

- the richness of dry-subhumid drylands in wild relatives of modern crops, often higher than in the semiarid ones (Zohary and Hopf 1993), and
- the potential role of dry-subhumid trees in climate change mitigation.

Current development reduces the capacity of the biosphere to sequester carbon and mitigate global warming by a mounting reduction of tree cover not only in the tropics but also in the dry-subhumid drylands. Temperate non-drylands are ecologically more profitable for afforestation than any dryland type but this is also true with respect to food production. Food production has priority over afforestation, hence on a global prospective it is optimal to use non-drylands for food production and drylands for afforestation. Of all dryland types, only in the dry-subhumid, afforestation is ecologically feasible and hence also economically viable. The role of ecology is to determine the afforestation potential of dry-subhumid drylands as a global warming mitigation strategy. However, due to global warming itself, the incidence of forest fires will increase. Afforestation with fire awareness, that incorporates pastoralism and accommodates the conservation of biogenetic resources and other types of development, is one of the most exciting challenges for ecological research and development in the drylands. The options for sustainable development of dryland types are summarized in Fig. 8.1.

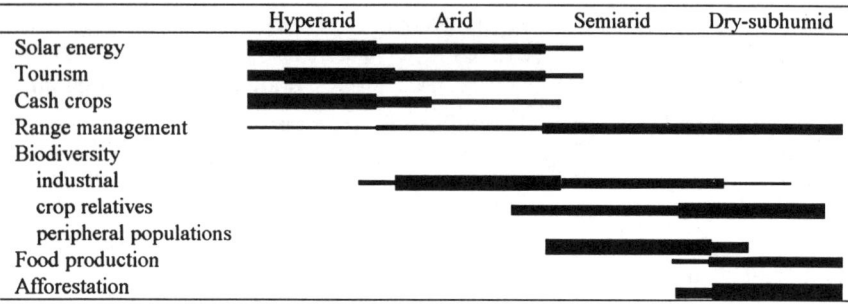

Fig. 8.1. Options for sustainable development

The figure indicates that hyperarid drylands are best suited for the development of solar energy uses, for tourism and cash crops, including algae and fish, that thrive on sunlight and saline water. For arid drylands, using the biodiversity for industrial production of potentially useful natural compounds would be a first choice, with use of solar energy and tourism also possible. Semiarid and dry-subhumid drylands afford more possibilities. For semi-arid drylands, the biodiversity of peripheral populations of plants can be used to mitigate detrimental effects of climatic change in other regions. Range management and using the biodiversity for industrial purposes for improving agricultural crops are also possible. Dry-subhumid areas can serve for afforestation, conservation of biodiversity, and progenitors and relatives of cultivated plants. Range management and food production are also possibilities for dry-subhumid areas.

References

Joseph JH, Ganor E (1986) Variability of climatic boundaries in Israel - use of modified Budyko-Lettau aridity index. J of Climatology 6:69-82

Safriel UN, Volis S, Kark S (1994): Core and peripheral populations and global climate change. Israel J of Plant Sciences 42:331-345

UNEP (1992) World atlas of desertification. Editorial commentary by Middleton NJ, Thomas DSC. Edward Arnold, London

Zohary D, Hopf M (1993) Domestication of plants in the Old World. Clarendon Press, Oxford

Part Two

CITIES OF COLD AND HOT DESERTS

9 Physical Environment and Social Attractiveness of Frontier Settlements: Cities of Siberia, Russia

Boris A. Portnov
J. Blaustein Institute for Desert Research, Ben-Gurion University of the Negev, Sede-Boker Campus, 84990, Israel

9.1 Introduction

The response of city residents to urban improvements, as well as the degree of their satisfaction with the urban physical environment are issues of special importance to urban-planners and decision-makers. Improvements in the urban physical environment (renewal of historical districts, upgrading of engineering infrastructure, expansion of the city's road system, etc.) may not be evaluated positively because they do not adequately reflect expectations and needs of city-dwellers. When local financial resources are limited, future improvements in the urban physical environment must, therefore, *be planned and prioritized* according to their real importance to city residents.

The *social attractiveness* of the urban physical environment may also inform decision-makers about future *property values*. During the period of transition from a planned to a market economy when isolated market transactions do not reflect actual property values, social evaluations may temporarily substitute genuine market assessments.

To understand how social evaluations of the urban physical environment may affect planning policies, three important questions should be answered:

1. How accurately does the social attractiveness of the urban physical environment (UPE) represent genuine qualities (i.e. amenities and disamenities) of this environment?
2. Which factors determine the relative attractiveness of urban territories in settlements of different sizes and geographic locations?
3. How do city-dwellers' evaluations of the quality of UPE differ from those of professionals, i.e. city-planners, engineers, local officials, etc.?

The research was carried out between 1990 and 1994 under the present author's supervision by the Laboratory of Regional Problems of Urban Planning at the Krasnoyarsk Civil Engineering Institute of Russia. The research covered five cities of both East and West Siberia, Russia: *Novosibirsk* (1.5 million resi-

---

[1] The chapter is based on a shorter article published by this author in Annals of Regional Science (1998) 32(4): 525-548

dents), *Krasnoyarsk* (1.0 million residents), *Barnaul* (600,000 residents), *Norilsk* (180,000 residents), and *Lesosibirsk* (50,000 residents). The cities were selected as to represent Siberian settlements of different sizes which are located in all the main climatic and geographical zones of the region. The following geographic areas of Siberia were covered by the analysis: South-East (Krasnoyarsk), South-West (Barnaul, Novosibirsk), the Center (Lesosibirsk) and the North (Norilsk) (Fig.9.1).

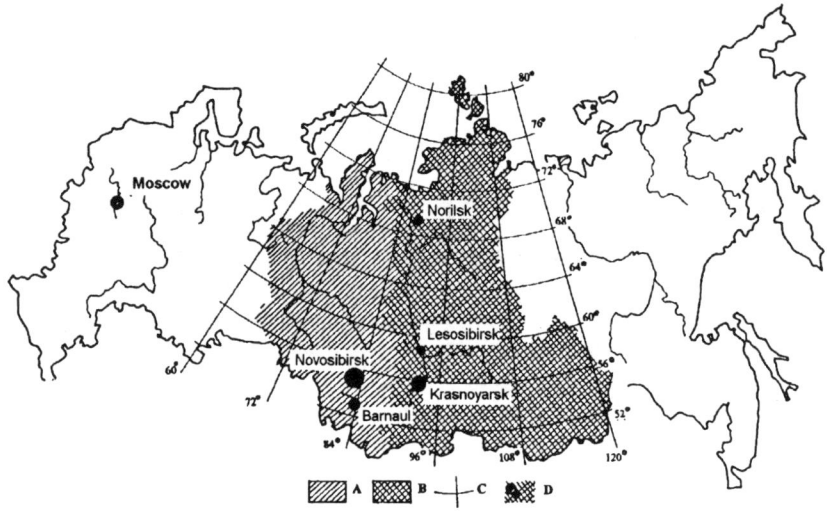

Fig. 9.1. Geographic location of the cities surveyed

A - West Siberian region of Russia; B - East Siberian region of Russia; C - longitude and latitude; D - the cities surveyed

The research indicated that social evaluations of the urban environment do not reflect adequately physical qualities of this environment. The comparative analysis of Siberian cities of different size and location demonstrated that social attractiveness, evaluated by sociological polls and surveys, tended to reflect a combination of different environmental factors (i.e. the average level of urban development achieved in a given city, the range of intra-city divergences, and social composition of urban population) rather than the actual physical qualities of the environment. It was also revealed that priorities and assessments of ordinary city residents and city officials (urban planners, designers, city engineers) differed prominently in many respects. While specialists evaluated spatial qualities of the urban environment (access to a city center, and access to recreational areas) as the most important ones, ordinary city-dwellers paid more attention to functional issues (services and facilities in neighborhoods, structural condition of buildings).

9.2
Previous Research

The interrelationships between the urban environment and social behavior is a key issue for urban psychology and sociology (Lynch 1960; Glass et al. 1977; Timmermans 1991). The syntax of urban space and the problems of morphological analysis in the context of the relationship between the urban environment and social structures or events are reviewed by Hillier and Harison (1984), and Teklenburg et al. (1993). In these studies, regression modeling is a commonly used technique for estimating the specific aspects of these relationships. Assessments of the urban environment through the use of computer techniques were also conducted by Tugnutt and Robertson (1987), Rahman (1992), Smardon (1986), and Gravetter and Wallnau (1992).

Specific forms of a community life in the relationship to urban spatial characteristics were examined by Keane (1991). The negative aspects of "urban environment – social behavior" interactions (the relationship between residential density, city-residents' environmental satisfaction and crime levels) have been examined by Bonnes et al. (1991) in Rome, and Lin and Hanlong (1990) in the Shanghai urban area.

The interactions between urban physical environment and its evaluation by city dwellers have been thus studied at both an entire city level and the level of particular districts and neighborhoods. However, infra-city inequalities in the level of *social attractiveness* of various city districts for settling and business activity in the interaction of this attractiveness with a range of urban physical qualities (i.e. development and planning patterns, social services, transport, shopping facilities, ecological conditions, etc.), which is the focus of the present study, have not yet received sufficient attention. In addition, the cities selected for the present study (i.e. urban settlements of both East and West Siberia of Russia) have been previously covered by very few urban and environmental studies.

9.3
The Region

Siberia is that territory of Russia which lies to the east of the *Ural* Mountains and to the west of the Russian Far East. It falls into two unequal parts - *East* and *West Siberia* (Fig.9.1). The total territory of the region amounts to some 6.2 million km^2 or 36.5% of the entire Russian territory. The total population of Siberia is 23.1 million people or 16.4% of the country's population.

The entire region is frigid in winter. Air temperatures are below -30° (C) for approximately 80 days annually in *Norilsk* and *Dudinka* (East Siberian cities situated above the polar circle) and for 25-30 days in Krasnoyarsk and Irkutsk (the southern part of the region). In *Dixson* (the north of East Siberia), winds blow at a speed of 15 km per hour 240 days a year, while winds of more than 40 km per hour blow for an average of 61 days.

Two-thirds of Siberian population live in 107 urban settlements. Nearly all the major population centers of Siberia (Novosibirsk, Krasnoyarsk, Irkutsk, Kansk and Achinsk) are located along the Trans-Siberian Railroad (the *Transsib*) in the southern part of the region.

The physical layout of Siberian settlements differs significantly from that of settlements located in other regions of Russia. It is especially different from urban settlements situated in the European part of the country (*Moscow, St.Peterburg, Nizhniy Novgorod*, etc.).

First of all, the *specific natural landscape* of the region (presence of large rivers, steep slopes, and other natural restraints) led to a considerable spatial dispersal of residential and other functional areas in Siberian cities. This trend has been reinforced by the *extreme climatic conditions* (long period of low winter temperatures, snow-drifts, strong and continuing winds). Combined with a lack of utilities and transport infrastructures, these conditions called for the location of residential and industrial development along all-weather arterial roads. This tendency was accompanied by extensive development of the spaces between the roads that are less accessible in winter. As a result, some Siberian settlements are spread over considerable distances, up to 20-50 km, with a population density as low as 50-150 persons per km^2.

Second, during the Soviet era, the spatial development of all settlements – including, of course, those in Siberia – as determined by *long-term master plans*. These regulated the conditions for future land use stringently. If the sites designated for construction works were not developed (for instance, due to a lack of necessary financial resources), they remained unused or were put to extensive agricultural use. As a result, there was a scarcity of urban development and the density of the population remained low. This countrywide trend was even stronger in Siberia because of the a traditional notion about its *land abundance*.

Third, as Show (1987) points out, the *Soviets,* like the *tsars* before the 1917 *Bolshevik* revolution, were determined to exploit the remote natural resources of Siberia as *quickly and cheaply as possible*. To achieve this goal, they had a tendency to skimp on infrastructure where not absolutely necessary. A study of contemporary Siberian urban planning provides evidence that this tendency led to the concentration of extensive cottage areas around industrial enterprises scattered at considerable distances from each other. These mixed industrial-residential structures had limited utilities and deficient infrastructures. They came into existence because this type of development allowed local and central governments to house the labor force without diverting state resources from industrial needs. Furthermore, this type of development could be accomplished without expensive investments in land preparation and upgrading infrastructure (Portnov 1994).

9.4
The Cities

The cities included in the analysis -- Novosibirsk, Krasnoyarsk, Barnaul, Norilsk and Lesosibirsk – differ in respect to many development indicators including population size, location and environmental conditions.

Novosibirsk is the largest city of the region, 1.5 million residents, and the official capital of Russian Siberia (see Fig. 9.1). The city was founded at the turn of the century simultaneously with the construction of the Trans-Siberian railroad. A comparatively low level of air pollution, beautiful natural surroundings, and highly developed (at least for the region) social and cultural infrastructures (universities, theaters, museums, etc.) make this city more attractive and desirable for settling and living.

Krasnoyarsk is one of the oldest, 370 years, and largest industrial centers of the region. Its size approaches 1.0 million residents. Despite the extremely favorable natural conditions (the Yenissei river, forests, lakes, spurs of the Sahyany Mountains, etc.), the city is aesthetically poorly developed, and highly polluted. There are huge territorial differences in development patterns, and the level of social facilities and commercial services in the city's residential neighborhoods. Large areas of physically deteriorated single-family housing form 'the exterior' of the city.

Barnaul is an industrial city of medium size, 600,000 people, located in the southern part of the region. It is characterized by relatively new housing and industrial development. The social facilities and commercial services in the city are not as advanced as they are in Novosibirsk, Krasnoyarsk and other large cities of the region.

Norilsk is the central city of the Greater Norilsk Metropolitan Area (GNMA). The city was founded in the mid-1930s as a part of Stalin's *Gulag*.[2] Norilsk is heavily polluted and has an extremely high average density of development (up to 350 residential units per hectare), and the flat landscape of Arctic latitudes. It is believed that GNMA is one of the coldest and windiest inhabited areas on Earth.

Lesosibirsk is a small and remote urban community located in the central part of the region. Its population amounts to 50,000 residents. The city is aesthetically poorly developed but comparatively less polluted. The city is formed by a conglomerate of small villages grouped around two large timber-industry enterprises. It stretches along the *Yenissei* river for over 50 km from north to south. It is isolated from the rest of the region, and has few social, recreational and cultural facilities in its residential neighborhoods.

---

[2] GULAG is a chain of forced labor camps established by the Soviet Government for political prisoners in Siberia, Russian Far East and other remote and climatically harsh regions of the country

9.5
Economics of Transition

An intrinsic aspect of economic reforms in Russia and other countries of the former Soviet Union is the transfer of publicly owned urban land to long-term private ownership. This process requires the estimation of initial land prices for municipal sales and for taxation purposes. However, under conditions of a developing land market, when comparative land prices in different city districts are not available, only indirect methods and techniques can be employed for such assessments.

An analysis of a current Russian land market was carried out under the World Bank's supervision (Kaganova 1996). The results indicate that very few urban land parcels had been privatized to date. Only in 1995, privatized enterprises were allowed to purchase the land they occupied in full ownership. For example, in St.Peterburg, which is the second-largest Russian city, as few as *several dozens* privatized enterprises had obtained land titles by the end of 1995. In addition, an insignificant amount of urban land had been allocated in ownership to families for single-family houses, private garages, and seasonal houses (*dachas*), or privatized by families currently living in existing single-family homes.

The pace of urban land privatization in other cities and regions of the country was even slower. According to the Krasnoyarsk State Land Committee's data, only *seven* land parcels were sold before 1996 in this largest urban center of East Siberia, and some additional 20 lots had been prepared for forthcoming municipal sales.

Golovatskaya et al. (1994) argue that the stagnant situation on the country's land market is caused to a great extent by a lack of *adequate approaches to land appraising which can adequately reflect the conditions of a "thin" land market*. Local municipalities are not only afraid to loose their control over urban land that is currently under their jurisdiction, but also concerned that the potential gains from sales of municipal land could be artificially low with respect to its future value. These municipalities prefer not to rush into aggressive land marketing without *reliable assessments of the current and prospective values* of so scarce resources as good land (Portnov and Maslovskiy 1996).

Meanwhile, when a number of property transactions is limited, the process of land and other property valuation is not straightforward. This task requires use of indirect methods and indicators, while such assessments are often hindered by a lack of reliable data.

One of the most common techniques, used under such conditions for urban land appraisal, is based on the *Average Standard Fee* (ASF), stated annually or quarterly by the federal government for cities of different size as an average value of a unit of land. ASF is subsequently distributed among city districts and forms a base for the valuation of land and other types of properties as well as for taxation purposes. However, the allocation of ASF across urban areas is frequently based on arbitrary assessments of local officials and often leads to fairly heated discussion between the municipality, tenants and property owners.

These specifics of the transition clearly call for developing of alternative approaches to land valuation, and this need was considered as one of the main incentives for the present study.

9.6
Research Method

The research survey was carried out in five phases.

The first phase included designing questionnaires and disseminating them among the cities' residents *via* a newspaper survey. The survey posed five main questions. The first question was illustrated with a map (see Fig. 9.2) to help evaluate social attractiveness of the city districts. In each city surveyed, the map showed a generalized city plan with the location of main city districts accompanied by their conditional names. For example, it showed the *Center, Pokrovka, Nikolaevka*, etc.

The districts were chosen according to three criteria:

1. Similarity in development (the same or similar pattern of land use and planning lay-out within a district);
2. Similar size (from 20,000 to 40,000 people);
3. Familiarity to the general population.

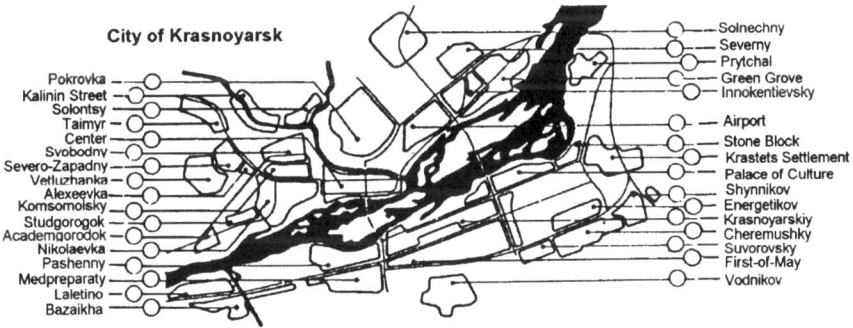

Fig. 9.2. A generalized plan of the city of Krasnoyarsk analogous to that used in the newspaper survey

Spaces in which to rank each district's attractiveness were provided alongside the maps. The respondents were asked to evaluate each city district according to three criteria:

- general attractiveness for residence;
- quality of the physical environment;

- attractiveness for business activity.

The respondents were also asked to evaluate their own districts separately or simply underline its name on the map. They were asked to rank each district on a scale of *1 to 10* (10 being the highest rank).

Other items of the questionnaire (questions #2-#5) were designed to determine residents' evaluations of some specific aspects of their city's development: priorities in urban renewal, and neighborhoods amenities and disamenities. These questions served primarily an auxiliary function. Basically, they were included in the questionnaire in order to draw the residents' attention to the main question (#1). Since the survey was anonymous, the readers were asked to indicate their gender, age, and education. This information was required to trace any bias in the representation of particular social groups in the sample. It would also allow due adjustments to be made.

The questionnaires were published in major newspapers of the cities surveyed (Novosibirsk, Krasnoyarsk, Norilsk, Barnaul and Lesosibirsk), with a total circulation of some 470,000 copies. The readers were asked to fill out the questionnaires and to send them back, either to the editorial board or to the respective city council. As a result of this procedure, we obtained approximately *8,500* completed questionnaires (1.8% of the newspaper circulation). Almost 80 per cent of these questionnaires were obtained within two weeks after publication.

This approach to dissemination of questionnaires deserves a comment.

Using the newspapers to distribute the survey, as well as the design of questionnaires and, particularly, that of spatial diagrams to evaluate the city districts' attractiveness was our original idea. As far as we know, there have been no similar surveys in the practice of urban physical planning and land assessment. An obvious advantage of such an approach lies in its ability to collect a considerable amount of sociological data representing various groups of the city population within a very limited time span. However this approach is insufficient, in making a valid representation of particular groups of the general city population in the sample, because the survey used a self-selected sub-set of the population. The analysis of the questionnaires showed, for instance, that the social structure of the sample had some deviations from the general city population (within 10-15 percent across some socio-demographic groups). Any differences were sequentially adjusted to reflect the proper proportion of the population. To this end, some questionnaires received from "over-represented" groups were excluded from the final set. A procedure of adjustment, using specially designed computer software, was also employed. This procedure was aimed at weighting the assessments of the various groups of respondents according to the share of these groups in the general population.

In addition, to test the homogeneity of population, a comparative analysis of the population's socio-demographic structure was carried out (see Appendix 1). This analysis indicated that the disparities between the samples and populations are rather minor, and these disparities generally fall within a 10-percent error margin established for the survey. Moreover, in order to secure *homoscedasticity of errors*, the procedure of logarithmic transformation was also applied to the original research variables.

Independently the same questions were directed towards a group of experts, i.e. city-planners, urban designers, the local government officials. As a result, approximately 60 additional questionnaires were received in each city.

The second phase of the research included the statistical analysis of the questionnaires. During this phase, the respective *averages* for each city district were computed separately according to four parameters:

1. The attractiveness for settling (the index of a district's prestige or IP);
2. The index of quality of the residential environment, as assessed by the inhabitants for their district of residence (the RE index);
3. The specialists' evaluation of city districts (the SE index), and
4. The evaluation of the attractiveness of districts for business activity (the BA index).

Finally, *the third and fourth phases* of the research dealt with evaluating the physical qualities of city districts *(third phase)*, and synthesizing the regression models between the social evaluations and the districts' physical qualities *(fourth phase)*. The following factors were included in the study area as explanatory variables:

- F_1 - Average Access Time (AAT) to a city center in minutes;
- F_2 - AAT to main city industrial districts (workplace accessibility);
- F_3 - AAT to nearby natural amenities, i.e. parks, lakes, riverfront, etc. (recreation accessibility), in minutes;
- F_4 - the level of the district's functional development measured as a total number of social facilities and businesses available in neighborhoods, such as schools, shopping facilities, healthcare, sports facilities, etc.;
- F_5 - the level of air pollution, a relative index reflecting annual average of major air pollutants;
- F_6 - the structural condition of buildings measured as an average percentage of physical depreciation;
- F_7 - available infrastructure in neighborhoods (a number of the available engineering utilities such as water supply, sewage, electricity, natural gas, etc.);
- F_8 - the architectural and aesthetic qualities of the development ranked for a particular district on a scale from 1 to 10;
- F_9 - aesthetic qualities of the natural landscape also ranked on a scale from 1 to 10, and
- F_{10} - density of development in m^2 of gross building area per hectare of land.

This list of factors thus covered the main categories that physical planning traditionally deals with: *district location, utilities, services, physical conditions and aesthetic appearance of development, and the quality of natural landscapes.*

These factors, in various combinations, are also most frequently mentioned in the numerous studies as contributing to residential land value and the social perception of the quality of the urban environment (Hester 1984; Rapoport 1977; Wittick et al. 1977).

In order to evaluate the districts' qualities, the following indicators were used: time and distance measurements (F_1, F_2, F_3), an analysis of the city councils' statistical data (F_4, F_5, F_6, F_7, F_{10}), and expert assessments (F_8 and F_9).[3]

9.7
Spatial Patterns of District Attractiveness

The analysis of the index of prestige's (IP) mean values allowed us to classify all districts in the cities surveyed into four *typological groups of territories* (TGT).

I TGT included the districts with the highest score of social attractiveness (above six points on a ten-point scale). This group covers both central city districts and the new residential areas of external suburban fringe, situated in the most favorable landscape and ecological conditions (beautiful nature, clean air, etc.).

II TGT embraces districts with a high level of attractiveness. IP values, estimated for these districts, vary from 5 to 6 on a ten-point-scale. In particular, this TGT includes neighborhoods built in the 1940's - the early 1960's. As a rule, these neighborhoods are well established and provide numerous amenities (schools, shopping facilities, health, and sport edifices) to their residents.

III TGT is formed by the districts with an average level of IP (3 through 5 points on a ten-points scale). The districts of this group are represented by both low-density neighborhoods of single-family houses, located within external fringes of central city districts, and by suburban neighborhoods of four- and five-story buildings constructed in the late 1950s - the early 1970s according to considerably low construction standards of these years in comparatively unfavorable areas of poor landscape and ecology.

Lastly, IV TGT incorporates urban districts having a lowest rating of social attractiveness These "backward" districts are graded below three points on a ten-point-scale. In the main, this TGT includes districts located far away from a city center (especially those whose ecological conditions are extremely unfavorable: high level of air pollution, plain landscape, etc.) and small suburban settlements distanced from a metropolitan center.

---

[3] The expert judgments of the aesthetic attractiveness of the landscape and development (F_8 and F_9) were used as a general indicator of respective aesthetic qualities of the urban environment, as is commonly done in a number of urban studies (see, for instance, Rapoport 1977). To minimize possible bias, different groups of experts were used to grade overall attractiveness of the districts and the districts' aesthetic qualities.

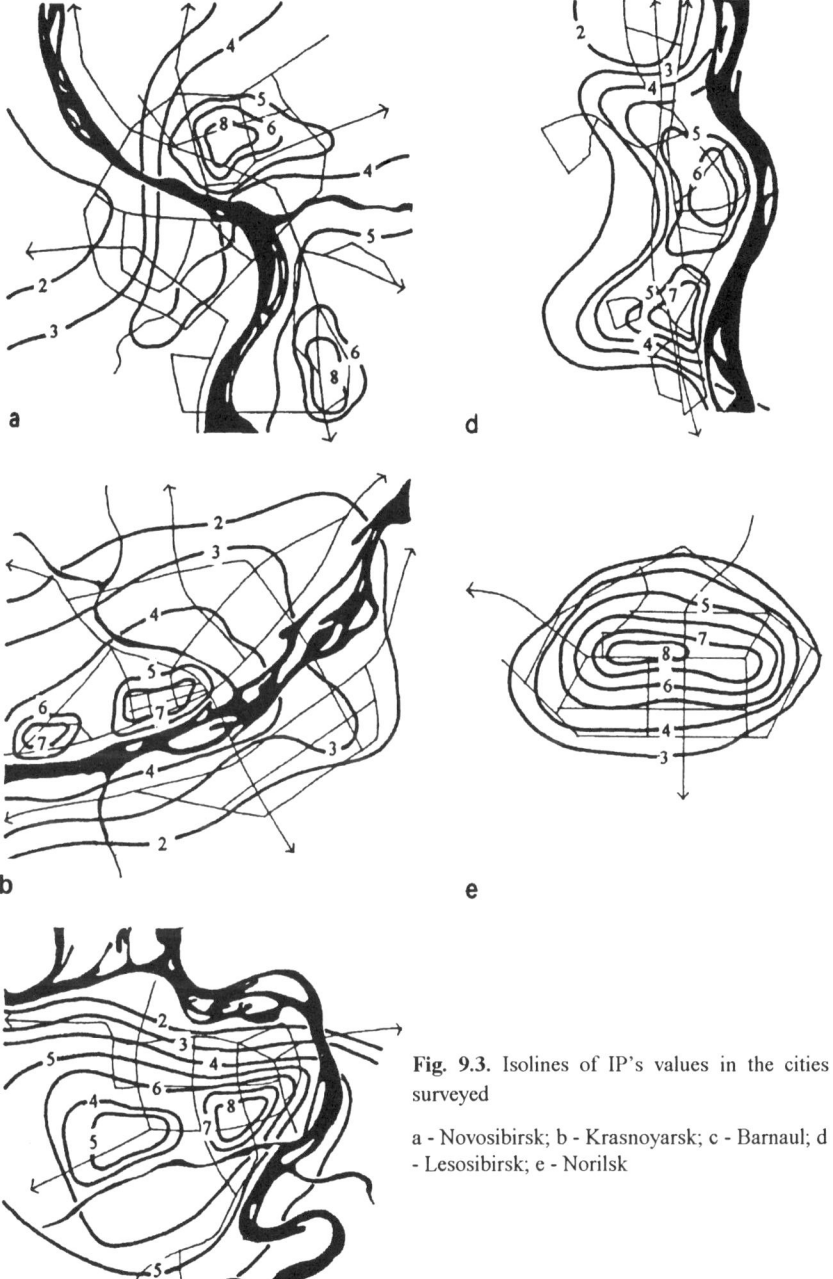

Fig. 9.3. Isolines of IP's values in the cities surveyed

a - Novosibirsk; b - Krasnoyarsk; c - Barnaul; d - Lesosibirsk; e - Norilsk

As Fig. 9.3 shows, the geographic distribution of TGT in the cities surveyed exhibit three types of spatial patterns:

1. The "two peaks" pattern which characterizes the situation in the poli-center cities with multiple urban disamenities located throughout the urban territory (Novosibirsk, Krasnoyarsk and Lesosibirsk);
2. The "one peak" pattern of a compact one-center-city with industrial and landscape disamenities located outside the city borders (Norilsk), and
3. The "peak-hollow" pattern of a one-center-city with major urban disamenities (unattractive industrial buildings, sources of air pollution, etc.) scattered across the city territory (Barnaul).

The percentage of districts included in each TGT is given in Fig. 9.4.

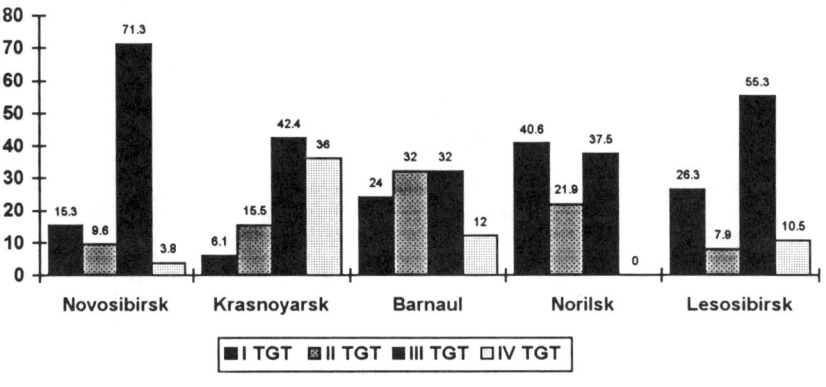

Fig. 9.4. Percentage of districts included in different TGT

As we can see, in large cities of the region (Krasnoyarsk and Novosibirsk), the number of less attractive districts *(III and IV TGT)* exceeds that of more "prestigious" areas *(I and II TGT)*. In Novosibirsk, for instance, this ratio between 'unattractive' and 'attractive' districts is 75.1 per cent vs. 24.9 per cent, respectively. In Krasnoyarsk, this ratio equals 78.4 vs. 21.6 per cent. In the smaller cities surveyed (Barnaul, Norilsk), this relationship is reversed: the quantity of highly attractive districts prevails that of less attractive areas: 56 and 44 per cent for Barnaul and 62.5 and 37.5 per cent for Norilsk.

To explain this phenomenon, *the effect of a city's size on its districts' physical diversity should be taken into account:* Residents of a small city do not have enough "varieties" to choose from. Therefore, they may appreciate even insignificant positive differences among the city's neighborhoods. Meanwhile, in a large territorially sprawled city (Novosibirsk, Krasnoyarsk), there are more spatial patterns for comparison, and this provides residents with a broader scale for differentiating among various city areas.

9.8
Components of Attractiveness

The relationships between the index of prestige (IP) and some physical indicators of city development are represented in Tables 9.1 and 9.2.

As Table 9.1 shows, IP's range (the difference between IP's maximum and minimum values) does not appear to have *any direct correlation with city size (population) and territory*. Indeed, small Lesosibirsk does not exhibit smaller differences of the districts' attractiveness than Krasnoyarsk whose population is far greater. Although Barnaul is half the size of Novosibirsk, the former has a greater range of IP's values than the largest city of the region (7.0 and 5.9, respectively). This preliminary conclusion is further validated by the result of the analysis of variance (Table 9.3) which shows that only the impact of *climatic harshness* on the IP's range is statistically significant (P<0.05).

Table 9.1. Physical parameters and the dispersion of IP in the siberian cities surveyed

City	Population, millions	Territory, km^2	Climatic harshness[4]	Dispersion of IP			
				Maximum	Minimum	Range	Mean
Novosibirsk	1.50	467.5	2.5	8.3	2.4	5.9	4.5
Krasnoyarsk	1.00	317.5	3.0	6.4	1.7	4.7	3.7
Barnaul	0.60	245.0	2.0	8.3	1.3	7.0	5.0
Norilsk	0.20	7.5	7.0	7.6	3.5	4.1	5.6
Lesosibirsk	0.05	47.5	5.0	7.2	2.8	4.4	4.8

Table 9.2. Dispersion of values of some factors influencing IP's values in the Siberian cities surveyed

City	F_1 (distance to a city center)		F_4 (level of functional development)		F_5 (level of air pollution)		F_9 (aesthetic quality of natural landscapes)	
	Range	Mean	Range	Mean	Range	Mean	Range	Mean
Novosibirsk	55.0	33.7	88.0	61.3	2.8	1.1	9.0	5.1
Krasnoyarsk	50.0	35.1	166.0	44.3	9.2	2.6	6.5	4.4
Barnaul	55.0	25.8	80.0	38.6	7.0	2.8	6.0	4.1
Norilsk	15.0	11.0	196.0	103.5	1.5	6.7	0.2	3.5
Lesosibirsk	95.0	24.8	38.0	24.4	0.2	0.2	4.3	3.2

However, IP's range correlates firmly *with the range of the physical environment's qualities* in the cities (Table 9.2). A relatively high range of IP in Leso-

---

[4] The indicator of climatic harshness used in the analysis is suggested by Kruslinskiy (1986) to grade climatic differences in Siberian cities. The indicator is based on the integral assessment of various climatic features including the average wind speed, relative humidity, a number of days with discomfort air temperatures, etc. The high values of this indicator correspond to harsh climatic conditions, while low values indicate relatively mild climate.

sibirsk therefore, may be caused by the extremely high range of "the location factor" F_1 (distance to a city center). It is also likely that the high range of F_4 (level of functional development) may be responsible for considerable differences between IP's maximum and minimum values observed in territorially compact Norilsk.

Table 9.3. The impact of selected exogenous factors on IP's range and values (analysis of variance)

Factor	IP's mean value		IP's range	
	F-ratio	F-probability	F-ratio	F-probability
Population	2.089	0.244	0.506	0.528
Territory	2.465	0.288	0.422	0.703
Climatic harshness	2.278	0.228	19.273	0.021

Note: A logarithmic transformation was applied to the original variables. The Hartley's F-Max test for violation of the homoscedasticity assumption confirmed that the variances are indeed similar.

There is also a correlation between *the IP's mean values and mean values of some environmental factors*. The high average attractiveness of Norilsk's districts may, for instance, be explained by an extremely low value of "the location factor," F_1 and by a relatively high level of the city's functional development F_4 (see Table 9.2). On the other hand, the low attractiveness of Krasnoyarsk's districts is likely to result from unfavorable access conditions to the city's central area (F_1), a comparatively low level of functional development of the city neighborhoods (F_4), and high level of air pollution in city neighborhoods (F_5). In other words, the differences in the social attractiveness of city districts appear to be caused by *the range of physical conditions, and the level of urban development, achieved at an entire city level, rather than by the city's physical size and territory*.

However, the social attractiveness of various city districts is only partially explained by their objective physical qualities. Indeed, the cities of the region that are less desirable for living (*Norilsk* and *Lesosibirsk*) exhibit a higher average level of social attractiveness than *Novosibirsk*, which is more developed and favorable in this respect (see Table 9.1).

9.9
Relative Importance of Influencing Factors

To explain the relationships between IP values and their physical determinants (see Section 9.6), the following linear regression model was used:

$$IP = B_o + B_1 \times F_1 + ... + B_{10} \times F_{10} + \varepsilon, \quad (1)$$

where $B_o, B_1, ... B_{10}$ are the regression coefficients; $F_1, ... F_{10}$ are the respective factors (explanatory variable), and ε is a random error term.

9 Physical Environment and Social Attractiveness of Frontier Settlements

As in the analysis of variance (see the previous section), a logarithmic transformation was applied to the original variables. Hartley's F-Max test for violations of the homoscedasticity assumption confirmed that variances are indeed homoscedastic (Appendix 2).[5]

Table 9.4 portrays the factors that found to be the most statistically significant and were selected the *stepwise multiple regression* (SMR) procedure. Each of the factors is significant *at least* at a 0.05 level.

First of all, the level of statistical significance of the models, estimated by R^2 values appears to be extremely high in four cities of the scope: 0.834 (Krasnoyarsk), 0.847 (Barnaul), 0.816 (Norilsk), and 0.821 (Lesosibirsk). Although for Novosibirsk, the regression fit is somewhat lower, it is also reasonably high (R^2 =0.465). This indicates that the social attractiveness of the urban physical environment, measured by IP, reflects physical parameters of this environment relatively well.

The variables included in the models are not of the same level of significance in each city. The variable named "*the distance to a city center*" is statistically significant in four cities of the sample (Krasnoyarsk, Norilsk, Barnaul and Lesosibirsk), while it is extremely weak in Novosibirsk. Another variable named "*architectural and aesthetic qualities of development*" is statistically significant only in Novosibirsk and Lesosibirsk.

When compared with the dispersion of the corresponding factors' values (see Table 9.2), the nature of this phenomenon seems to be clear: *a factor "works" in the cities whose urban environment provides sufficient diversity in the respective physical phenomenon.* For instance, in Norilsk, which is highly and evenly polluted, the variable of air pollution is not statistically significant, while in far less polluted Krasnoyarsk and Barnaul, the factor in question is of great importance because these cities exhibit significant intra-city differences in respect to the indicator in question (see Table 9.2).

The main factors that determine the IP's level in the cities surveyed are: 1) the distance to a city center (this factor is included in four out of five regression models as statistically significant); 2) the level of functional development (two cities of the sample); 3) architectural and aesthetic qualities of development (two cities); 4) aesthetic qualities of the natural landscape (three cities), 5) structural conditions of development (two cities), and 6) the level of air pollution (three cities of the scope).

A more detailed analysis of these variables helps to explain some additional aspects of the IP's phenomenon.

F_1 *(the distance to a city center)*. As expected, this factor has a negative influence on the districts' social attractiveness (IP). It means that as the distance from

---

[5] The test examines research variables in order to avoid extreme departures of their variances, which may, theoretically, cause biased estimations of regression coefficients. Particularly, the test verifies that the ratio between the maximum and minimum values of the variances does not exceed a certain critical threshold which depends on the overall number of research variables, the size of a sample, and a given level of statistical credibility

a city center increases, the IP's level generally declines. Unsurprisingly, this factor is of prime significance in the north cities surveyed (Lesosibirsk and Norilsk) because waiting time for public transport is vital for the general city population. In more mild climates (Krasnoyarsk, Novosibirsk, and Barnaul), this factor loses its prime importance.

F_4 *(the level of functional development)* is included in two regression equations (Krasnoyarsk and Norilsk) as statistically significant one. This factor tends to be of prime significance in the cities with a high level of functional development (Norilsk), while it is weak in relatively underdeveloped settlements (*Lesosibirsk, Barnaul*). This fact thus further validates our initial conclusion that a factor's contribution to IP is determined by the combination of its range and mean values achieved in a given settlement.

F_5 *(the level of air pollution)* is statistically significant in three cites of the sample (Krasnoyarsk, Barnaul and Lesosibirsk). This factor is of negative correlation with the dependent variable (IP) and tends to be most important in the cities in which city territory is big enough to provide essential variation in the level of air pollution among city districts. It explains why this factor is not statistically significant in extremely polluted Norilsk which has similar levels of air pollution citywide (Table 9.2).

F_6 *(the structural conditions of development)* is primarily important in the cities of highly depreciated housing and industrial buildings (Krasnoyarsk and Lesosibirsk), and is not statistically significant in the cities with newer development (Novosibirsk, Norilsk, Barnaul).

F_8 *(the architectural aesthetic qualities of spatial surrounding)* and F_9 *(the aesthetic qualities of natural landscape)* are statistically significant in three cities surveyed (Novosibirsk, Barnaul and Lesosibirsk). The importance of these factors is especially notable since they are traditionally underestimated by local urban and regional planners and regarded as supplementary ones compared to physical parameters of urban environment.

9.10
"Experts" and "Residents": Different Visions

There are essential differences among city districts in respect to the level of general attractiveness (IP), the evaluation of the districts by their residents (RE), and experts' evaluations (SE) (Table 9.5).

The following formula describes the relationships between IP's, RE's and SE's values as of the city of Krasnoyarsk:

$$SE = 1.37IP - 1.31 \quad (R^2 = 0.777) \qquad (2)$$

$$RE = 0.94IP + 0.95 \quad (R^2 = 0.452) \qquad (3)$$

Table 9.4. Regression coefficients (B) and their t-statistics in Siberian cities surveyed

Variable	Krasnoyarsk		Novosibirsk		Norilsk		Barnaul		Lesosibirsk	
	B	t	B	t	B	t	B	t	B	t
F_1 (distance to a city center)	-0.921	-3.7[a]	-	-	-4.687	-7.8[a]	-1.594	-2.4[b]	-1.413	-3.3[a]
F_2 (workplace accessibility)	-	-	-1.036	-2.4[a]	-	-	-	-	-	-
F_4 (level of functional development)	1.285	4.6[a]	-	-	0.988	2.2[b]	-	-	-	-
F_5 (level of air pollution)	-3.178	-8.1[a]	-	-	-	-	-5.600	-6.8[a]	-0.579	-1.9
F_6 (technical conditions of development)	-1.045	-4.0[a]	-	-	-	-	-	-	-1.231	-2.6[a]
F_7 (engineering infrastructure)	-	-	2.197	2.6[a]	-	-	-	-	-	-
F_8 (architectural and aesthetic qualities of development)	-	-	1.576	2.1[b]	-	-	-	-	1.680	3.1[a]
F_9 (aesthetic qualities of natural landscape)	-	-	1.321	1.7	-	-	2.535	5.4[a]	1.056	2.8[a]
Bo (Constant)	6.916	5.6[a]	5.279	3.9[a]	8.954	8.3[a]	.673	0.5	2.094	1.8
F-statistic	34.06		13.51		62.27		38.93		23.74	
R^2	0.834		0.465		0.816		0.847		0.821	

[a] indicates a two-tailed 0.01 significance level; [b] indicates 0.05 significance level

The mutual interaction between these variables is, however, more complex. The regression analysis indicates that the main factors influencing SE and RE indicators are different. The equations, exhibiting these relationships, follow (only factors significant at a 95% confidence level are included; initial values of the factors are logarithmically transformed):

$$SE = 17.5 - 1.29F_1 - 2.56F_3 - 3.15F_5 \quad (R^2 = 0.832) \quad (4)$$

$$RE = 3.76 - 4.76F_5 + 1.87F_8 \quad (R^2 = 0.545) \quad (5)$$

Compared to IP (Table 9.4, Krasnoyarsk), the SE index (Equation 4) has higher values of regression coefficients for F_1 and F_3 (location parameters of development) while it "underestimates" the significance of F_4 (level of functional development), and F_6 (structural conditions of development). The specialists thus tend to underrate functional qualities of the urban environment (the structural conditions of building and engineering services available in neighborhoods) that are important for the general city population, while they pay considerably more attention to location qualities of urban districts (distance to a city center). The comparison between the RE and the SE indices also indicates that the specialists tend to underestimate the level of air pollution which is extremely important to ordinary city-dwellers.

The significant differences between IP values and those of local residents (RE; (Table 9.5) reflect the fact that IP was evaluated by the residents of the entire city, whereas RE reflects a district's attractiveness *evaluated by the residents of a district*. It is thus not surprising that the RE index emphasizes the role of air pollution in the area (F_5).

The data in Table 9.4 and Fig. 9.5 allow us to introduce different visions of the *"perfect" urban environment,* as depicted by specialists and ordinary city dwellers. To make this classification, statistically significant factors were ranked in descending order according to their standardized *(beta)* values. A sign in parentheses indicates the direction of a particular factor's influence (i.e. either positive or negative):

- *Specialists:* remoteness from recreational areas (-); low level of air pollution (+); closeness to a city center (+) (the SE index).
- *City residents:* clean air (+), numerous community services in neighborhoods (+); new development(+); remoteness from a city center (-) (IP);
- *Local district residents:* low level of air pollution (+), attractive landscapes (+) (the RE index).

In comparison to the other "perception" indices (IP and SE), the RE index (residents' evaluation) primarily reflects basic functional qualities of urban surrounding (pollution, closeness to the city center, social facilities). These qualities are not so noticeable to outsiders. In contrast, the variables "SE" and "IP" primarily emphasize the more superficial qualities of the urban environment, such as visual attractiveness, and structural conditions of development. This orientation of RE towards "internal" functional factors makes this index useful for the evaluation of *long-term neighborhood and community development programs* designed (at least by definition) to serve the residents' real needs rather than to enhance market profit, land values, etc.

Table 9.5. Evaluation of different districts of Krasnoyarsk by the city-residents (the IP and RE Indices) and by specialists (the SE Index)

District	IP	RE	MV	SE	District	IP	RE	MV	SE
Center	1.00	1.00	1.00	1.00	Bazaikha	0.50	1.44	-	0.55
Academgorodok	0.97	1.51	0.75	0.86	Solnechny	0.48	1.47	0.40	0.32
Vetluzhanka	0.89	1.54	0.55	0.60	Severny	0.47	0.53	0.55	0.33
Nikolaevka	0.87	1.30	-	0.54	Vodnikov	0.47	0.53	0.41	0.20
Severno-Zapadny	0.85	1.32	0.55	0.66	Pokrovka	0.44	1.12	-	0.45
Airport	0.82	1.14	0.71	0.50	Palace of Culture	0.44	0.65	0.54	0.25
Svobodny Avenue	0.81	1.26	0.75	0.65	Green Grove	0.44	0.76	0.49	0.25
Studgorodok	0.73	1.04	0.62	0.79	Energetikov	0.42	0.40	0.34	0.17
Komsomolsky	0.73	1.02	0.67	0.69	Laletino	0.42	0.36	-	0.53
Pashenny Island	0.73	0.88	0.53	0.31	Solontsy	0.40	0.80	-	0.36
Alexeevka	0.64	1.54	0.66	0.57	Cheremushky	0.39	0.73	0.37	0.19
Krasrab	0.62	1.03	0.56	0.45	Stone Block	0.35	0.61	-	0.22
Kalinin Street	0.59	0.64	-	0.30	Suvorovsky	0.34	0.40	-	0.18
Taimyr	0.58	1.04	-	0.39	Shynnikov	0.34	0.51	0.38	0.16
Innokentievsky	0.54	0.91	0.59	0.35	Prytchal	0.29	0.50	-	0.14
First-of-May	0.51	0.78	0.47	0.26	Krastets	0.26	0.28	0.37	0.13
Medpreparaty	0.51	0.46	0.54	0.19					

Notes: The values of the indices for the district "Center" were conditionally assumed as 1.00 while the values in other city districts were re-computed correspondingly; "-" means lack of data. IP = the index of prestige (residents' evaluation index); RE = index of quality of the residential environment, as assessed by the inhabitants for their district of residence; MV = market value index (location component of housing market transactions); SE = specialists' evaluation index

9.11
District Attractiveness to Business Activity

An additional variable should be introduced. This is the attractiveness of urban territories for *business activity* (the BA index). The regression model of this index was calculated for the city of Krasnoyarsk (only factors statistically significant at a .05 level are included):

$$BA = 25.54 - 8.75F_2 + 12.92F_4 - 9.17F_7 \quad (R^2 = 0.792) \quad (6)$$

Comparing to the general social attractiveness of urban territories (the IP index), this indicator is more focused on variables reflecting the functional and location qualities of city districts (workplace accessibility, the level of functional development and engineering utilities). The influence of the aesthetic qualities of the urban environment (F_8 and F_9) and the level of air pollution (F_5) on this aspect of the social attractiveness is not significant.

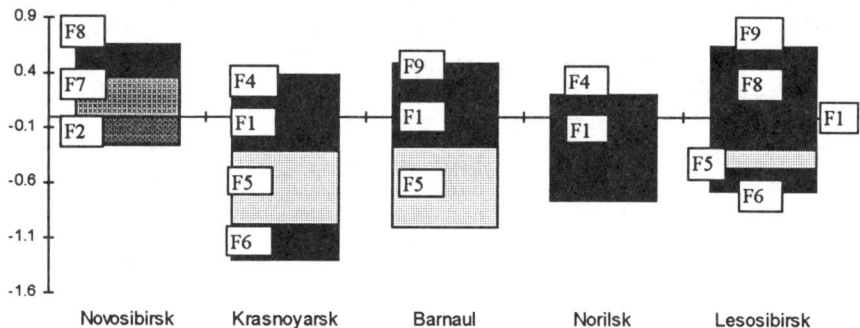

Fig. 9.5. Relative importance of the factors influencing the attractiveness of UPE in Siberian cities surveyed (see text for the factors' definitions)

9.12
IP and the Market Value of Residential Land

For the city of Krasnoyarsk, the values of the social indices (IP, SE and BA) were compared with the location component (spatial gradient) of the market value of residential land (the MV index). Since only housing market transactions were available at the time of the survey, we decided to extract *"the location component"* from the price of apartments sold on the open market. This task involved the following procedure.

The statistical data drawn from the city's major real estate agencies were analyzed and classified by the district, type of apartments, floor on which the housing unit was located, wall materials, etc. The survey covered the period between July and October 1995. In total, 150 market transactions representing 23 out of 33 city districts were analyzed. Approximately 98 percent of dwelling units sold were located in multi-storey buildings. Thus, it was necessary to narrow down price differences related to the size of the unit, the floor of its location, etc. To this end, one particular type of dwelling (a studio apartment located on a first floor of a five-storey building with brick walls) was taken as a *standard unit (SU)* for each city district. This type was chosen because it was the most popular one amid market transactions studied. The prices of other housing units were then adjusted to the "common denominator" of *SU* in a corresponding district. The prices of other housing units were then adjusted to the "common denominator" of *SU* in a corresponding district (see Table 9.5, Fig. 9.6).

As Table 9.5 shows, the *index of prestige* (IP) gives the best equivalent of the city districts' housing market values (MV) in 20 out of 23 cases. The specialists' evaluations (the SE index) turned out to be in this respect more approximate. The SE index coincides with MV better than IP in only three cases out of 23. Compared to the SE index, IP is also more precise in *reproducing spatial patterns of MV* (see Fig. 9.6). As for the RE index (the index of quality of the residential environment, as assessed by the inhabitants for their district of residence), it does not coincide with MV's value at all in most cases (Table 9.5).

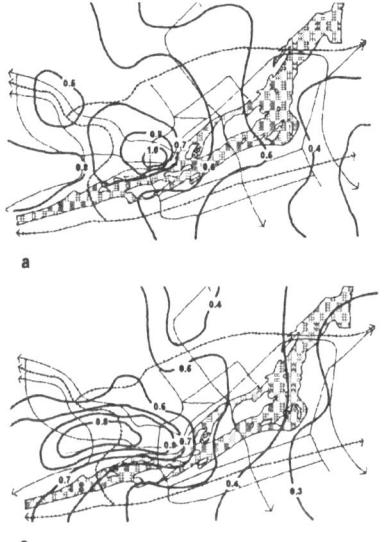
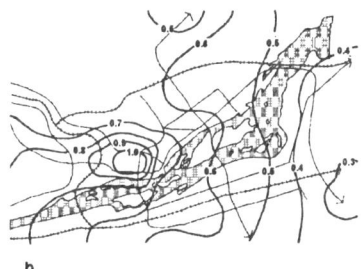

Fig. 9.6. Relative attractiveness of residential land in the city of Krasnoyarsk according to different estimation indices

a – location component of housing prices (the MV index); b – residents' evaluation index (RE); c – specialists' evaluation index (SE).

To explain the substantial differences observed among the indices studied, it was suggested that *MV and social "equivalents" (SE, IP and RE) respond differently to a particular combination of urban physical qualities represented in certain city districts*. The regression model obtained for the MV index is in line with this conclusion:

$$MV = 1.784 - 0.154F_1 - 0.223F_3 \quad (R^2 = 0.896) \qquad (7)$$

As the comparison of Equations (4) and (7) shows, the SE index is more closely related with factors F_1, F_3 and F_5 (access to the city center and workplaces, and the level of air pollution) than MV. At the same time, the IP index matches MV relatively well with respect to the *b*-value of its location component: F_1 - access to the city center (see Table 9.4). This seems to be essential for proper understanding of the close correspondence between two indices (IP and MV), since F_1 "contributes" heavily to MV's values.

9.13
Social Factors

City-dwellers living in a same district evaluate its physical quality differently. These divergences are especially visible across different social and demographic groups (Table 9.6).

Table 9.6. The effect of social factors on IP's evaluations

Socio-demographic group	Evaluation of IP by residents of different urban areas		
	City center	Central area's fringe	Suburban neighborhood
Gender:			
male[a]	1.00	1.00	1.00
female	1.04	1.11	0.98
Age, years:			
14-16	1.08	1.18	0.93
17-26	1.04	1.23	0.91
27-36[a]	1.00	1.00	1.00
37-60	0.92	0.94	1.07
above 60	1.10	1.50	1.08
Education:			
elementary school	1.10	0.71	1.27
high school[a]	1.00	1.00	1.00
college, university	1.06	0.90	1.08

[a] Conditionally assumed as 1.00

Since the detailed analysis of the relationships in Table 9.6 does not fall within the scope of the present analysis, only general observations will be offered. Apparently, central city districts and the central districts' fringe are more attractive to women than because these districts are saturated with shopping and cultural facilities. Suburban neighborhoods, located closer to the natural surroundings, are more attractive to male residents. The city center and the center's fringe are more attractive to young adults (14-26) and seniors (above 60), while suburban districts are evaluated higher by people of middle-age (37-60), as well as by seniors (above 60). The peripheral neighborhoods are also especially attractive to respondents with elementary education. The latter group of population apparently represents a farming-oriented category of "new city dwellers" who recently moved to the urban area from the country-side. They are not fully accustomed to city crowdness and tend naturally to more quite peripheral areas.

9.14
Applications in Planning

The results of the present case study can be of use in two areas: a) the evaluation of residential land under conditions of an emerging property market; and b) long-term physical planning.

During the beginning phase of developing of a residential land market, when both housing and land transactions are scarce and comparable sales data are not available, the sampling method developed and used in the present case study (publication of specially designed questionnaires in major city newspapers) may be used to rank the spatial differences in residential land attractiveness among residential districts. The results of this survey can also be used for the subsequent computation of the downward or upward coefficients of the *Average Standard Fee* (ASF) discussed in Section 9.5.

The hierarchy of the factors influencing IP and other indices of district attractiveness considered in this chapter (IP, SE, and RE) may also be used for *urban physical planning*. For instance, using these indicators, we can determine basic planning strategies which are conducive to positive changes in the social attractiveness of urban areas (for more details see Portnov and Maslovskiy 1996).

As an example, in order to increase social attractiveness of residential districts in, for instance, Krasnoyarsk (Fig. 9.5), four main strategies can be employed: 1) the reduction of the level of air pollution in city neighborhoods (F_5); 2) the saturation of the residential neighborhoods with community services and facilities (F_4); 3) the renovation of obsolete buildings and neighborhoods (F_6), and 4) the upgrading of the urban road system by introducing new highways, access streets, an advanced transportation system which can reduce the access time from remote city districts to the downtown area (F_1).

Fig. 9.7 illustrates the expected changes in the city of Krasnoyarsk's residential land value due the planned construction of the city subway system. These changes are forecasted using the MV model (Equation 7). As Fig. 9.7 shows, the existing large differences in residential land value between the city center and some remote peripheral districts (especially those located in eastern part of the city – Green Grove, Airport) are significantly reduced. In addition, the average increase in residential land value citywide may be expected to be roughly 30 percent.

These research results were positively regarded by the local city officials in the cities surveyed. Since 1992, the IP and the RE models that we computed have been used for the evaluation of residential buildings and underlying land in the course of privatization in *Krasnoyarsk*, *Barnaul* and *Lesosibirsk* as upward and downward coefficients for the average standard fee based upon the above *Average Standard Fee* (ASF), stated by the federal government for the respective settlements as a minimal price of an average land unit. The differential indexes thus derived have allowed city authorities to differentiate land use fees for particular city districts and establish initial land auction prices, even when genuine market assessments for all city districts were not available.

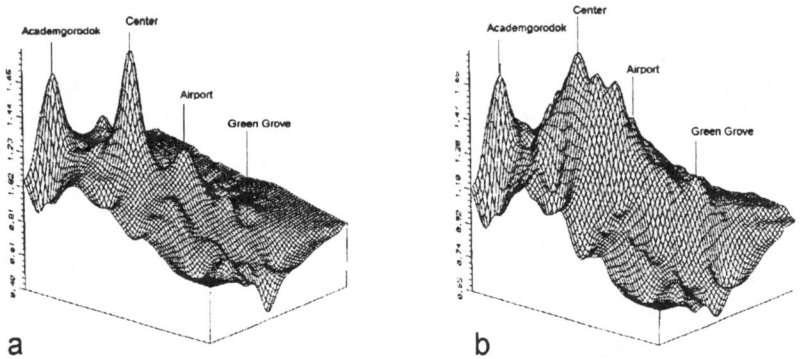

Fig. 9.7. Diagrams show the exsisting differences in the city of Krasnoyrsk's residential land value (a), and those forecasted as a result of the planned construction of the city subway system (b)

9.15
Conclusion

The comparative analysis carried out in Siberian cities of different size and location demonstrates that the indicators of social attractiveness evaluated by sociological polls and surveys reflect the combination of different environmental qualities (i.e. an average level of urban development achieved in a certain city, a range of city environmental qualities, social composition of urban population) rather than the objective physical qualities of the environment. As a result, less developed and less desirable for living cities, as indicated by their physical qualities, may exhibit a higher mean level of social attractiveness than more prestigious and developed urban communities.

The differences between environmental values of specialists and general city population are also important. While specialists (city-planners, local city officials, etc.) estimate spatial qualities of the urban environment (access to a city center, recreation accessibility) as the most important ones, ordinary city dwellers pay more attention to the ecological and functional issues (the level of air pollution, structural conditions of development, social facilities and services in the neighborhood). Thus, planning improvements may not reflect people's real needs as much as they meet priorities and ambitions of "planners-in-charge" and decision-makers.

A set of indices introduced in the analysis (the Index of Prestige, the Residents' Evaluation Index, the Business Attractiveness Index) and their explanatory models can be used both for long-term planning and urban land valuation. One potential application of these models is to forecast the impact of planning physical improvements on the urban land attractiveness. Another potential use is to assess the effectiveness of comprehensive plans, general development programs, and community development programs in urban communities.

References

Bonnes M, Bonaiuto M, Ercolani AP (1991) Crowding and residential satisfaction in the urban environment. Environment and Behavior 23(5):531-552
Glass D, Singer J, Pennlaker J (1977) Perspectives on environment and behavior. Plenum, New York
Golovatskaya N, Lasurenko S, Novitskiy I, Fedorovskay E (1994) Privatization of lands and regulation of land rights in a city. Stroyizdat, Moscow (in Russsian)
Gravetter FJ, Wallnau LB (1992) Statistics for the behavioral sciences. West Publishing Company, St.Paul
Hester RT Jr (1984) Planning neighborhood space with people, 2nd Edition. Van Nostrand Reinhold Company, New York
Hillier B, Harison J (1984) The social logic of space. Cambridge University Press, Cambridge
Kaganova O (1996) Are contemporary fundamentals of appraisal applicable to undeveloped lands in Russian cities. Paper presented at 5th International Research Conference of AREUEA. Orlando, Florida, May 23-25
Keane C (1991) Socio-environmental determinants of community formation. Environment and Behavior 23(1):27-46
Krushlinskiy V (1986) The cities of Siberia. The Krasnoyarsk University Press, Krasnoyarsk (in Russian)
Lin N, Hanlong L (1990) A structural model of social indicators and quality of life: a survey of life in the urban Shanghai. Social Science in China, 2:180-211
Lynch K (1960) Image of the city. Technology Press, Cambridge
Portnov BA (1994) Increase in urban land effectiveness under changing socio-economic conditions. Modeling, Measurement & Control D, 10(1):1-28
Portnov BA, Maslovskiy VP (1996) Residential land attractiveness in an emerging property market. Netherlands J of Housing and Built Environment, 11(2):107-130
Rahman OMA (1992) Visual quality and response assessment: an experimental technique. Environment and Planning B: Planning and Design, (19):689-708
Rapoport A (1977) Human aspects of urban form: towards a man-environment approach to urban form and design. Pergamon Press, London
Show D (1987) Siberia: geographic background. In: Wood A (ed) Siberia: prospects for regional development. Croom Helm, London, pp 9-34
Smardon RC (1986) Review of agency methodology for visual project analysis. In: Smardon RC (ed) Foundation for visual project analysis. John Wiley, New York

Teklenburg JAF, Timmermans HJP, Wagenberg AF (1993) Space syntax: standardized integration measures and some simulations. Environment and Planning B: Planning and Design (20):347-357

Timmermans H (1991) Decision-making process, choice, behavior, and environmental design: conceptual issues and problems of application. In: Environment condition and action: an integrated approach Garling T, Evans G (eds) Oxford University Press, New York

Tugnutt A, Robertson M (1987) Making townscape. Mitchell, London

Wittick A (1977) Encyclopedia of urban planning. McGraw-Hill, New York

Appendix 1
The comparative analysis of socio-demographic parameters of the population of the cities surveyed

Parameter	Percentage by settlements				
	Novosibirsk	Krasnoyarsk	Barnaul	Norilsk	Lesosibirsk
Gender:					
male	49.7	49.8	49.8	50.1	49.4
female	50.3	50.2	50.2	49.9	51.6
Age, years:					
14-16	3.9	3.8	3.9	4.2	3.7
17-26	22.0	21.9	22.3	24.0	23.3
27-36	16.7	16.9	16.2	20.8	18.2
37-60	24.8	25.0	24.7	24.8	24.9
above 60	10.4	10.6	10.8	3.8	7.6
Education:					
elementary school	38.3	39.3	39.0	30.6	40.1
high school	24.1	24.8	26.7	36.9	28.3
college, university	37.6	35.9	34.3	32.5	31.6

Source: Statistical Bureaus of the Krasnoyarsk, Barnaul and Novosibirsk region.

Appendix 2
Homogeneity-of-variance test for violation of the equal variance (homoscedasticity) assumption using Hartley's F-max method

Factor	Variance after logarithmic transformation				
	Novosibirsk	Krasnoyarsk	Barnaul	Norilsk	Lesosibirsk
F_1	-	0.21	0.09	0.04	0.08
F_2	0.12	-	-	-	-
F_4	-	0.16	-	0.07	-
F_5	-	0.08	0.05	-	0.16
F_6	-	0.16	-	-	0.06
F_7	0.05	-	-	-	-
F_8	0.07	-	-	-	0.11
F_9	-	-	0.12	-	0.11
F-Max	2.40	2.62	2.41	1.82	2.74
F-Critical ($\alpha=0.01$)	3.01	3.37	3.81	2.64	3.52

10 Planning Theories versus Reality: A Desert Case Study[1]

Isaac A. Meir
J. Blaustein Institute for Desert Research, Ben Gurion University of the Negev, Sede Boker Campus, 84990, Israel

10.1 Introduction

Theories and models are by definition of an abstract, general nature. Thus, they refer to no place in particular and, consequently, are disconnected from landscape, climate and culture. This is especially so with theories based on economics and statistics, or with different design fashions, such as those we have witnessed in the last century.

The spreading of such theories and their internationalization is not a new trend. The Hippodamic grid, the *castrum romanum*, the *forum* and many building and open space types may be seen in places as different from each other as Asia Minor, North Africa and Western Europe. However, this expansion of theories, practices and styles has become easier with the development of modern telecommunications and fast, cheap transportation. New tendencies and theories pass from source to consumer even before the source has had the time to implement the theory and check it thoroughly. Thus, similar built environments pop up in different places almost simultaneously and the analysis and assessment of their suitability occurs sometimes only in the form of post-occupancy evaluation.

The way and extent that such processes influence the built environment vary according to the scope of the theory or fashion. A purely decorative wave of cut-and-paste forms may affect built forms, but not necessarily their functioning. On the other hand, a movement or school that deals with the overall design process and end product will certainly affect forms and functions and their inter-relations. For example, although the statement "a house is a machine for living in" (Le Corbusier 1927) certainly refers to the functional aspects of design, it nevertheless affects forms and aesthetics to a great extent. The way the end product interacts with its natural environment and its constraints is more-often-than-not the result of uncalculated processes governed by chance.

All of these claims may be well illustrated in the case of Be'er Sheva, a desert city functioning as the administrative, cultural, economic and political center of the Negev Desert, an arid region constituting over 65 per cent of the area of Israel (within the 1967 borders). At an altitude of about 300 meters above sea level, the climate in the vicinity of Be'er Sheva is characterized by wide variability of temperatures and relative humidity, both diurnally and seasonally. Summer days may reach well over 35°C with relative humidity often being lower than 30 per cent,

---
[1] Partly based on: Meir I.A. (1992) Urban space evolution in the desert – the case of Be'er Sheva. Bldg & Environment 1(27): 1-11

while summer nights are considerably cooler, with temperature sometimes reaching below 18°C and relative humidity over 95 per cent. Winters are cold, with daytime temperatures sometimes lower than 8°C, and nightly minimum close to or even below 0°C. Yearly rainfall average is approximately 300 mm; daily global radiation in June is about 680 $Cal\,cm^{-2}\,day^{-1}$. Prevailing winds blow from northwest (except for night winter winds blowing from the east) reaching average speeds of 8-10 $m\,sec^{-1}$ (Stern et al.1986).

Be'er Sheva may serve as a convenient case study because its modern history is relatively short (it was founded in the beginning of this century), the town was built without any constraints of existing tissue, past history or historical buildings and clusters,[2] and it belongs to a time and place which idealized the modern and was in constant search of new ways and theories.

10.2
A Short Note on Settlement in the Past

Be'er Sheva can be traced back to the Book of Genesis. It is referred to in connection with Abraham and Isaac (Israeli 1979). Archaeological finds date back to the Chalcolithic period (4000 BCE). However, the history of the settlement is not continuous. After an interval of about 1800 years in the Bronze period, the area was resettled until the time of the Kings of Judea (586 BCE), only to be abandoned again. The Roman and Byzantine periods are marked by renewed settlement which continued until the 7th century CE. After the Arab occupation and until the end of the 19th century, the whole Negev area was devoid of permanent settlements and served mainly as the dominion of the local nomads, who visited the area periodically, mainly because of a number of water wells located there.

This discontinuity is attributed to the harsh environmental conditions of the desert. Human settlement in the area was made possible only at times of strong central government, which initiated settlement and supported the settlers (Herzog 1979).

10.3
The Modern Era

10.3.1
Ottoman Period and the European Influences (1900-1917)

Under Ottoman rule, the Negev desert started changing. A strong central government influenced the local level again and attempts were made to settle the nomad tribes of the area. It was decided to found a new town which would function as the administrative, commercial and military center of the region. The site

---

[2] Remains dating from the Chalcolithic to the Byzantine and Early Islamic periods were largely ignored during the first years of design and construction, thus creating an uneasy *status quo*. The Ottoman part of the city has been built on Byzantine ruins, whereas the *castrum romanum* lies underneath newer buildings.

chosen for this town was an area located at the center of the tribal territories of the three major Bedouin groups. It included the rail tracks and a number of water wells, and coincided with the site of ancient Be'er Sheva. As a first step toward the foundation of the new settlement, some 223,000 dunam[3] were purchased from the local Bedouin tribes.

The settlement was founded in 1900. The plan for the new town was prepared by two architects/engineers, a German and a Swiss, the former serving as a survey engineer in the Turkish army. The European spirit of the plan is clear: the town is based on a rectangular grid of lots measuring 60x60 meters, divided by streets 15 meters wide. The first stage of development was to include 60 lots. The main street was 20 meters wide and divided the town into two equal parts. The grid was rotated by 45° in relation to the north, the main street being positioned northwest to southeast. This was the beginning of a controversial evolution process. The main reason for the selection of the specific site was its proximity to the existing wells and the Be'er Sheva stream. The whole plan was superimposed onto the site regardless of the local topography. The concept of grid is alien to the local, Middle Eastern (and also the Mediterranean) concept of urban space, which is mainly based on high density, narrow public open spaces and introvert buildings facing small, weather controlled, easily manageable courtyards.

The public buildings constructed in the town (council house, town hall, mosque, school) were extrovert buildings with sloped tile roofs, set in the middle of the plot, surrounded by garden. On the other hand, the houses built by the first settlers are mostly one storey flat roof houses closed to the environment, opening to protected courtyards. According to testimonies of travelers who visited the town during that period, most of the private courtyards included cisterns, fountains and fruit trees, ameliorating the microclimate. Although trees were planted along the streets, use of public open space was not taken into account and the plan did not provide for a center or even for a market, even though the town was supposed to serve as the commercial center for the local tribes (Ben-Arieh and Sapir 1979). Be'er Sheva remained under Ottoman rule until WWI, when it was occupied by the British, together with the rest of Palestine.

10.3.2
Colonialism and British Mandate (1917-1948)

The British seem to have left their marks on most of the settlements developed under their rule, as well as on the architecture of the period. Cities like Jerusalem, Tel-Aviv and Haifa include parts designed and built either in a colonial style, or in the spirit of the modern European movements evolving at the time.

However, Be'er Sheva was only marginally influenced by the British. The master plan of the town (prepared in 1937 by Kendall, the town planning adviser to the Government of Palestine, who also prepared the master plan of Jerusalem in 1944) remained in essence very similar to the Ottoman plan, although certain important additions (such as industrial areas and public open spaces) were made to the original. One major intervention in the urban fabric was the design of

---

[3] 1 dunam = 1,000 m^2

curved streets at the edges of the settlement, which were supposed to create a feeling of compactness. The plan also suggested the construction of arcades along the main commercial streets, but the parts actually covered are negligible. Apart from a few functionalistic, rather insignificant public buildings (such as those of the police compound), most of the private houses constructed at the time are similar to the ones built under the Ottomans.

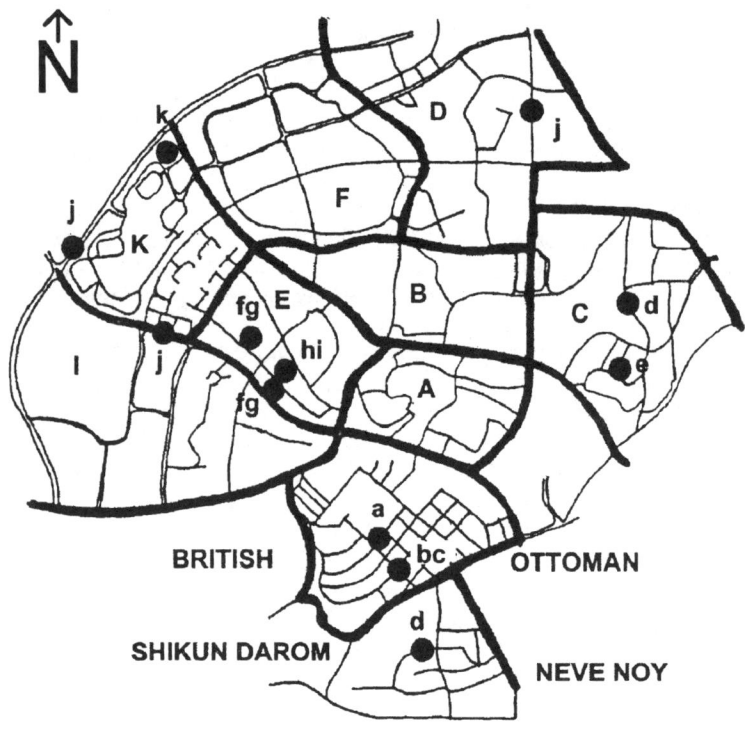

Fig. 10.1. Diagrammatic map of Be'er Sheva with measurement locations indicated by lower case letters

10.3.3
Suburban Agriculture of the First Israeli Period (1948-1950)

Plans for the foundation of a Jewish city alongside the Ottoman one existed already in the early 1940s, but Jewish settlement began from the existing part (mostly in infill housing similar to the existing patio houses of the previous period). However, since plans for new neighborhoods already existed, the tendency of the authorities was to establish new, "western" type clusters. These plans included not only the probable location of new neighborhoods, but also detailed plans for the use of local stone and other materials for construction.

The main planning concepts, which influenced these new neighborhoods were two: garden cities and suburban agricultural units. The former concept stemmed from the European reaction to the poor living conditions created in the urban centers at the time of the industrial revolution, while the latter was developed in Israel as a settling pattern for remote areas lacking economic infrastructure.[4] The main purpose of this second pattern was to enable the creation of urban settlements, development towns, which would be based on family agriculture for self-consumption as a secondary source of support. Whatever their roots, the physical implication of these two patterns was similar. Since they both dictated the allocation of relatively large plots of land per dwelling unit, the clusters they formed were of low density. Wide open spaces were created which could not provide any protection from either the hot summer sun, or the strong desert winds, be they hot or cold, carrying dust and sand, or accompanying a sudden down-pour.

Two suburban agricultural neighborhoods were planned and built according to these principles (Neve Noy and Shikun Darom), both of them to the south of the Ottoman town. The suburban or rural character of the built environment, formed by single family houses in plots of an average of 1.5 dunam, was certainly alien to the environmental conditions of the desert.

10.3.4
Garden City and Neighborhood Unit (1950s)

The concept of the garden city, created by Howard, but also appearing under different names in the works of other European architects motivated by romanticism as well as by humanism, such as Garnier, may be considered as the theoretical basis for the suburban development in Europe and, primarily, in England (Howard 1965; Wiebenson 1969). These and other concepts and theories were imported to a newly formed society by architects who were educated in Europe. Furthermore, the concept of a settlement formed by small houses set in green gardens seems to have appealed to those architects since it resembled conceptually the image of an oasis. The fact that soil types, water availability and intense radiation might prove problematic for their western concepts of gardening and landscaping, seems to have been disregarded.

An additional concept which was imported from abroad and has had a definite influence on the cityscape of Be'er Sheva, is the concept of the neighborhood unit, according to which each neighborhood is a self-contained urban unit, including services and facilities (clinic, movie theater, shops). This concept enabled - may even have dictated - the non-continuous, fragmentary development of the city. Neighborhoods A, B and C, which were built at the time, have no functional, visual or other connection between themselves. Each one started from the core outwards. Thus, the borders among the different urban units were reduced merely to transportation routes. It is claimed that the concept stemmed from the

---

[4] An extensive discussion of the European roots of such prototypes has been presented by Shilony (1998), especially in Chapter One: The European Background, pp.23-36.

fact that the mobility of the people at the time was relatively low and private car ownership was negligible. The self-contained neighborhood unit was supposed to provide basic and immediate needs and goods for the neighborhood population. It may be argued, though, that especially because of this low mobility level and because of the harsh climatic conditions of the region, it would have been much more practical and logical to compact the developing areas and try to create a built continuum.

Construction was undertaken by large contracting companies under the authority and supervision of the Ministry of Housing. Housing units were standard. Most of the buildings were two storey high and included four apartments of up to 32 m^2 each. Later on, the small size of the unit, and the relatively large average size of the families housed there, created a wave of additions to the houses by the tenants, many of them illegal. Furthermore, the large open spaces between buildings, meant to function as public, green areas, were left unattended and were used for infill housing which was different from the existing units. Both of these processes resulted in a very confused functional and visual agglomeration, which, in turn affected the image and, consequently, the development of the different neighborhoods in a negative way.

10.3.5
The New Master Plan and Design Experimentation (1960s)

The late 1960s in Israel were characterized by a general euphoria and accelerated development in all fields. Master plans were prepared for the different settlements, most of them taking into account the most optimistic statistics. Be'er Sheva was included in the list of settlements to be properly designed. The new master plan set the target population to 250,000 in the year 2000. In addition, it attempted to finally move the existing central business district (CBD) from inside the dense old city (the part built at the time of the Ottomans and the British) to the main intersection of roads connecting Be'er Sheva with Tel-Aviv to the north, Eilat to the south, Hebron and Jerusalem to the east, and Ashdod and Ashkelon on the coastal plain to the west. The CBD and civic center, situated at the southeastern part of the city would spread along the main avenue of Be'er Sheva and would create an axis of services and facilities, which would connect it to the university campus, and the hospital and medical center compound situated near the north entrance to the city. The neighborhood unit concept guided the master plan to divide the city in squares of roughly 700x700 meters framed by wide roads. Each one of these units was supposed to contain 4-7,000 dwelling units which, in turn, would enable the development of neighborhood services (Golani and von Schwarze 1970: 3.36-3.41). Guided by the rapid development forecasts, the planners left large land reserves inside the neighborhood units, as well as along the area which was to become the new CBD and the civic center.

Cluster and building design during this period were marked by experimentation. Patio houses arranged in low rise/high density patterns; long apartment buildings forming either protective edges for the low rise clusters, or forming enclosed public open spaces; "streets in the air"- open corridors for apartment buildings; passages for pedestrians enclosed in multi-functional buildings; and high rise apartment buildings with units shading the neighboring ones, are but a few of the prototypes built at the time. Although important in themselves, most of these prototypes remained isolated incidents in the urban landscape of Be'er Sheva, either because they proved to be unpopular or because they were too costly. In any case, the impact they have on the urban landscape is that of a building type museum, in which one may find one example of each type of building or cluster.

Out of this inventory of building patterns, two had a great impact on the urban landscape. Long apartment buildings of four storeys on pilotis may be found either arranged along traffic arteries, or arranged around semi-public open spaces. The idea of enclosed public open space seems quite attractive at first glance. However, since the supposedly protective edges are on pilotis, air movement through the open space is free and uncontrolled, thus, turning the open space into an unpleasant area for most of the year. As for the buildings arranged along the traffic arteries, they have a relatively small impact on the urban landscape as perceived by the pedestrian, either because they are disconnected from the public pavement or because they are oriented regardless of wind and sun directions and thus provide little or no climatic protection on the street level (Meir 1989).

The most promising cluster type of this period, the low rise/high density cluster of patio houses, protected by long apartment buildings positioned at the edges, was implemented once, in the "model neighborhood" in neighborhood E. Although construction was of a relatively poor quality (uninsulated envelope, leaking joints, poor ventilation), there was, and still is, demand for the houses. Mobility in the neighborhood is relatively low (Golani and von Schwarze 1970: 4.72-4.82). The reason why the pattern was not repeated seems to lie in a misconception of the suitability of open spaces for a region such as the Negev. It was assumed that stacking the dwelling units into apartment houses would free the ground area from construction and would enable the creation of large and continuous open spaces. These, however, proved to be anything but an asset, since they are exposed to unfavorable climatic conditions and are very expensive to maintain, demanding a lot of irrigation and work.

Whatever the reservations may be today, the master plan of Be'er Sheva was considered a major breakthrough in professional thought in town planning, and the interdisciplinary team responsible for the plan was awarded the Reynolds Prize in 1969.

10.3.6
Prefabrication, High Rise Buildings and Traffic Separation (1970s)

It is hard to really distinguish between the different styles of the 1960s and the 1970s, since the two intermingle. Nevertheless, one may identify the 1970s by three patterns: prefabrication, high rise buildings and traffic separation.

Prefabrication seemed to be the proper solution to a number of problems, such as housing solutions for massive immigration, demand for bigger apartments and lowering the cost of construction through standardization. The fact is that in many cases, prefabrication resulted in an unimaginative repetition of elements and units. The technical solutions were poor and no special attention was paid to the environmental constraints of the region, since the solutions were uniform for the whole country. This may be explained by the small size of the overall housing market in a small country, a fact which may be conceived as not justifying the creation of different prefabricated types of buildings for the different climatic zones (Etzion 1985).

High rise buildings would enable the creation of public open spaces bigger than those achieved through the use of the four-storey buildings. Furthermore, it was assumed that the higher an apartment, the more pleasant the environmental conditions would be, since there would be better ventilation and more dust-free wind. The truth, though, is that wind at the height of a few storeys above the ground becomes a problem because of its high velocity, and causes high heat losses through the building envelope, both by accelerating infiltration through fenestration, and by accelerating convection and conduction losses. Even the problem of airborne dust is not so easily solved, since dust particles are easily raised to a height of 2-3 kilometers above ground and may settle slowly over a period of days or weeks after a storm (Erell 1997; Golany 1978). Nevertheless, the higher densities (and the resulting economic benefits) reached through the construction of high rise buildings, brought about the design of a large number of point blocks 8-14 storeys high, both in new neighborhoods (F, I, K) and as infill in existing ones (B).

In some cases, such as along one of the main (excessively wide) avenues of the city, clusters of these point blocks create visually perceptible edges and define well the public and semi-public/semi-private open space. From the climatic point of view, most of these buildings and clusters have created unpleasant conditions in their vicinity, mostly by creating wind turbulence around their base.

An additional concept, which guided town planning and urban design at the time, was that of absolute differentiation between pedestrian and vehicular traffic systems. Professional literature of the period includes a number of publications advocating this tendency (Breines and Dean 1974; OECD 1974; Woods 1975). Naturally, this theory, too, found its way into Israeli town planning and urban design (Mertens and Golani 1973). In practical terms, this resulted in a multiplicity of open spaces devoted to traffic, as well as in the location of parking space at the fringes of the dwelling clusters. Both of these results proved to be of negative effect, functionally and climatically. The distances between parking and house are long (by today's standards, which demand parking as close to the house as possible) and in many cases, the hierarchical division of vehicular systems proved to be unnecessary, since the traffic volume in many of the secondary (intra-neighborhood) and tertiary (cluster) roads proved to be very small.

10.3.7
Satellite Rururban Development (1980s)

One of the most serious problems created in the older neighborhoods was that of social stagnation and consequent disintegration. Since most of the apartments built until the 1980s were relatively small, and since the prosperity of the 1980s brought about the rise of living standards, most of the economically stronger strata of the population started abandoning the older parts of the city. At the same time, the overall population of the city, as well as that of the whole Negev region, was reduced by emigration to the center and north of the country. As a consequence, neighborhood services and facilities became obsolete or their level and quality dropped. Apartment prices dropped and the whole process became a repeating cycle. This process was accelerated and encouraged by the new land policy of the government, which enabled for the first time the massive creation of single family housing neighborhoods and suburbs by leasing state owned land to private individuals. Such satellite settlements appeared around Be'er Sheva, in a process similar to that which occurred in the rest of the country. Although this process started already in the late 1970s, the 1980s were the period of real "suburbanization." However, the establishment of suburban-rururban quarters could not suffice the high demand, while, at the same time, the authorities tried to divert part of the socio-economically stronger strata back to the city. The Municipality of Be'er Sheva, in co-operation with the Ministry of Housing, revised the master plan of the city and decided to create zones for low rise/low density housing inside the plan's area. Such neighborhoods developed relatively fast both as separate quarters, and inside existing neighborhoods, be they in the core of the city or at the fringes. The average plot area allocated for such projects ranged from 400 to 500 m^2. The density thus created, is closer to the "garden city" concept of the city's first years, than to the urban image the master plan was supposed to create. Furthermore, the definition of the settlement edges is very weak since the low density neighborhoods placed at the fringes seem to dissolve into the desert rather than differentiating the built, treated environment from the natural, exposed one.

An additional conceptual change in the design of some of the low density quarters, especially those inside the urban fabric, is that of public open space definition and design. In many cases, the concept of absolute differentiation and separation between vehicular and pedestrian traffic has been abandoned and attempts are made to return to a "traditional" street, in which different types of activities take place in the same space. This, in turn, certainly reduces the size of public open areas and has a high potential in creating weather protected, manageable open spaces. Most of these streets are based on the prototypes developed in the 1980s in various countries such as Holland and the United States (Royal Dutch Touring Club 1980; Appleyard 1981).

10.3.8
Emergency Planning for Immigrants (the Early 1990s)

The beginning of the 1990s was marked by a new wave of massive immigration to Israel, mainly from East European countries, but also from Ethiopia, South America and elsewhere. A significant portion of the thousands of people that arrived each month reached the Negev region (mainly because no housing remained available in the vicinity of the large urban centers). Lack of housing had already become a great problem and the authorities were trying to deal with the situation as with an emergency. To do so, temporary housing was provided for the new immigrants until permanent one was constructed.

A first neighborhood, Nahal Beqa, with some 2,300 units in such temporary houses (imported and locally manufactured mobile and other lightweight structures) was added to Be'er Sheva, along one of the streams to the southwest of the city. Since the structures were "temporary," the neighborhood, or at least its design, were not to be considered permanent. A second one, Ramot, designed by the emergency planning processes adopted at the time, and including largely, but not only, lightweight houses, is located on the hills to the northeast of the older neighborhoods. Although this location has certain advantages (cooling summer breezes and better views), the whole project suffers from a number of problems. It is hard to estimate the impact this new massive construction will have on the overall urban environment of the city. However, it is clear that the site selected for development is disconnected from the existing urban tissue. Furthermore, the distance between it and the city may result in a feeling of isolation.

The light structures, be they mobile or permanent, are unsuitable for climates with extreme fluctuations such as those in the desert. In addition to that they were not even required to comply with various Israeli standards, such as those concerning materials and insulation (although stricter requirements were adopted for electricity, gas piping and plumbing).[5] Simulation and monitoring results of such structures vis-a-vis conventional, heavy ones with external insulation, have shown that under the climatic conditions of the Negev, the former will suffer of internal conditions by far inferior to those of the latter. When optimally operated in summer, lightweight buildings monitored were uncomfortable for over 10 hours per day (over 27°C at 30 per cent relative humidity, reaching peaks of 34°C). In winter, they suffered of wide fluctuation reaching minima of about 6°C at night (Pearlmutter and Meir 1995, 1998). This sudden massive addition resulted in extensive consumption of energy for air conditioning and heating, which, combined with a general steep rise in air conditioner purchases, reached absolute highest-ever maxima during the winter of 1997-98 and the summer of 1998 (Meir and Messinas 1998).

10.4
Microclimatic Variability

To verify some of the assumptions and statements made in the presentation of the different development stages and neighborhoods, indicative measurements were performed. These included maximum ambient temperatures and wind velocities in the public and semi-public/semi-private open spaces. The collection of data was undertaken during the months of April, May and June 1991. Data collected included ambient (and sometimes surface) temperatures and wind velocities.

The purpose of the measurements was to obtain information on thermal comfort conditions in the open spaces. Readings were taken at street level, at an average height of 1.5 to 1.6 meters above ground. Characteristic points for each one of the urban space types dealt with in the paper were identified (Fig. 10.1) and an average thirty measurements for each site were made. Both temperature and wind measurements were made between 14:00 and 15:00 hours, during which time both reach their maximum. Measurements were made in spaces of different orientation.

---

[5] The official regulation was published by the Minister of Interior, and was announced in the Building Centre of Israel Bulletin (1991, no.20, p.29). The rationale behind this decision was that such buildings have a short life span, therefore building standards relating to conventional construction should not necessarily apply to lightweight "temporary" structures.

10 Planning Theories vs. Reality   197

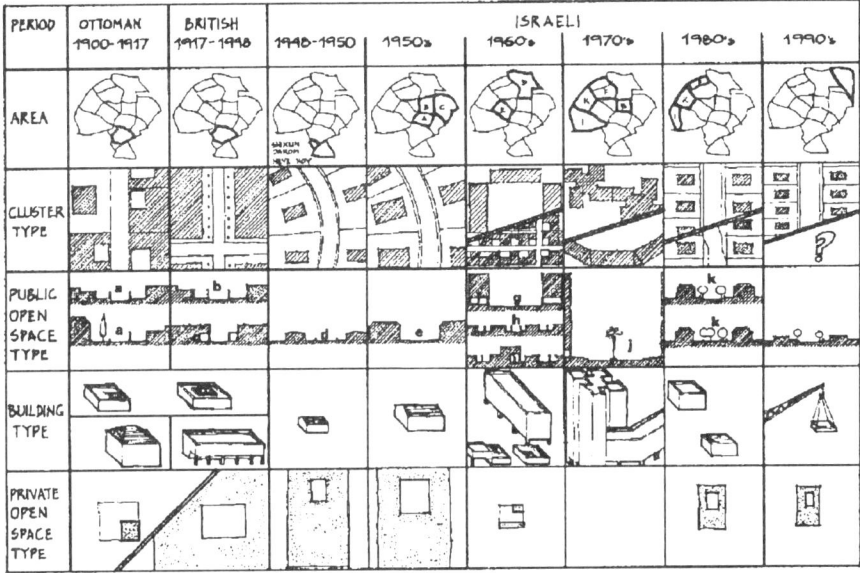

Fig. 10.2. Urban forms: cluster, building and open space types (measurement locations indicated by lower case letters)

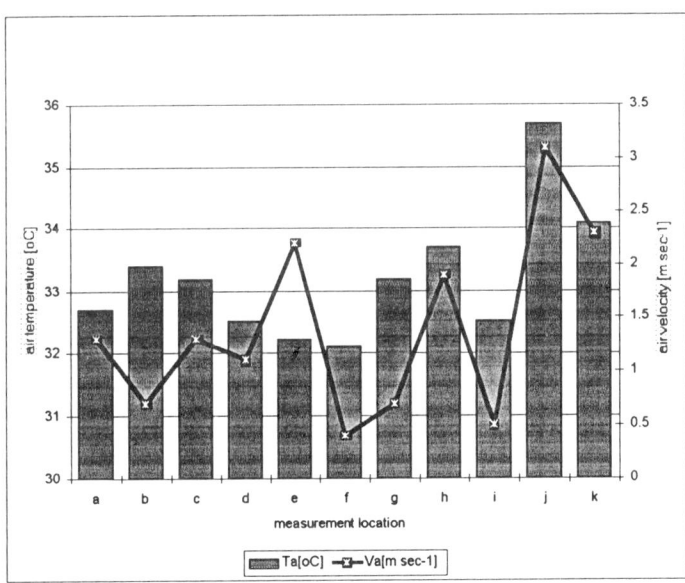

Fig. 10.3. Dry bulb and wind velocity average values measured (letters correspond with those in Fig.10.2 and Fig.10.3)

The typology of each site measured appears in Fig.10.2. Lower case letters correspond to those indicating the location of the sites in the city, as shown in Fig.10.1. Average ambient temperature and air velocity values for each site and space type are shown in Fig. 10.3.

The average maximum ambient temperature measured was 33.9°C, which is almost identical with the average maximum for the month of June. Deviations from the average ranged from −1.8 to +1.8°C. Air velocity in the different spaces varied widely (as expected) from an average lowest of 0.3 to an average highest of 3.1 m sec^{-1}. The average air velocity for the months of April, May and June at 14:00 is 5.7 m sec^{-1} and is the highest monthly average of the year. However, it is important to add that this value corresponds to air velocity measured at the meteorological station, at the top of a 10 m mast, situated in an open space. It is also important to add that absolute maxima of air velocity reach up to 11 m sec^{-1} in summer and may reach up to 18.5 m sec^{-1} in winter. Similarly, instant maxima measured during data collection were much higher than the average values mentioned here.

The lowest temperatures (32.2°C, or 1.7°C below average) were measured in two storey house clusters of the 1950s (or similar to those) (Fig. 10.2e). Average maximum air velocity in these areas was 2.2 m sec^{-1}. The next lowest temperatures were measured in the low rise/low density areas built in 1948-50 (32.5°C or 1.4°C below average) (Fig.10.2d), and were similar to those measured in covered streets of low rise/high density areas of the 1960s (Fig.10.2i). However, air velocity in these two cluster types was very different (1.1 m sec^{-1} in the former compared to 0.5 m sec^{-1} in the latter). The similarity in ambient temperature may be attributed to the high proportion of shaded open space provided by mature vegetation and building overhangs respectively. This may also explain the difference in air velocity, with the latter, dense cluster type providing better protection from wind.

The next group of temperatures was measured in the Ottoman and British period parts of the city. Temperatures of 32.7°C (1.2°C below average), 33.2°C (0.7°C below average) and 33.4°C (0.5°C below average) were measured in the Ottoman, British covered and British open spaces respectively (Fig. 10.2a; Fig. 10.2c; Fig. 10.2b). Air velocity in these areas reached a maximum of 1.3 m sec^{-1} in the Ottoman part and a minimum of 0.7 m sec^{-1} in the British part. The ambient temperature differences in these three cluster types has to be attributed to faster heat dissipation to the environment in the less dense Ottoman sector, rather than heat accumulating in the denser British sector. The pattern of air velocity in the British period sector is not very clear since the complicated morphology of the arcades and overhangs creates locally varying air movement patterns. Somewhere between these two cluster types, lie the open spaces of the 1960s defined by three to four storey buildings on three sides (33.0 to 33.2°C or 0.9 to 0.7°C below average, depending on the relation of vegetation to hard paving, the latter being the hottest) (Fig. 10.2g). Air velocity in these spaces reached an average of 0.7 m sec^{-1}. It is interesting, however, to note that conditions of the adjacent pilotis areas (Fig. 10.2f) are much better, with ambient temperature reaching 32.1°C (1.8°C below average) and average air velocity reaching 0.4 m sec^{-1}. (This air velocity was measured in pilotis areas of buildings in the cluster core, while air velocity in similar building areas at the cluster fringes reached more than 1.5 m sec^{-1}.)

The next cluster group is characterized by average maxima just above the average ambient temperature, and includes the uncovered open spaces of the low rise, dense clusters of the 1960s (33.7°C, or 0.2°C above average) (Fig. 10.2h) and the new suburban type clusters of the 1980s (33.7 to 34.1°C or 0.2 below to 0.2°C above average, depending on the density and height of the buildings) (Fig. 10.2k). The two cluster types were exposed to air velocities of 1.9 m sec^{-1} and 2.3 m sec^{-1} respectively.

The worst environmental conditions measured were those in the wide open spaces of the 1970s characterized by three to four lane roads and wide pavements, empty land reserves and high rise buildings (Fig.10.2j). The average maximum temperature there reached 35.7°C (1.8°C above average) and air velocity reached 3.1 m sec^{-1}. Combined with the intense solar radiation common at this time of the year, and the airborne dust, these conditions become highly uncomfortable.

From occasional observations, it seems that the environmental conditions in summer nights and in winter will follow symmetrical patterns. Dense clusters will cool slowly during summer nights, and will heat up slowly in winter days (only a small proportion of the covered open spaces enjoys solar access). On the other hand, wide open spaces will cool fast during summer nights and, although at least partly exposed to the winter sun, will suffer during the cold months from thermally uncomfortable conditions because of their exposure to strong winds.

The data collected point toward a number of definite conclusions. It is obvious that clusters similar to these from the 1970s (wide open spaces defined by high buildings) suffer of most unfavorable conditions. It is also logical to assume that the growth of vegetation in the new suburban type areas of the 1980s will probably bring conditions in them closer to those in the clusters of the late 1940s and the 1950s, namely will cause the reduction of ambient maximum temperatures.

The cluster types enjoying the most favorable conditions were those with low rise/high density construction. Although ambient maxima there were not the lowest measured, it may be assumed that thermal comfort conditions exist throughout most of the year due to the existence of mild air movement. These conclusions are correct provided that the relative humidity during heat peak hours is close to the average value of about 30 to 35 per cent (average relative humidity in April, May and June at 14:00). However, higher relative humidity values (40 to 50 per cent) are not an uncommon phenomenon.[6] Higher temperatures are also common, either in the summer or during the transition periods. In such cases of high ambient temperature and relatively high humidity, intense solar radiation and hot dusty winds, being in the open spaces becomes unbearable.

Air velocity measurements in the built environment showed wide variations among the different space typologies. The winding form of the streets and the fabric of the urban mass complicate the analysis of air movement patterns. It is

---

[6] Such high relative humidity may be caused by regional changes, such as wind blowing from the sea or barometric changes, sometimes combined with local humidity generators, such as air conditioning and evaporative cooling, irrigation and heavy traffic. It is hard to estimate the exact effect these sources have on local conditions, since no such systematic research has been undertaken so far in the area, but residents have been complaining of a rising humidity.

obvious that in most cases measured, air movement near the ground was not necessarily the direct result of regional winds, but rather an indirect creation of local air movement stirred by strong winds at higher altitudes. In any case, measured values of air velocity near the ground were by far higher from those measured or calculated for similarly defined areas (urban settlements, city centers) (Davenport 1965; Poreh and Paciuk 1980). These conclusions explain to a great extent the fact that no differences were observed in air movement velocity or pattern in open spaces of different orientation.

10.5
Conclusions

The modern city of Be'er Sheva has a history of some ninety years. In this period the population increased from a few tens in the early 1900s to over 150,000 inhabitants (to enhance the booming growth of population, it is enough to mention that from 1964 to 1972 the population of Be'er Sheva grew by some 132 per cent) (Golani 1973). Most of the construction and development were undertaken under pressures of sudden immigration waves or other political and economic constraints. Most of the first planners and architects had very little or no acquaintance with the special environmental constraints of the desert. Lack of previous building history in the recent past enabled relating to the area as *tabula rasa* and made possible the implementation of new, innovative theories and practices. As a consequence, experimentation was undertaken on a relatively large scale, something that would have been impossible in any existing settlement.

The controversy starts at the point where this attitude extends from the design theories and architectural heritage to the environmental conditions. This lack of relation to the special constraints of the desert enabled the implementation of theories and fashions, which were developed in environments vastly different from the one in discussion. Low density neighborhoods, dispersed development and large land reserves left in the core of the settlement for future development created a fragmented urban tissue. Furthermore, the inflated statistics on which the master plan was based, and the sudden halt in population growth sometime in the late 1970s, postponed the infill process which was supposed to create an entity out of the existing fragments.

The socio-political developments of the 1990s and the way the newly created pressure for massive housing construction was dealt with, point toward the possibility that past and present trends will continue. According to recent reports, the city, in the development stage of the late 1980s and early 1990s, included 634 sites categorized as public open spaces, reaching a total of 4,576 dunam, or, approximately, 20 per cent of the built area of the city. Put otherwise, the average open space area per inhabitant at the time of the survey was 41.5 m^2, while the equivalent in Holland was 60 m^2 and in the United States, 50 m^2. Out of this area, about one fifth is unattended (Haussmann 1988). To that, one should add an estimated 5,000 dunam of undeveloped land reserve situated in the core of the settlement, thus reaching an almost 1:1 ratio between built and open/empty areas.

Except for some special cases, such as the "example neighborhood" mentioned previously, in which most of the vegetation used in the open space landscaping was drought resistant, most of the public gardening has non-indigenous plants and, thus, the amounts of water needed to irrigate these areas in the specific environment may reach up to 1,000 m^3 per dunam per year for lawn, and about 12 m^3 per year for a tree. Considering these facts, proposals for temporary coverage of land reserves by vegetation become unreal. Consequently, these unattended areas become dust generators, increase the albedo of the urban environment, increase the amount of heat which is absorbed by the bare surfaces and later dissipated, and enable the strong desert winds and storms to cross through the built environment unobstructed. The discontinuity of built environment makes walking in the street a very unpleasant necessity, both under the burning summer sun and in the cold winter winds.

In addition to these problems, it is hard to disregard the lack of cohesiveness among different building types spread throughout the city, creating a visual discord and accentuating the city's fragmented and non-continuous character.

10.5.1
Positive Intervention in the Existing Fabric

A general revision of priorities is of utmost importance for the future development and image of the city of Be'er Sheva. There is vital need for a number of measures to be taken in order to improve the present situation and prevent development and construction from following chance rather than planning. These measures should include:

- Revision of the existing master plan and its adaptation to present needs of reality as have evolved in the city.
- Revision of the land reserve policy, which will enable the empty sites to be either developed now or properly environmentally protected until permanent development.
- Definition of the edges of the built environment and those of the neighborhood units.
- Revision and redefinition of the cluster, building and open space types appropriate for the desert environment of the city.
- Proper evaluation of the effect low density quarters have on the overall urban environment, functionally, climatically and visually.
- Preparation of detailed guidelines for the design of climatically protected open spaces.

These guidelines should dictate and enable the creation of a weather protected continuum in the urban space (Meir et al. 1998). The massive amount of new construction required by large-scale immigration provides a unique opportunity to achieve these goals. It should be exploited in order to change the existing urban fabric into a cohesive entity. This is also true for a number of other settlements in Israel.

10.5.2
Creating Desert Responsive Urban Forms

In a more general, universal sense, the following conclusions may be derived from the case of Be'er Sheva:

- Development should be continuous rather than leap-frogging.
- Each stage in the development of a desert settlement should be an entity in itself.
- Land reserves in the built environment should be avoided as much as possible.
- Low rise/high density clusters should be preferred, provided solar access in winter and proper ventilation in summer are ensured.
- Wide open spaces, high point blocks and low density clusters should be avoided as much as possible.
- The concept of thick green belts providing protection from wind and dust should be reassessed, especially in countries such as Israel, the water resources of which are limited and where drought is a periodical occurrence. The implementation of this concept seems possible only through the use of indigenous plants, the introduction other drought resistant ones, and the exploitation of runoff and treated sewage for irrigation.

10.6
Theory and Implementation

In parallel with the emergency planning processes of the early 1990s, the design of a new area for Be'er Sheva was undertaken. The site selected is located to the north of the city, stretching from the Be'er Sheva - Tel Aviv road to the west, to the Be'er Sheva - Hebron road to the south, and embracing the Ramot emergency neighborhood. This new area, Ramot Proper, includes some 8,000 housing units, as well as educational, cultural, recreational and commercial facilities, and open spaces. The area was divided into five neighborhoods, each one designed by a different architect, but all of them coordinated by the planner of the general master plan.[7] The five plans, initiated by the Israel Ministry of Housing and Construction were undertaken by five interdisciplinary teams, each one including consultants from the various relevant fields (water, sewage, roads, transportation, electricity, communications, landscaping, archaeology etc). The author of this paper was the energy and climate consultant of the project.

The overall design provided for minimum landscape damage where possible, as well as the restoration of already damaged areas. It also accentuated the importance of runoff utilization in green open spaces, appropriate vegetation for water conservation, shaded paths for pedestrians and bicycles, and sewage treatment and reuse. The various clusters and layouts were considered from their climatic and energy aspects, redesigned and adapted. Design guidelines, theoretical

---

[7] The Ramot Proper master plan was designed by Arie Rahamimoff Architects and Urbanists, Jerusalem, who also supervised and coordinated the five neighborhood master plans and interdisciplinary teams.

and technical information, recommendations and economic analysis of various features were included in a number of working papers and reports presented to the Ministry and the designers (Meir 1996a, 1996b). Although not incorporated as compulsory instructions, many of the guidelines developed through the design process have found their way into the design guidelines accompanying each separate master plan. These include cluster, building and fenestration orientation; open spaces treatment, including orientation, proportions, finish materials and vegetation; passive cooling and heating of buildings; recommended sections for walls and roofs, encouraging the adoption of insulation standards higher than the minimal ones required by law; and proper treatment of external surfaces for maximum comfort (thermal, visual etc). It is hoped that the energy conscious features of the Ramot Proper area will provide a climatically better environment for the residents, and will raise higher public awareness to, and expectations from, environmentally conscious and responsive design. The presence of climate and energy consultants in master plan design teams (as well as those of detailed plans) is becoming a common practice for all projects initiated by the national planning and development authorities. Although still far from ideal, it may be said that planning in the Israeli desert is taking a positive step toward sound environmental adaptation and sustainable practices.

References

Appleyard D (1981) Livable streets. University of California Press, Berkeley
Ben-Arieh Y, Sapir S (1979) The beginning of Be'er Sheva at the end of the Ottoman period (in Hebrew). In: Gradus Y and Stern E (eds) Beersheba. Keter Publishing House, Jerusalem, pp 55-68
Breines S, Dean WJ (1974) The pedestrian revolution: streets without cars. Vintage Books, Random House, New York
Davenport AG (1965) Wind loads on structures. Technical Paper #88, National Research Council, Ottawa
Erell E (1997) An experimental evaluation of strategies for reducing airborn dust in desert cities. Bldg & Envir 32(3):225-236
Etzion Y (1985) Desert architecture - the architecture of the extremes. In: Gradus Y (ed) Desert development - man and technology in sparselands. D Reidel Publishing Company, Dordrecht, pp 81-102
Golani Y (ed) (1973) Planning and implementation 1971-72 (in Hebrew). State of Israel, Ministry of Housing, Planning & Engineering Division, Jerusalem, pp. 134-136
Golani Y, von Schwarze DG (eds) (1970) Israel builds 1970. State of Israel, Ministry of Housing, Jerusalem
Golany G (ed) (1978) Urban planning for arid zones. John Wiley, New York, p 26
Haussmann Architects Consultants (1988) Public open spaces survey of Be'er Sheva (in Hebrew). State of Israel, Ministry of Housing, Town Planning Division, Tel Aviv
Herzog Z (1979) The Chalcolithic, Canaanite and Israelite periods (in Hebrew). In: Gradus Y and Stern E (eds) Beersheba. Keter Publishing House, Jerusalem, pp 27-37
Howard E (1965) Garden cities of tomorrow. Faber & Faber Ltd, London
Israeli A (1979) Ancient Be'er Sheva and its status (in Hebrew). In: Gradus Y and Stern E (eds) Beersheba. Keter Publishing House, Jerusalem, pp 21-24

Le Corbusier (1927/1970) Towards a new architecture. The Architectural Press, London

Meir IA (1989) Climatic sub-regions and design contextualism. Bldg & Envir, 24(3):245-251

Meir IA (1996a), Be'er Sheva - Ramot Proper: climatic analysis and recommendations for climatically responsive design. Working Papers I-V, submitted to the Israel Ministry of Housing and Construction. Center for Desert Architecture and Urban Design, Sede Boker Campus

Meir IA (1996b) Be'er Sheva - Ramot Proper: climatic analysis and recommendations for climatically responsive design. Final Report, submitted to the Israel Ministry of Housing and Construction, Center for Desert Architecture and Urban Design, Sede Boker Campus

Meir IA, Etzion Y, Faiman D (1998) Energy aspects of design in arid zones. State of Israel, Ministry of Energy & Infrastructure, Division of Research & Development, Jerusalem

Meir IA, Messinas EV (1998) Retrofiting of existing housing stock: a feasibility case study. In: Maldonado E and Yannas S (eds) Environmentally friendly cities. Proceedings of PLEA 98, pp 283-286

Mertens H, Golani Y (1973) The influence of physical structure of a residential quarter on its environmental quality. In: Harlap A (ed) Israel builds. State of Israel, Ministry of Housing, Planning & Engineering Division, pp 63-68

OECD (1974) Streets for people. Organization for Economic Cooperation & Development, Paris

Pearlmutter D, Meir IA (1995) Assessing the climatic implications of lightweight housing in a peripheral arid region. Bldg & Envir, 30 (3):441-451

Pearlmutter D, Meir IA (1998) Lightweight housing in arid zones: thermal comfort and energy use. In: Bruins HJ and Litwick H (eds) The Arid Frontier. Kluwer Academic Publishers, Dordrecht, pp 365-381

Poreh M, Paciuk M (1980) Criteria for identifying wind problems in initial planning stages (in Hebrew). Building Research Station, Technion, Haifa

Royal Dutch Touring Club (1980) Woonerf: a new approach to environmental management in residential areas and the related traffic legislation. Royal Safety Directorate, ANWB, The Hague

Shilony Z (1998) Ideology and settlement: the Jewish National Fund, 1897-1914. Magness Press, The Hebrew University, Jerusalem

Stern E, Gradus Y, Meir A, Krakover S, Tsoar H (eds) (1986) Atlas of the Negev. Dept. of Geography, Ben-Gurion University of the Negev, Be'er Sheva

Wiebenson D (1969) Tony Garnier: the cite industriele. George Brazilier, New York

Woods S (1975) The man in the street: a polemic on urbanism. Penguin Books Ltd, Harmondsworth, Middlesex

11 An Experimental Evaluation of Strategies for Reducing Airborne Dust in Desert Cities[1]

Evyatar Erell and Haim Tsoar

J. Blaustein Institute for Desert Research, Ben Gurion University of the Negev, Sede Boker Campus, 84990, Israel

11.1 Abstract

The effect of buildings on the dry deposition of dust was investigated in Be'er-Sheva, a desert city in southern Israel, and at two reference points in the surrounding countryside. The mineral and chemical composition of dust sampled at all sites was similar, reflecting the composition of the local loess soil, its likely origin. However, dust deposited in the traps set up in the vicinity of buildings in the city was significantly coarser than the dust which accumulated in similar traps at exposed sites in the countryside. The amount of dust in the urban dust traps was on average more than twice the amount deposited in the rural area. The differences in grain-size distribution and quantity of dust are accounted for by the disturbances to the natural environment caused by the presence of buildings and by human activity in the city.

This study suggests that strategies commonly employed in the design of buildings and urban space to reduce exposure to dust, such as the construction of walled courtyards, are not effective. A significant reduction in the concentration of dust in the vicinity of buildings in desert cities may require a comprehensive approach which deals with the entire urban area and its immediate surroundings, particularly with a view to reduce the availability of erodible particles by means of planting or paving all exposed land surfaces.

11.2 Introduction

Apart from the problems of physiological comfort generated by adverse climate, by far the most significant irritant in arid lands is air pollution by dust and sand (Saini 1980). Architects and designers practicing in desert areas are generally aware of the importance of dealing with the discomfort caused by dust, yet this awareness is rarely translated into appropriate design solutions, due to misconceptions concerning the environmental processes involved. In recent years, concepts imported from temperate climates have influenced the design of desert

---

[1] Reprinted from: Building and Environment 32(3) Erell E, Tsoar H "An experimental evaluation of strategies for reducing aiborn dust in desert cities," pp 225-236, Copyright 1997, with permission from Elsevier Science

towns, which often incorporate features more appropriate to "garden cities." Design guidelines for planners working in arid areas are based on the personal experience of their authors, but studies designed to provide the theoretical basis for understanding the effect of buildings on the deposition of dust are still lacking.

11.3
Background

11.3.1
The Transport and Deposition of Dust in the Urban Environment

The term "dust" may be used to describe fine particles which are transported in suspension by turbulent eddies in the wind. The term "surface dust", refers to dust found on the ground, in streets etc., and "atmospheric dust" to particles found in suspension in the air for relatively long periods of time (Fergusson 1992). Dust particles are smaller than 100 μm in diameter. The movement of dust particles through the air may be described in three phases (Pye 1987): entrainment, dispersion and deposition.

Entrainment: Particles may be brought into the air stream by any of the following forces, or a combination of them: aerodynamic forces acting on the individual particles, such as lift and drag; transfer of momentum caused by the impact of other moving particles, rain drops or hail stones; or disturbance by pedestrians, animals and vehicles.

Dispersion: Particles may move through the air in a number of modes, depending on their mass and the strength of the aerodynamic forces acting upon them (Fig. 11.1). The largest particles are rarely lifted above the surface, and their movement caused by the impact of smaller grains is described as "surface creep." Grains of fine sand, 125-250 μm in diameter, move in a bouncing motion, through repeated impact with the surface of the ground, known as "saltation," and rarely reach a height of more than 1 meter (Bagnold 1941). Fine particles (<100 μm in diameter) may be present in the air in "suspension" for varying periods of time, depending on their mass and the properties of the airflow, such as turbulence and atmospheric stability. Particles whose diameter is less than 20 μm may be transported thousands of kilometers from their source, and even cross oceans (Tsoar and Pye 1987).

The grain size distribution of dust particles in the air is a function of height above the ground. Large particles - >20 μm in diameter - are found only close to the ground, their concentration decreasing exponentially with height (Goosens 1985). The concentration of very fine particles - <10 μm in diameter - is almost independent of height.

Deposition - particles are deposited through the action of one of four mechanisms (Pye 1987):

- The particles are washed out of the atmosphere by precipitation.
- The particles settle due to the force of gravity. When the mean vertical component of the local airflow is lower than the terminal settling velocity of a particle, it will eventually settle to the surface.
- Very fine particles in long-term suspension may form aggregates as the result of electro-static forces or due to Brownian motion. The resulting particles have a higher settling velocity and are deposited through the force of gravity.
- Particles may be trapped when they come in contact with vegetation, water, etc.. The accumulation of dust resulting in the formation of loess soil occurs when there is vegetation in the area where particles are deposited (Tsoar and Pye 1987).

Patterns of particle entrainment and deposition may be also be affected by the topography, though the mechanisms involved are still in dispute (Pye 1987; Goosens and Offer 1988, 1990).

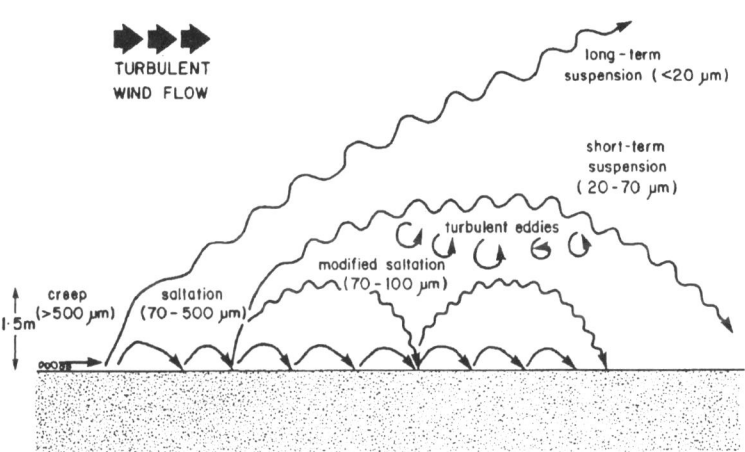

Fig. 11.1. Modes of particles transport by wind (Pye 1987)

11.4
The Urban Climate

The accumulation of dust depends on a supply of airborne particles and on the action of forces resulting in their deposition. The urban environment may have an effect on the deposition of dust because, under certain conditions, it may be an enhanced source of erodible particles, and because the urban micro-climate differs in several important ways from that of the surrounding rural area (Landsberg

1981; Oke 1987). The differences between the urban climate and that of the surrounding countryside are affected by the extent of the built-up area; by the spatial organization of the city, especially with respect to the natural topography; by the morphology of the urban fabric - the density of the built-up area, the height of buildings and the extent of uniformity of building size and form; by the amount and types of vegetation in the city; and by the existence of local sources of pollutants, such as industry and traffic.

11.4.1
Temperature

The air above a city is usually warmer than that above the surrounding countryside. The magnitude of this phenomenon, known as urban heat island, depends on the size of the city (Oke 1973). The urban heat island is a result of two main differences between the city and the surrounding countryside: differences in the physical qualities, and differences in the topography. The first factor refers to material that covers the city surface (concrete, asphalt) that has high thermal capacity, while the second factor refers to city buildings that create roughness and strong turbulence of the wind relative to the open areas outside the city (Chandler 1965; Skibin 1979). This effect is very salient during nights with thermal inversion that causes air circulation patterns that enfeeble the inversion and create a dust dome over the city. Measurements of the heat island of Be'er-Sheva done in December 1975, revealed an increase of up to 5°C at the city center (Skibin 1979), which is compatible with the inferences of Oke (1973).

11.4.2
Rainfall

The amount of precipitation measured in large cities is often greater than that measured in the surrounding rural area. Landsberg (1981) attributed this finding to the effect of the urban heat island, which results in thermal convection and atmospheric instability; and to the presence of condensation nuclei formed as a result of air pollution in the city.

11.4.3
Wind Regime

The urban wind field is affected by the aerodynamic roughness of the city and by the thermal forces resulting from the urban heat island. It is characterized by large variations in wind speed and direction, which are mostly due to the disruption in the airflow caused by the presence of buildings. These local variations may be more significant than the regional differences between urban and rural conditions.

The *mean* wind speed in the city is lower than in the surrounding countryside (Landsberg 1981). However, under certain conditions, the frequency of calm conditions may be lower. In London, the thermal convection associated with the heat island may result in updrafts near the city center and air being drawn in from the nearby rural areas, particularly at night (Chandler 1965).

The *vertical profile* of the airflow over large urban areas is affected both by thermal convection and by the aerodynamic roughness of the city surface. The atmospheric boundary layer is thicker over cities, and the rate of increase in airspeed as a function of elevation is slower (Davenport 1965).

An isolated building causes a disruption to the airflow which is expressed in changes of wind speed and direction at various points around it, and in the creation of turbulence in its vicinity (Givoni and Paciuk 1972; Aynslety et al. 1977; Paciuk and Poreh 1982). The wind speed on the windward side of such a building may be 30 to 60% of the velocity measured at roof height, while at its sides wind velocity may be 95% of the free flow. The lee side of the building is characterized by generally low airspeed but great turbulence.

The presence of adjacent buildings affects the airflow in more complex ways. The ratio between building height and the distance between adjacent buildings - H/W - is one of the most important parameters affecting the airflow (Oke 1987). For values of H/W less than 0.3 in the case of buildings of approximately square plan form, and 0.4 in the case of row houses, the airflow around the buildings will be similar to that found near free-standing buildings. When buildings are grouped more closely, up to a value of H/W of about 0.7, each building will be affected by turbulence due to the disruption caused by windward structures. As buildings are grouped ever closer, most of the airflow is deflected above roof height, and the spaces between them are exposed to turbulent eddies.

11.5
Experimental Sampling, Dust Deposits

The research was based on the sampling of dust deposited at a number of sites in the city of Be'er-Sheva, and at two reference sites in the adjacent rural area (Fig. 11.2.).

Be'er-Sheva is located on the edge of the Negev Desert in the southern part of Israel, approximately 45 km from the Mediterranean Sea. Mean annual precipitation is 203 mm, with rains occurring only between October and May. The mean daily temperature in winter (January) is 11°C, with a diurnal range of 10 degrees. In summer (July), the mean daily temperature is 25°C, with an average maximum of 33°C and a minimum of 18°C. Relative humidity is low, averaging 45-50% at mid day in winter and less than 30% in the summer (Bitan and Rubin 1991)

The wind pattern in summer is generally stable. The winds are light easterly in the morning, swinging to the northwest and increasing in speed to about 20km/h in the afternoon, and dying down completely toward midnight. In winter, the degree of wind constancy is lower than in the summer, but the magnitude of the wind is higher.

Dust storms occur in Be'er-Sheva mainly in winter and spring. Low pressure systems moving along the north coast of Africa are typical of the spring "Hamsin" depression. Dust storms in winter are the result of the "Cyprus depressions" similar to the one that occurred during the storm of November 3, 1991. As the focus of such a system approaches, the area is affected by strong easterly winds, which swing to the southwest and then to the northwest following the passage of a cold front. Dust storms of this type often extend over a front hundreds of kilometers wide, ranging from North Africa over the Sinai Peninsula and through the southern part of Israel (Ganor 1991; Ganor et al. 1991).

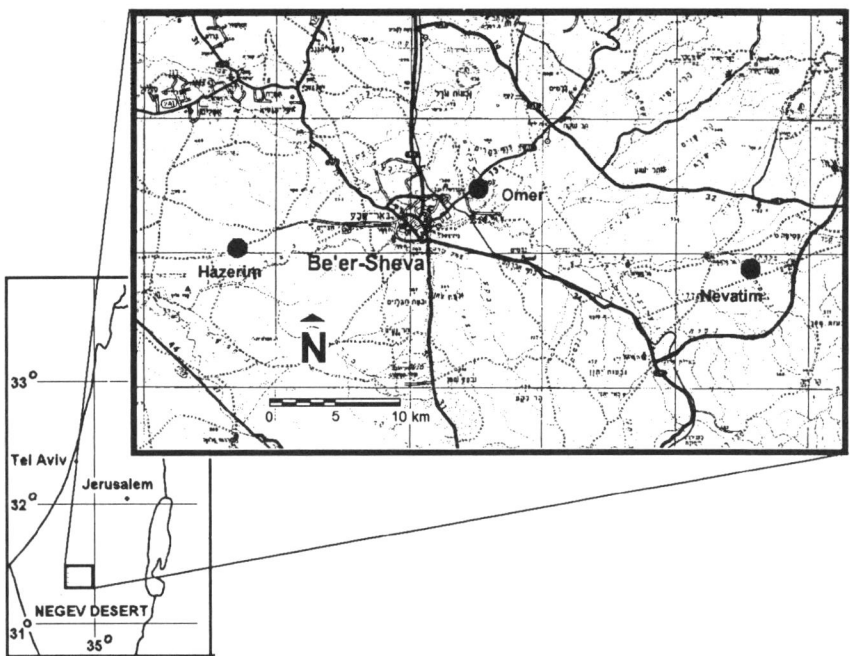

Fig. 11.2. Map of Be'er Sheva and the surrounding area, showing location of dust sampling sites

Dust deposition in summer is due to local effects. Concentrations of dust in the air are generally higher than during the winter. Following prolonged periods without rain, the topsoil is dry and often barren of vegetation, and is a potential source of erodible particles. High air temperatures and intense solar radiation result in elevated surface temperatures, which in turn cause thermal convection. Under suitable meteorological conditions - i.e., light breezes and local atmospheric instability - dust devils may be formed due to the thermal convection created over local hot spots on the surface. In the presence of erodible material, dust may be raised in plumes of 2-3 meters in diameter to a height of up to 200 meters, moving rapidly across the surface and dissipating in a short space of time.

Local soils are mostly desert loess, composed of silt (70-95%), clay (0.2-27%) and sand (0-9%). Natural ground cover is fairly sparse, consisting mostly of grasses and low shrubs.

11.5.1
Description of Sampling Locations

The sampling locations within the city of Be'er-Sheva were selected so as to provide data from sites with different morphology, random exposure to prevailing winds, and varying distance from the edge of the built-up area or from known local sources of dust such as construction sites. Five urban sites were selected (Fig. 11.3):

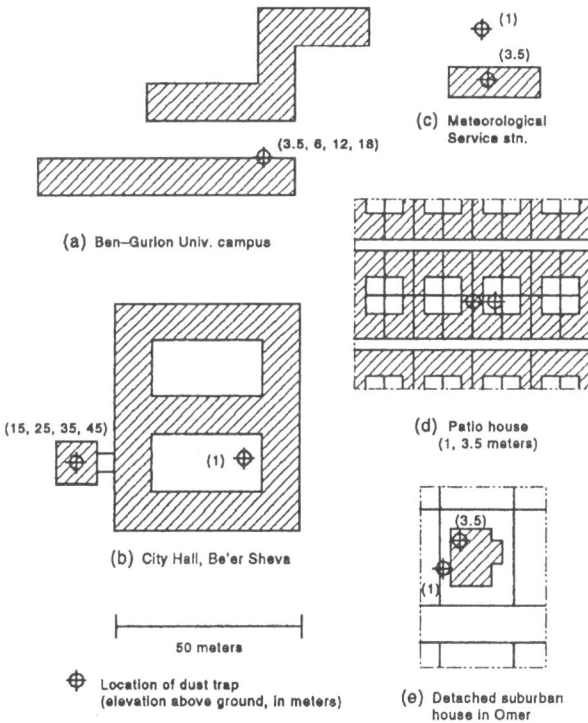

Fig. 11.3. Diagrammatic block plans of the five urban sampling sites, drawn to approximate scale, showing the location and elevation of the dust traps

Campus of the Ben-Gurion University: The dust traps were set up on window sills on the north-facing wall of an elongated building 4 stories high, 3.5 meters, 6 meters, 12 meters and 18 meters above the ground surface. A building of similar height faces it across the street. For part of the period when dust was sampled, construction work was being carried out on a building site about 100 meters west of the sampling site.

- *City Hall, Be'er-Sheva:* The City Hall building is located near the center of the urban area. However, its immediate vicinity is vacant of buildings, and the land surface is exposed for most of the year, being neither paved nor landscaped. It is 3 stories high, and has two rectangular courtyards completely enclosed by the perimeter of the building. An observation tower about 60 m high is adjacent to the main building on the west. Dust traps were placed near the center of one of the enclosed courtyards, and on the open landings of the observation tower, 15 meters, 25 meters, 35 meters and 45 meters above the ground.
- *Israel Meteorological Service Station:* The station is located about 200 meters north of the City Hall building. The nearest building west of the station is about 500 meters away, across a 6-lane main road. Adjacent buildings in other directions are 1 storey high, and are separated by vegetation. The dust traps were placed on the roof of the station, about 3.5 meters above the ground, and on a metal stand 1 meter above the ground.
- *Patio House:* The sampling site is in a low-rise, high-density neighborhood of buildings designed around small interior courtyards 15-50 square meters in area. All pedestrian and vehicular access is paved, but there is very little vegetation in the public spaces. This type of neighborhood layout, known as "carpet construction", leaves no exposed ground surfaces, creates very few exposed perimeter walls, and provides light and ventilation to the dwelling units through the enclosed patios which also function as attached, semi-outdoor space. The neighborhood is situated near the western edge of the city, surrounded by 4-storey buildings standing on pilotis. The dust traps were placed on the roof of the building, about 3.5 meters above the ground, and inside the small courtyard on a stand 1 meter above the ground.
- *Private Residence in Omer:* The sampling site is near the center of a suburb of Be'er-Sheva about 5 kilometers southeast of the city. The neighborhood is characterized by dense vegetation, well-kept public space and very little exposed land. The dust traps were placed on the flat roof of the building about 3.5 meters above the ground, and on a metal stand 1 meter above the ground, near the house and beneath a large tree.

Two rural sites, several kilometers away from the nearest buildings, were selected as a reference. The dust deposited in them is taken to be representative of the undisturbed, rural periphery of the city. The dust traps in each of the sites were placed on 6 meter high poles erected for this purpose, on struts 1 meter, 2.7 meters, 4.4 meters and 6 meters above the ground.

1. The site near *Hazerim*, about 13 kilometers west of Be'er-Sheva, is characterized by fairly sparse perennial vegetation. The pole was erected on a ridge rising about 30 meters above the surrounding terrain to the south and east.
2. The site near *Nevatim*, about 27 kilometers east of Be'er-Sheva, is in an area of gentle hills. Ground cover is sparse, consisting mostly of annual shrubs and grasses, which wither and dry out in early summer.

11.5.2
Field Methods

The research involved the sampling of dust simultaneously at a number of locations, over extended periods of time. Most of the sampling sites were unattended, requiring the use of cheap, expendable equipment. The research plan did not require the compilation of absolute measures of the amount of dust deposited, since the emphasis was on obtaining comparative data.

The dust traps used, following Ganor (1975), consisted of shallow plastic bowls, with a diameter of 22 centimeters and depth of 5 centimeters. The bottom of each bowl was covered with a layer of glass marbles 16 millimeters in diameter. The bowls had a plastic cover and were sealed upon removal from the site to prevent contamination of the samples. At the rural sampling sites, the dust traps were placed at several heights on metal struts extending from 6 meter high towers erected for this purpose. In the city, dust was sampled at several heights in each location, according to the characteristics of each site. In all cases, the lowest trap was set at a height of 1 meter above the ground.

11.5.3
Laboratory Methods

Dust was brought to the laboratory in sealed plastic bowls. The dust was washed from the container with distilled water, and leaves, dead insects, etc., were removed using a 63 µm sieve. Grain size distribution of the dust was analyzed using a Coulter Counter. The chemical composition was determined by SEM, using a dispersive X-ray spectrometer; elements lighter than sodium are not detected using this method, and the results are deemed to be accurate to within 5%. Mineralogical composition was established by X-ray diffraction. The dust was then filtered onto Whatman GF/C paper, oven-dried, and weighed to measure the amount of non-soluble material in the sample.

11.6
Results

The results in the following section are based on data compiled from four separate sequences of samples, taken between April, 1991 and September, 1992. One of the sequences displays data from samples taken during a dust storm which extended throughout the southeastern basin of the Mediterranean on November 3 1991. For the other sampling sequences, dust traps were placed in the field for periods of approximately one month at a time, during the spring, summer and early autumn. Due to various problems, dust samples were not available at some of the locations for at least part of the duration.

Data concerning the quantity of dust deposited are of most interest to designers and to the general public. Other data, such as the grain size or the chemical and mineralogical composition of the dust, are of value since they may help to shed light on the processes of transport and deposition involved, as well as indicating possible sources of the particles found in the dust traps. The data are presented here in summary form, in order to establish the experimental basis for the discussion that follows.

11.6.1
The Dust Deposition Rate

The amount of dust deposited at a height of 1 meter above the surface was used as the basis for comparison between the different locations investigated. Table 1 compares the average rate of dust deposition in the urban locations with the rate of dust deposition in one rural site (near Nevatim, 27 kilometers east of Be'er-Sheva). During the dust storm, dust was also sampled in the second rural site, 13 kilometers west of the city. The data given for this sampling period is the average for the two sites. Later studies established that conditions in the two rural sites are in fact similar.

The amount of dust deposited in the sites monitored in the city of Be'er-Sheva was on average 108% greater than the amount of dust deposited in the rural reference sites. While there were large variations in the amounts of dust deposited near buildings in the city, both on a temporal and spatial basis (see Table 11.1), the amount of dust deposited in the rural areas was always smaller than that deposited in the least exposed urban site in Omer, a suburb of Be'er-Sheva.

Table 11.1. The rate of dust deposition in traps at 1m height [$g/m^2/day$]

	spring 1991	summer 1991	autumn 1992	average for normal weather	dust storm, 3/11/91
City Hall	0.240	0.132	0.155	0.176	3.948
Meteorological Service station	0.194	0.202	0.207	0.201	3.553
Omer	0.186	0.127	0.086	0.133	2.500
patio house	0.333	-	0.168	0.251	5.264
Ben-Gurion University *	0.317	0.152	0.311	0.260	-
mean - urban	0.254	0.153	0.154	0.204	3.816
Hazerim (13 km west)	-	-	-	-	1.448
Nevatim (27 km east)	0.093	0.081	0.095	0.090	1.842
mean - rural	-	-	-	-	1.645

It should be noted that due to the limitations of the sampling technique, the rates of dust deposition reported here give only a rough approximation of the actual rates of deposition. They do, however, enable comparison between the various sites.

11.6.2
Grain Size Characteristics

The differences in grain size characteristics of dust deposited in urban and rural dust traps are displayed in Fig 11.4.

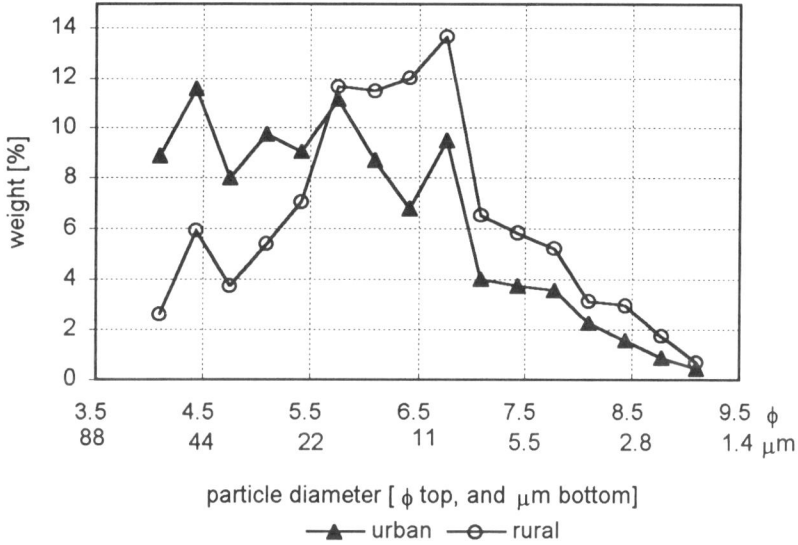

Fig. 11.4. Particle-size distribution of composite samples representing rural and urban dust

Composite samples were assembled representing dust deposited in rural and urban areas, using data from all urban and rural sites. Urban dust sampled at a height of 1 meter above the ground was found to be coarser, having a mean diameter of 19-23 μm, compared with a mean diameter of 13-16 μm for rural dust sampled at the same height. The frequency distribution for rural dust was approximately log-normal, and the dust was fairly well sorted. Urban dust was relatively poorly sorted, a possible indication of the effects of more complex air flow patterns. The frequency of particles with a diameter in excess of 20 μm was fairly high, unlike rural dust, but there was no preponderant size group.

11.6.3
Chemical and Mineralogical Composition

The chemical and mineralogical composition of the dust collected from all traps, in the city as well as in the rural areas, was similar. The dust particles were composed mostly of quartz and calcite, with smaller amounts of dolomite, feldspar and clay minerals (mostly kaolinite). These minerals are found in similar proportions in the local loessial soil, indicating that the likely source of the dust collected in the dust traps is the local unvegetated soil surface.

The chemical analysis of the dust was carried out by SEM (Scanning Electron Microscope), using the energy dispersive technique. The major constituents, in descending order, were silicon and calcium; minor constituents were aluminum, iron, magnesium, potassium and sodium. Table 11.2 shows the relative abundance of the most common elements in the dust sampled and in local soils, compared with average global values for soil and urban surface dust, adapted from Fergusson (1992), and the Israel Background Reference Standard proposed by Ganor (1991).

Table 11.2. The composition of dust deposited in Be'er-Sheva and in the rural hinterland, compared with the proposed Israel Background Reference Standard (Ganor 1991) and mean global values for soil and urban surface dust, adapted from Fergusson (1992)

element	Be'er-Sheva - urban		Be'er-Sheva - rural		Israel	global	
	dust	soil	dust	soil	dust	dust	soil
Na	1.9	0.3	1.5	<0.1	1.1	3.7	1.0
Mg	3.9	3.6	3.5	4.0	5.3	2.2	1.0
Al	10.2	9.8	9.9	10.0	9.6	12.0	14.6
Si	42.6	47.4	45.2	41.5	42.7	48.0	67.7
P	0.3	0.9	0.6	1.1	1.0	-	0.2
S	0.4	0.2	0.2	0.5	0.5	-	0.1
Cl	0.4	0.1	0.4	0.1	0.4	0.2	<0.1
K	2.3	3.1	2.2	2.7	1.7	4.5	2.9
Ca	28.7	25.7	27.4	30.3	29.9	18.0	3.1
Ti	1.2	0.8	0.7	1.0	0.9	0.6	1.0
Mn	0.3	0.3	0.3	0.1	0.1	0.2	0.2
Fe	7.8	7.9	8.1	8.8	6.7	10.5	8.2

The chemical composition of the dust deposited in the traps shows a very high correlation with the chemical composition of local soils, sampled in each of the locations and analyzed using the same technique (Fig 11.5a and 11.5b, showing the correlation for urban and rural sampling sites, respectively.)

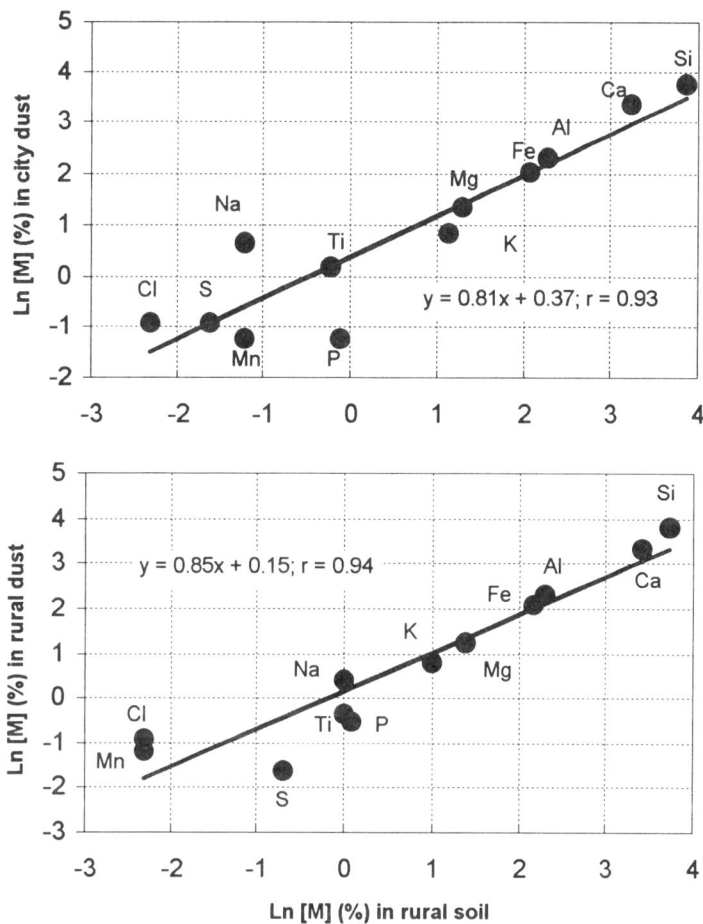

Fig. 11.5. Correlation between the chemical composition of local soil and airborne dust deposited in Be'er Sheva (a) and in the rural sites (b)

11.6.4
The Effects of a Major Dust Storm

Dust deposition was measured during a major dust storm which occurred on November 3, 1991, by means of traps which were exposed on November 1 and sealed on November 4, immediately after the storm. Extremely high dust concentrations - over 4,000 µg/m^3 were measured during this storm in Sede-Boker, 50 kilometers south of Be'er-Sheva (Offer and Azmon 1994) - occur on the average only 1-2 times a year (Offer and Zangvill 1992). The airborne dust concentrations in Be'er-Sheva under normal weather conditions are 100-200 µg/m^3 (Ganor 1975).

The Amount of Dust:

The rate of dust deposition during the storm was about 20 times that recorded during normal weather conditions in three other sampling sequences (Table 1). The difference in deposition rates between urban and rural locations found during normal weather conditions was also observed during the storm: The amount of dust deposited in the urban sites was on the average 123% greater than the average for the two rural sites.

Grain Size:

The mean grain size of dust deposited in the rural traps during the dust storm was smaller than that of dust deposited during periods of normal weather conditions - 12.7 μm vs. 15.1 μm. In urban dust traps, the difference -19.7 μm vs. 20.2 μm - was not significant, in spite of 15 meter per second winds and the high concentration of dust particles in the air.

Chemical Composition:

The composition of dust deposited in the two rural sites was very similar to that sampled concurrently by Offer and Azmon (1994) in Sede-Boker, about 50 kilometers south of Be'er-Sheva, using a High Volume Sampler. This indicated that that the composition of dust deposited during the storm over a wide area of the Negev desert around Be'er-Sheva was uniform.

The composition of dust deposited in the rural sites during the storm was similar to that deposited during normal weather conditions in spring and summer of the same year, as well as to that of the local soil. This indicates that a large proportion of the dust deposited during the storm was probably from local sources, while the contribution of Saharan dust, which has relatively less calcium and more aluminum (Ganor et al. 1991), is less significant.

11.7
Discussion

11.7.1
The Sources of Urban Dust in Desert Cities

The origin of dust particles affects the strategies adopted to reduce dust deposition. It is therefore important to establish whether most of the dust deposited originates within the city, or in the surrounding countryside. In a study conducted in the Tel-Aviv metropolitan area, near the Mediterranean Sea, the large variations in deposits of aerosols were attributed to patterns of land use, especially the presence of local dust sources such as heavy industry, and to the distance from the coast (Mamane et al. 1982). In the case of Be'er-Sheva, the differences in the quantity of dust deposited and in the grain size characteristics of the dust indicate that the city does in fact have an effect on the processes of dust transport and deposition. However, the composition of the dust particles sampled in the city and at the two rural sites was similar, and did not indicate whether the difference in quantity and grain size distribution were due to the presence of inner-city sources of dust, or whether the city was acting as a more efficient dust trap than undisturbed rural terrain.

Establishing the provenance of airborne dust particles is extremely difficult. While the presence of rare minerals is an indication of the possible source of the particles containing them, fine silt and clay particles may be transported thousands of kilometers, from any of several possible sources. However, coarse particles (diameter > 20 μm) are unlikely to be transported more than several kilometers from their source when exposed to moderate winds and a neutral atmosphere (Tsoar and Pye 1987).

Offer and Goosens (1990) reported that in a field study conducted near Sede Boker, a rural area about 50 kilometers south of Be'er-Sheva, no correlation was found between wind speed near the surface (up to mean hourly values of 7 m/s) and dust concentrations in the air. This was taken to indicate that most of the dust particles were from a distant source (thus implying that there were few particles in the local region susceptible to entrainment by wind of low to moderate speeds. High dust concentrations measured while local wind speeds were low were described as "dust haze," rather than "dust storms." There were only two occasions during the six-month study period when high wind speeds (>11 m/s) were accompanied by dust storms associated with local deflation. Offer and Goosens also indicated a marked increase in the concentration of dust in the air during the daytime (compared with the night hours) and during the summer (compared with winter). This was attributed to turbulence and atmospheric instability caused by thermal convection.

The relatively low concentration of coarse particles in the dust sampled at the rural sites (Nevatim and Hazerim) is thus in accordance with the results given by Offer and Goosens (1990). The supply of readily erodible dust particles in the rural areas surrounding Be'er-Sheva depends on the extent to which the natural ground cover has been disturbed. Where the microphytic crust is unbroken or where larger plants are sufficiently abundant, the availability of erodible dust particles is limited. The amounts of dust deposited at rural locations in the vicinity of Be'er-Sheva may be viewed as the natural "background" rate of deposition. The relatively low rate of dust deposition in rural areas is accounted for by the nature of the natural ground cover. While virtually none of the land around the city is in a pristine state, the area enclosed within a number of large military installations in the vicinity is largely undisturbed, and its condition is an indication of the natural state of the terrain. In the absence of human activity, the landscape is characterized by a matrix of crusted soil with microphytic communities and distinct patches of perennial plants (Boeken and Shachak 1994). Soil covered with a well-developed microphytic crust has a tightly structured surface with very few loose particles. Airborne dust is trapped by the perennial plants and forms a typical mound underneath them.

Human activity often results in the destruction of the fragile microphytic crust, and the loess soil may then be eroded by the wind. In the absence of the crust, surface runoff is smaller during rain events, and the existing shrubs are able to draw upon less rainwater for their survival. The result is often the desertification of areas lying on the fringe of existing deserts, even though they may have sufficient rain to maintain a fairly rich cover of vegetation.

The relative abundance of coarse particles sampled in dust deposited in the city of Be'er-Sheva, but not in the rural sites, suggests that these particles, a large proportion of the dust, originated within the urban area. The availability of coarse dust particles in the city is explained by the effects of human activities, notably construction work, but also motorized traffic. The sources of dust within Be'er-Sheva are mostly associated with disturbances to the soil - building sites, vacant lots serving as temporary parking spaces, etc.. In all such areas, the topsoil is broken up by vehicles and ground to a fine dust. Any disturbance, such as that caused by the passage of vehicles or even pedestrians, results in the ejection of dust particles into the air. The turbulence and strong vertical flows typical of the urban wind field result in the entrainment of larger particles than would be caught in the smoother airflow typical of rural conditions.

In conclusion, the processes of dust deposition in Be'er-Sheva and the surrounding rural areas may described as follows: Fine particles (< 8 µm in diameter), which may originate from long-distance sources, were probably found in suspension in similar concentrations throughout the region, and were deposited in the dust traps in comparable quantities. Coarser particles were deposited in greater quantities close to their source - within the urban area. The increase in dust deposition in the urban area is attributed to two factors:

- The abundance of erodible particles, due to the effects of human activities
- The unique properties of the urban wind field, particularly increased turbulence and vertical air flows.

11.7.2
The Effect of Common Design Strategies for Reducing Exposure to Airborne Dust

Givoni (1989) distinguished between two types of dust storms:

1. *Regional* storms, in which dust extends to heights of several hundred meters or more, and covers large areas - hundreds of square kilometers. Such storms occur several times a year, and are not a daily phenomenon.
2. *Local* dust waves, in which dust particles travel only several hundred meters from their source, at relatively low altitudes. Dust waves of this type occur daily in many parts of the world.

Little can be done to prevent or even minimize the impact of regional dust storms on outdoor spaces (Givoni 1989). However, it is often claimed that well adapted design can reduce the occurrence and minimize the impact of local dust waves (Konya 1980; Meir et al. 1990; El-Shakhs 1994).

The effect of some of the most common strategies for reducing exposure to airborne dust was investigated by placing dust traps in locations purportedly affected by those measures.

Solid Boundaries:

Solid boundaries, such as masonry walls, are often claimed to provide protection from dust in a certain area on their lee side (Saini 1980; Konya 1980; Meir et al. 1990). This claim was examined by placing dust traps in two adjacent locations near the center of Be'er-Sheva: the central courtyard of the City Hall building and the grounds of the local meteorological station, about 200 meters away.

A comparison of the dust samples taken in these sites indicates that a fairly large courtyard such as the one investigated has little effect on exposure to dust:

- The average *quantity* of dust sampled inside the courtyard was slightly smaller than in the exposed grounds of the meteorological station - 0.31 grams and 0.35 grams, respectively - but the difference is statistically not significant. In two of the four sampling periods, including the dust storm, the amount of dust sampled inside the courtyard was greater than that sampled outside it.
- The *mean diameter* of the dust particles sampled inside the courtyard - 18.1 µm - was slightly smaller than that measured at the adjacent meteorological station - 21.3 µm. This difference was found to be significant at the 0.01 level, and is explained by the smaller proportion of coarse silt particles in the courtyard, due to the effect of the 12 meter high perimeter walls.

Small Patios:

Small patios are regarded as the ultimate "protected" outdoor space and are claimed to provide good protection from airborne dust (Saini 1980). The effect of small patios in creating dust-free zones was investigated by sampling dust inside a small courtyard approximately 3 meters by 3.5 meters wide, enclosed on one side by a two-storey building and on the other three sides by two-meter high walls (the "patio house").

- The amount of dust collected in the dust trap placed in the patio was greater than the average for all of the urban locations during each of the sampling periods. In two of the three periods when this site was accessible, including the dust storm on November 3, 1991, the amount of dust in this sampling site was greater than in any of the other sites sampled (Table 11.1).
- The quantity of dust sampled inside the patio, at a height of 1 meter above the ground, and its grain size distribution, were identical to that of the dust sampled on the roof of an adjacent shed, at a height of 3.5 meters above the ground. This is explained by the fact that there are no local sources of dust within the neighborhood; Thus, any dust particles deposited within the courtyards originated some distance away. They were first raised to roof height, at least 3 meters above the ground, then deposited evenly all over the neighbor-

hood. The small courtyard, rather than preventing the accumulation of dust, acted as a dust trap, preventing the re-entrainment of particles deposited in it.

Vegetation:

The accepted approach to protection from airborne dust advocates the creation of dust free zones by "filtering" the air, generally by means of vegetation. Systems of shrub and tree windbreaks are recommended to reduce the amount of dust carried from one part of the town to another (Orev 1979). Low hedges are considered an effective means of lowering wind speed and filtering dust near the ground, while allowing unobstructed flow at higher elevations (Givoni 1989). The basis for this approach is that vegetation is indeed a very efficient dust trap. Yet there is no experimental evidence to support the contention that local filters of airborne dust have any effect on the deposition of dust on an urban scale, or indeed at any definite area in the lee of such barriers.

The effect of vegetation on the deposition of dust was investigated in Omer, a garden suburb of Be'er-Sheva.

- Dust deposition in Omer was consistently lower than in the other urban locations - 17% to 53% less than the average for all urban sampling sites - yet was still 40% greater than the average for rural sites (Table 11.1). The reduction in the amounts of dust deposited relative to the other urban sites may probably be accounted for by the scarcity of local sources of dust: Omer is an established and affluent neighborhood, and public spaces are generally landscaped and well-kept. The completion of construction done at a building site on the southern part of the community is reflected in a decrease in the rate of deposition during autumn of 1992 relative to the rate of deposition in spring and summer of 1991. This decrease was accompanied by a reduction in the mean particle size from about 20-21 μm to about 16 μm - only slightly more than the average for the rural sites, 14 μm.

- Omer was the only site where the amount of dust deposited at a height of 1 meter above the ground surface was *smaller* than that deposited at roof height (Table 11.3). This may be explained by the following mechanism: the dust trap near the ground was in a protected spot shaded by the thick canopies of foliage above. The dust trap on the roof was therefore more exposed to dust bearing winds. Dust particles deposited in the trap 1 meter above the had been in suspension at roof height, and, upon settling, were filtered by the leaves of the trees. This mechanism also accounts for the fact that the grain size distribution of dust deposited in both traps in Omer was nearly identical, and did not display the reduction in particle size usually associated with an increase in height above the ground.

In conclusion, the filtering effect of vegetation is evident mainly in the confined areas where particles settling vertically are intercepted by foliage. The main benefit to be gained from widespread planting appears to be the reduction of readily erodible particles which are entrained by the wind. A comprehensive study involving the sampling of dust at a large number of points may be necessary to establish the effect of vegetation - if any - on the spatial distribution and overall levels of dust deposition in a community such as Omer.

Table 11.3. The effect of vegetation on dust deposition in Omer

	Dust deposition (g)	
	Yard 1m	Roof (3.5m)
Summer 1991	0.24	0.29
Summer 1991	0.25	0.27
Dust storm	0.19	0.24
Autumn 1992	0.20	0.29

11.7.3
Reducing Dust in Desert Cities - A Comprehensive Approach

Reducing exposure to wind has generally been equated with protection from dust. Thus, the following design strategies have been recommended for desert settlements (Meir et al. 1990):

1. Neighborhood layouts incorporating a sequence of tall buildings forming a "wall" along the perimeter of the built up area.
2. The use of earth berms as shelter along the windward side of the settlement.
3. Planting belts of trees along the perimeter of the urban area, to protect against the wind and to filter the dust.
4. Construction on the lee side of hills with respect to the prevailing dust-carrying winds.

The findings of this research cast doubt on the effect of such strategies, which are designed to create dust-free areas by intercepting airborne particles by various means. This is because fine particles - 20 m or less in diameter - are found in similar concentrations up to great heights above the ground. Since the supply of airborne dust particles is practically infinite with respect to a unique location, "filtering" the air flowing through vegetation or over projections may create dust-free zones only if such areas can be sealed to prevent unfiltered air from penetrating - a strategy obviously unsuited to outdoor spaces. Thus, strategies commonly employed to reduce exposure to coarser particles moving along the ground or close to it (Buchman 1979), are not effective in dealing with smaller particles, which are transported in suspension. The only benefit which may be realized by the use of walls, vegetation etc., is the reduction of wind speed in their lee, which may reduce the discomfort caused by the impact of airborne particles.

The most effective means of reducing exposure to airborne dust is by avoiding the creation of sources of dust within the urban area or in close proximity to it. The means of reducing exposure to dust in desert cities are therefore mostly in the hands of urban and regional planners, and not the designers of individual buildings.

The strategies which should be borne in mind when planning desert cities are:

- *Preserve Natural Ground Cover:* Land on the windward side of desert towns should be kept in its natural condition as far as possible, so as not to destroy the natural ground cover, thus reducing the erosion of dust particles. Restoration of vegetation is essential where the microphytic crust has been broken. Experiments are now being conducted aimed at reversing this process through the artificial creation of man-made mounds, on which perennial trees and shrubs are planted (Boeken and Shachak 1994). The restoration of ground cover to desert fringe areas by this means may be a very cheap means of producing a stable environment in which the supply of erodible particles is greatly reduced.
- *Avoid Exposed Land:* All open public (or private) space should be paved or planted. In desert climates, unused public space is often the source of dust. In humid climates, such land usually develops natural vegetation, or the soil is kept moist by rainfall for most of the year, thus preventing the formation of erodible dust particles.
- *Adopt Compact Plans:* Compact plans reduce the extent of land requiring treatment to avoid the formation of dust.
- *Plan for Contiguous Development:* Allow for future expansion without leaving unused land near the center of the urban area.
- *Reduce Plot Sizes:* Small plots are more likely to be fully landscaped, thus reducing the probability that dust bowls will be present within the urban area.

11.8 Conclusions

Urban areas appear to have an effect on the patterns of dust transport and deposition. Compared with undisturbed rural areas, dust deposited in the urban locations investigated in this research was coarser and found in larger quantities. The mineralogical and chemical composition of dust sampled reflected the composition of the soil in the Be'er-Sheva valley. The difference in grain-size distribution appears to indicate that a large proportion of the dust particles deposited in the city probably originated within the urban area. Measures often recommended to reduce exposure to dust in desert buildings, such as the design of enclosed courtyards, were not found to be effective. City planners must therefore avoid features of the urban landscape which are likely to become sources of erodible particles. However, dust sampled during a dust storm that affected the southeastern basin of the Mediterranean Sea indicates that there is little that can be done to reduce exposure to airborne dust during events of a regional scale.

References

Aynsley RM, Melbourne W and Vickery BJ (1977) Architectural aerodynamics. Applied Science Publishers, London

Bagnold RA (1941) The physics of blown sand and desert dunes. Methuen and Co, London

Bitan A, Rubin S (1991) Climatic atlas for physical and environmental planning (in Hebrew). Ramot Press, Tel-Aviv Univ., Tel-Aviv

Boeken B, Shachak M (1994) Desert plant communities in human-made patches - implications for management. Ecological Applications 4(4):702-716

Buchman M (1979) Urban design in arid zones. In: Golany G (ed) Arid zone settlement and planning - the Israeli experience. Pergamon Press. New York

Chandler TJ (1965) The climate of London. Hutchison, London

Davenport AG (1965) The relationship of wind structure to wind loading (Proc. Conf. Wind effects on structures Vol. 1) Natl. Phys. Lab., H.M. Stationery Office, London pp. 53-102

El-Shakhs S (1994) Sadat City, Egypt, and the role of new town planning in the developing world. Journal of Architectural and Planning Research 11(3):239-259

Fergusson JE (1992) Dust in the environment. In: Dunette DA and O'Brien RJ (eds) The science of global change: the impact of human activities on the environment. American Chemical Society, Washington DC

Ganor E (1975) Atmospheric dust in Israel - sedimentological and meteorological analysis of dust deposition. PhD thesis, The Hebrew Univ. Jerusalem

Ganor E (1991) The composition of clay minerals transported to Israel as indicators of saharan dust emission. Atmospheric Environment 25a (12):2657-2664

Ganor E, Foner HA, Brenner S, Neeman E, Lavi N (1991) The chemical composition of aerosols settling in Israel following dust storms. Atmospheric Environment 25a (12):2665-2670

Givoni B (1989) Urban design in different climates. World Meteorological Organization publication WMO/TD-No 346

Givoni B, Paciuk M (1972) Effect of high rise buildings on air flow around them. Building Research Station - Report submitted to the Israel National Council for Research and Development, Haifa

Goosens D (1985) The granulometrical characteristics of a slowly moving dust cloud. Earth Surface Processes and Landforms 10:353-362

Goosens D, Offer Z (1988) Loess erosion and loess deposition in the Negev desert: theoretical modeling and wind tunnel simulations (internal report). The J. Blaustein Institute for Desert Research, Ben-Gurion University of the Negev, Sede Boker Campus

Goosens D, Offer Z (1990) A wind tunnel simulation and field verification of desert dust deposition (Avdat Experimental Station). Sedimentology 37:7-22

Konya A (1980) Design primer for hot climates.The Architectural Press, London

Landsberg HE (1981) The urban climate. Academic Press, London

Mamane Y, Ganor E, Donagi AE (1982) Aerosol composition of urban and desert origin in the eastern mediterranean. Water, Air and Soil Pollution 18:475-484

Meir IA, Etzion Y, Faiman D (1990) Energy aspects of design in hot-arid climates (in Hebrew). Israel Ministry of Energy and Infrastructure, Jerusalem

Offer Z, Azmon A (1994) Chemistry and mineralogy of four dust storms in the northern

Negev desert, Israel (1988-1992). The Science of the Total Environment 143:235-243

Offer Z, Goosens D (1990) Airborne dust in the northern Negev desert (January-December 1987): general occurrence and dust concentration measurements. Journal of Arid Environments 18:1-19

Offer Z, Zangvill A (1992) Preliminary investigation of severe dust storms in the northern Negev desert of Israel. Meteorology-Climatology-Atmospheric Physics 1:443-449

Oke TR (1973) City size and the urban heat island. Atmospheric Environment 7:769-779

Oke TR (1987) Boundary layer climates. Methuen and Co, London

Orev Y (1979) Revegetation of the arid range. In: Golany G (ed) Arid zone settlement and planning - The Israeli experience. Pergamon Press, New York

Paciuk M, Poreh M (1982) The urban wind field (in Hebrew).The Israel Ministry of Construction and Housing and the Technion Research and Development Foundation. Haifa

Pye K (1987) Aeolian dust and dust deposits. Academic Press, London

Saini BS (1980) Building in hot dry climates. John Wiley and Sons, Chichester

Skibin D (1979) Meteorological aspects of life quality in Be'er-Sheva (in Hebrew). In: Gradus Y, Stern E (eds) Be'er-Sheva, Keter, Jerusalem, pp. 325-331

Tsoar H, Pye K (1987) Dust transport and the question of loess formation. Sedimentology 34:139-153

12 Planning in Desert Environments: Three Cases of Responsive Planning[1]

Yehuda Gradus
Department of Geography and the Negev Center for Regional Development, Ben-Gurion University of the Negev, Be'er-Sheva, 84105, Israel

12.1 Introduction

Since the early 1990s, Israel's settlement pattern has experienced significant growth as a result of the rapid absorption of over 400,000 new immigrants (between 1989-90). By the end of 1996, Israel's population reached 5.8 million inhabitants, having experienced annual growth of over five per cent during the years 1990 and 1991. While new settlements were not founded, new neighborhoods were built in nearly all existing communities, in some cases resulting in a significant physical expansion.

This chapter examines three man-built environment projects in the Israeli Negev desert in light of the urban and regional concepts applied in their development. I shall argue that preconceived urban models cannot simply be "transplanted" to arid zones; cultural and environmental considerations are essential for the implementation of such projects. A change in attitude is thus necessary in the planning process in desert areas: that is a move from the preconceived to the responsive planning process (Rahamimoff 1981).

The first example deals with the regional planning strategy of the entire settlement system in the Negev. The preconceived concepts derived from Zionist ideology will be examined and the impact of the arid environment on the evolution of the planned system and the readjustment of the concepts will be explored.

The second example deals with the urban planning concept of the internal structure of towns and neighborhoods in the desert, and how this concept has changed, due mainly to responsive environmental considerations.

The third illustration refers to planning for the indigenous nomadic Bedouin population of the desert and how planners perceived future urban-built environments for this population. The lessons learned from this experience will be discussed in the light of the readjustment of the original models to responsive planning, taking into consideration environmental and socio-cultural considerations.

---

[1] The chapter is partly based on a longer paper on this topic published by this author and E. Stern in: Y. Gradus (ed, 1985) Desert Development. Dordrecht, D. Reidel Publishing Company, pp 41-59

12.2
Israel: Ideology and Planning

Spatial planning in Israel is inseparable from ideology. Ideology is deeply rooted and influences development at all levels. Population dispersion, one of Israel's major national planning goals, is seen as the basis for claiming sovereignty over the land. In addition, making the desert habitable for humans has always been the dream of many Zionist leaders. The conquest and cultivation of desert areas were major objectives of the Zionist movement. Ben-Gurion, one of the founding fathers and first Prime Minister of Israel, expressed this vision very clearly in his writings:

> The small State of Israel cannot long tolerate within its bounds a desert that takes up half its territory. If the State does not put an end to the desert, the desert is likely to put an end to the State...These areas cannot be settled without the transformation of the facts of nature, an accomplishment not beyond the capacity of science in our day or the pioneering energy of our youth. Science and pioneering will enable us to perform this miracle (Ben-Gurion 1956).

In addition to the ideologies of population dispersion and the conquest of the desert, the Zionist movement looked upon agriculture as the foremost instrument for Israel's development. In early Zionist ideology, the city has traditionally been viewed as a necessary evil. As Cohen (1977) put it:

> Pioneering Zionism has been characterized by a strong pastoral or agricultural bias stemming mainly from the belief that the country will be conquered through the conquest of the soil, and socially rejuvenated by creating a healthy peasantry. The tendency of Zionist leaders, and particularly of the leaders of the pioneering socialist movement, which dominated the Jewish community, was to disregard the city completely.

However, successful absorption of the masses of new immigrants required that they be placed mainly in urban centers rather than in rural settlements, owing to the lack of opportunities for employment, and constraints such as shortage of water and fertile soil in the latter. The agricultural sector was not suited to realizing the Zionist dream and, therefore, small urban settlements with a rural atmosphere and linked to rural settlements were considered a second best plan - a compromise between ideology and reality.

12.3
Settlement System

Selecting a strategy of optimal spatial distribution of settlement in an unpopulated desert frontier is still a theoretical issue in most countries, where the desert environment is perceived as wasteland unsuitable for development. However, for the Israeli regional planners, this was an issue of real and immediate policy.

A preconceived concept derived from a ruralistic ideology has had a substantial impact on the evolution of the man-built urban system in the Negev desert. The dominant socialist Zionist agrarian ideology adopted a dispersed model for regional development. The plan was to create balanced and integrated regions throughout the country, each with a central urban core as a service center. These urban centers were intended to constitute a hierarchy of central places, ranging from small urban centers to metropolitan areas.

The planners, most of whom received their training in Europe, were obviously influenced by Christaller's (1933) Central Place theory; in this hierarchical system, each region is an integrated unit with a distinctive identity and character, almost self-contained in its service, but with a strong relationship between urban and rural settlement (Shachar 1971). An urban hierarchy of towns linked to rural settlements, based on this regionalistic concept, suited the rural bias of the Zionist ideology, but was entertained only as a second-best solution because of lack of opportunities for large-scale agricultural settlement. Thus, the idea was conceived of making the new urban centers to be as similar to agricultural settlements as possible. The population of these settlements had to be small, resembling large villages rather than towns, and the center had to be located in agricultural surroundings (Cohen 1977). The policy objective was population dispersal and spatial equality rather than concentration (see Chap. 7 of this volume). The application of this strategy exemplifies one of the few cases where development of an urban system was based on a theory of spatial organization (Shachar 1971; Alexander 1978). However, in the application of the concept, the physical desert environment was almost neglected. In certain cases, the planners realized that it would be difficult to establish towns in the arid environment solely as service centers to the rural hinterland. Therefore, they proposed an alternative - the towns southeast of Be'er-Sheva in the extreme arid area of the Negev would also provide housing for workers in the mining and other industries of the region. During the early 1950s, ten new towns were established in the Negev desert within a short space of time, based on this preconceived concept of the dispersed hierarchical central place (Fig. 12.1).

By the early sixties, it was evident that the economic development of the new towns could not be based to a large extent on providing central services to surrounding rural areas. Because of the scarcity of water, agricultural development in the hinterland did not proceed as planned. Studies conducted during the mid-sixties to evaluate the success of the towns concluded that they were not fulfilling the function for which they had been created, and that they displayed no significant interaction with the small agricultural settlements in the hinterlands which had developed northwest of Be'er-Sheva (Cohen 1967; Stern 1977; Krakover 1979). Some of the hinterland settlements interacting with these towns were veteran kibbutzim and moshavim, each with its own marketing and purchasing organizations. Therefore, they were by-passed as direct links developed between the nearest large towns and the agricultural settlements.

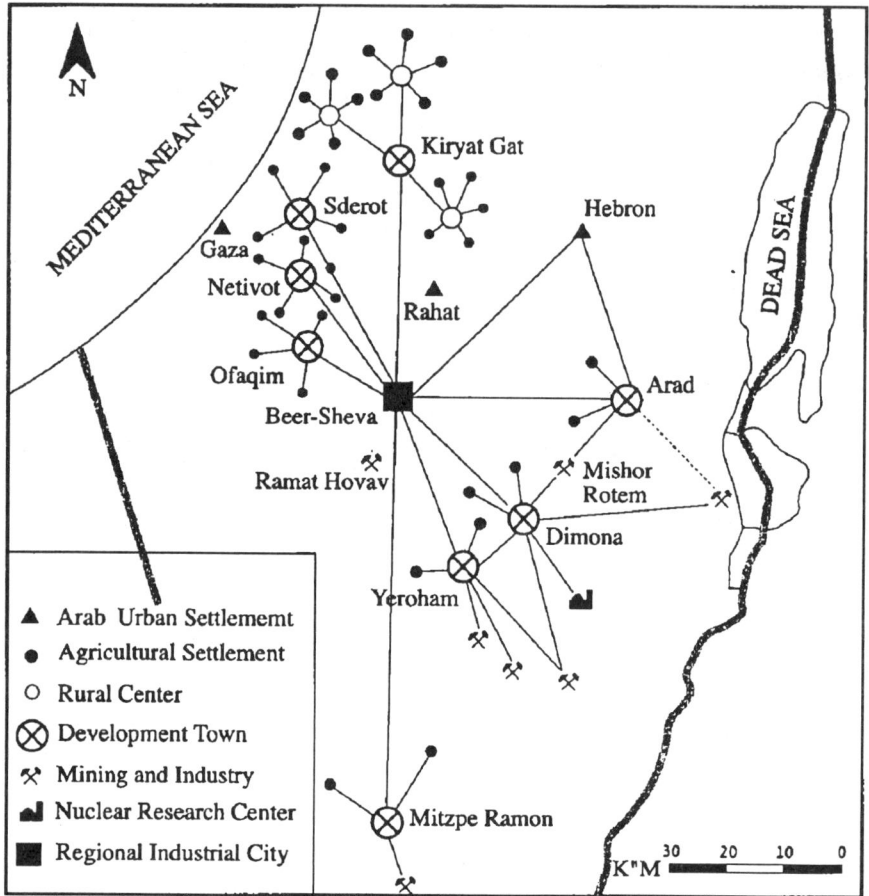

Fig. 12.1. The original Negev development concept of the 1950s

The towns, which had been designed to serve as centers, could not fulfill the expectations of their hinterland settlements in terms of the level of services. The level of health, education, and cultural services was low, and the physical urban planning of the towns inadequate. All these factors gave rise to an inferior quality of life, widespread poverty, and a high level of unemployment. The towns quickly turned into pockets of poverty, and the government was forced to provide direct assistance.

It became evident that the economic development of the new towns in the desert could not be based largely on providing central services to the surrounding region. A pressing need arose to create new job opportunities, and the government launched a massive industrialization program during the early sixties. By the early 1970s about half of all industrial manpower in the Negev was working in textiles and, within a short time, the towns which had been planned as service centers became typical company towns, vulnerable and dependent on outside decisions.

In conclusion:

The desert environment was thus neglected by the Israeli regional planners when they implemented preconceived rural concepts in the Negev. The Central Place theory as a strategy for regional development, as it was understood by Zionist planners, put the emphasis on dispersal, regional balance and hierarchical systems. They accepted this concept as a tool for regional development even though it was developed in an agricultural environment. An additional weakness of this theory is that its applicability is limited to the service sector and restricted mainly to consumer oriented activities. In cases such as that of the Negev desert, where the location of economic activity has been conditioned mainly by that of raw materials, and where there is a lack of widespread agricultural activities, the dispersed theory is virtually ineffective. It is evident from the Negev case that this strategy prevented the urban towns from becoming self-sustaining communities within their region. Only when a polarized growth strategy was chosen, could the region reach a self-sustaining stage. To achieve such a goal in isolated deserts, regional settlement development must be planned as an integrated and comprehensive unit, rather than as several small, isolated entities. A compact, functionally inter-related system with a major dominant growth center capable of providing such basic needs is desirable. Without such a center, self-sufficient regional development may be either very costly or ineffective.

12.4
Town-planning

The potential problems in applying a preconceived Western model of urban planning to non-western populations in an arid environment are demonstrated in the following example dealing with the designing of the internal structure of the city of Be'er-Sheva.

The school of thought prevalent among the planners of the new towns in Israel was that of garden-cities divided into neighborhood-units. The garden city movement emerged in the late 19th century as a reaction to the physical and social ills that industrialization had inflicted on many European cities. The purpose of the movement headed by Ebenezer Howard was to better the quality of life in British cities which were suffering from overcrowding and social and environmental pollution. In order to do this, the movement planned to create smaller urban units that would allow for social contact, while maintaining individual identity and obliterating the widening gap between city and country (Gradus 1978).

The idea of creating a rural atmosphere in an urban environment while keeping contact with the land, one of the major concepts behind the garden city movement, was readily embraced by leaders and planners of the socialist Zionist movement. In the early 1950s, this attitude was reinforced by the local leadership and founders of Be'er-Sheva, who were veteran members of the socialist agricultural sector.

European garden cities were planned for low density housing arranged in semi-self-sufficient neighborhood units. Each neighborhood would have in its center schools, shops, libraries, community centers, and so on. According to this concept, each unit was to be conducive to personal and community interaction in order to create a strong bond between the individual, the community and the urban environment. The aim was as much homogeneity as possible in the social structure of the neighborhood. There would be a uniform architectural style in each neighborhood to give a sense of unity to the entire community. Internal winding streets would be constructed, independent of and unrelated to the transportation system of other neighborhoods or to that of the entire city as a whole. Each neighborhood would be surrounded by a "green belt" in order to add to the rural character of the town and also to act as a buffer between the residential and industrial zones.

In the early fifties, most of these principles were indeed applied to Be'er-Sheva and other new towns in the Negev desert. Their application in Be'er-Sheva created a dispersed city composed of quarters remote and detached from one another, with no physical or social links between them (see Fig. 12.2).

In the 1950s-1960s, Be'er-Sheva's cityscape was rather monotonous. At the end of the 1950s, after the completion of most of the neighborhoods, the feeling was that the city suffered from lack of urban consolidation and from a great deal of dispersion, involving large expenditures to maintain infrastructural services such as sewage, water, telephone, electric systems and sanitary services. Inhabitants suffered from the great distances they had to walk in oppressive heat in search of services. The large spaces that had been left between the houses for gardens and cultivation of subsidiary farming remained unattended; residents made no attempt to improve their surroundings because of their indifference to the ornamental value of greenery, and due to the shortage of water. The vacant areas between neighborhoods which had been designated for public parks and greenbelts formed empty spaces in the urban environment, thus creating internal deserts within the city and lessening the urban feeling. The winding district roads unrelated to the general urban network created problems of communication and orientation.

The focusing of planning on the neighborhood also resulted in the absence of a main city center. By the early sixties, it had become imperative to re-examine physical planning. It had become evident that the garden city concept was inapplicable to Be'er-Sheva's desert environment and its socio-cultural reality. The planners realized that the imported plan which had been imposed on an area without regard to environmental conditions, had failed. The Be'er-Sheva plain had become a valley with a scattering of half a dozen unconsolidated and unrelated "villages."

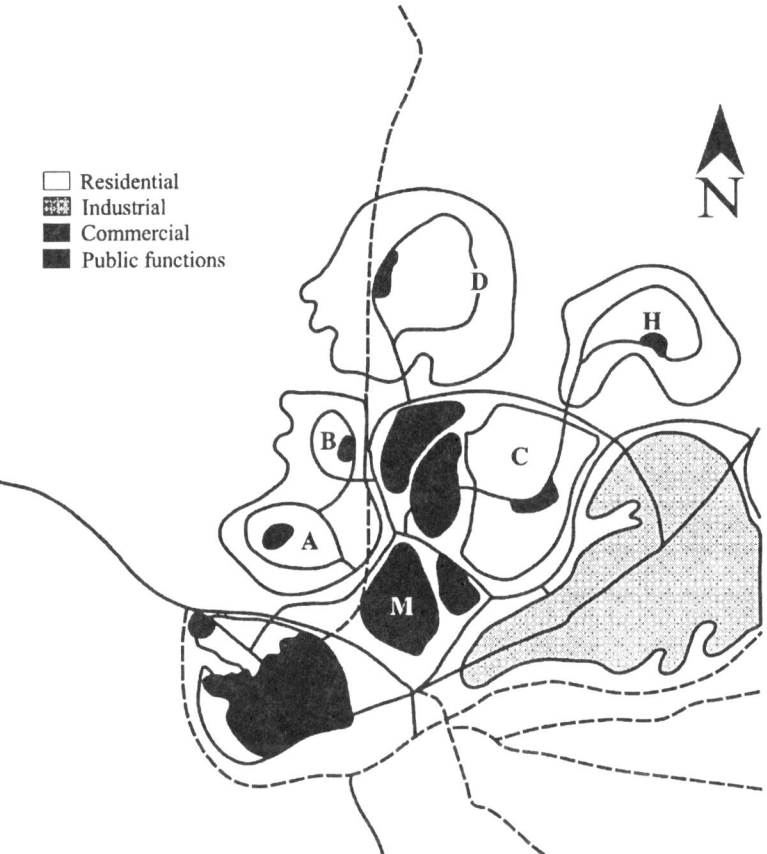

Fig. 12.2. Be'er-Sheva: comprehensive plan 1950

The new readjustment policies placed the emphasis on consolidating the city and condensing it in order to transform it into an organic unit functioning as a single economic and social entity. The awareness grew that, in a desert environment, an integrated planning approach is preferable to the autonomous dispersed-neighborhood units approach adopted earlier. In the replanning of neighborhoods, emphasis has been placed on condensing and consolidating the units and connecting them to each other; multi-storied houses were built in each neighborhood in the empty spaces which had earlier been earmarked for greenbelts or parks. This brought about a renewed flourishing of kindergartens and schools which had begun to decline. Shops and neighborhoods which had become run down revived. Stores were revitalized and some enlarged; houses and plots which had been neglected became valuable and, as a result, many residents began to improve and enlarge their homes. A hierarchical system of primary and secondary arteries was constructed between and within neighborhoods to ease traffic, and all roads were planned to aim at one axis channeling all the city's transportation (Fig. 12.3).

Fig. 12.3. Be'er-Sheva in the 1990s: physical layout and road pattern. (New residential neighborhoods, built after 1990 are shown in grey)

12.5
Bedouin Towns

Nearly 100,000 Bedouin living in the Negev belong to twenty-five tribes scattered over an area of 1000 km^2 in the north-central part of the region. Before 1949, a few Bedouin tribes practiced seasonal migration in this area, but most of them wandered with their flocks in the central and southern parts of the Negev. The tribes that moved to the area after 1949 are less attached to specific geographical territories within it and, therefore, are scattered over a larger territory, with different groups of the same tribe found in different locations. This spatial spread is also due to the process of spontaneous sedentarization which has been taking place over the last twenty years (Amiran et al. 1979; Shmueli 1976). Although the process probably started under the Ottoman rule (Musham 1970), it only recently became a planning problem. The scattered pattern of permanent and semi-permanent spontaneous settlements stood in conflict with development programs for the Negev, partly because of inefficient use of space. The municipal services required by a modern society, for example, are far more costly when they serve a widely dispersed population, particularly in an arid zone. In addition, the low level of habitation within the Bedouin settlements and the need for supplying regional services raised the necessity of controlling the spontaneous spatial spread of the Bedouin. Therefore the Israeli Government attempted to resettle the Bedouin in urban planned settlements forming an integral part of a regional development program. The first attempt to develop an urban settlement populated solely by former nomads failed because it implemented preconceived planning concepts, misinterpreting the effect of local environmental conditions on the Bedouin, as well as ignoring their socio-cultural needs and behavior.

Forming man-built environments for indigenous nomads is generally considered as a change more or less abrupt, and potentially disruptive to the whole system (Rapoport 1978). Efforts should therefore be made to reduce the incongruence between traditional lifestyles and settings and new ones, and to provide valid analogues of traditional forms when the latter cannot be used.

The first attempt to resettle the Bedouin in the Negev was undertaken in the early 1960s, when a new town of Tel-Sheva was planned to provide living space for 15,000 Bedouin residents. In 1966, the Ministry of Housing started the first phase, building 49 small houses (70 m^2) on 400 m^2 lots each (Lewando-Hundt 1979). The houses were built in a linear pattern, extending on both sides of a covered court accommodating a few shops and a clinic. No consideration was given to a spatio-social division or to possible interaction patterns. Both at the macro and the micro scales of design, inappropriate preconceived concepts, mainly related to the environmental arid conditions were practiced. The planners tried, for example, to adjust the buildings to the local arid conditions, by designing windows to account for only six percent of the buildings' outer-space. Plots were aligned with elongated backyards, supposedly for the use of husbandry animals. The city center was covered to create shady areas, and the total built up area was planned as a relatively high-density neighborhood.

The small houses offered in Tel-Sheva were unsuited to the large Bedouin families. The Bedouin, used to living in the open spaces, could not accept the idea of being compressed into small closed houses with few small windows intended to limit the penetration of dust during the frequent dust storms of the region. Actually, dust never bothered the Bedouin housewife, who accepted it as an unavoidable part of her surrounding environment. Consequently, the few Bedouin families that purchased houses in Tel-Sheva "solved" the problem by erecting their traditional tent in the back yard, while the brick house served their husbandry animals. Apart from the small family property, the high density of the town itself stood in conflict with the dispersed type of spontaneous settlements. High density obviously had an economic rational but evidently did not appeal to the Bedouin. Experience in Mexico and Botswana had already shown that the value of large plots and traditional areas at both the micro neighborhood and the neighborhood level may be very great. When traditional large plot sizes are significantly reduced, the change in lifestyle, activities, family, privacy, food habits, sociability patterns, and social relations is too abrupt.

Another misconception is reflected in the failure of the city center to function. Only one of the several shops built by the authorities continued to operate. The planners failed to consider the shopping practices of the Bedouin and the social attributes attached to them. The Bedouin are accustomed to doing their shopping once a week, mostly on their market day which takes place in an open plot in Be'er-Sheva. The traditional practice, and the proximity of Tel-Sheva to the business district of Be'er-Sheva itself, eliminated the potential for a relatively large shopping center in the planned settlement. Following the experience gained from the implementation of preconceived concepts in Tel-Sheva, a responsive planning process was successfully tried out in a second settlement, Rahat, the planning of which bears many of the theoretical considerations presented earlier in this section. The town, located 16 km north of Be'er-Sheva, was originally one of the largest concentrations of spontaneous sedentarization of Bedouin and ex-fellahin in the northern Negev. Its master plan (Fig. 12.4) was based on a population forecast of 21,000 in 1992, and 35,000 in 2000.

The basic principle of the plan was to set up a relationship between urban structure and Bedouin socio-cultural traditions (Stern and Gradus 1979). Thus, each social element in the Bedouin society was spatially identified with an urban element. The first stages of development were characterized by a widespread pattern of large neighborhoods inter-connected by a network of roads. Each neighborhood, however, served as an independent social framework, allowing for tribal territoriality. The internal structure of the neighborhood accommodated the hierarchical structure of traditional Bedouin society. Each street or alley was identified with an extended family, its households concentrated on adjacent lots. Such hierarchical coincidence between the urban and the social structures contributed to the individual's feeling of belonging, the Bedouin being able to maintain both tribal and spatial identity. Moreover, tribal territoriality is especially important in such a multi-tribal conglomerate which, within a defined geographical area, accommodates tribes traditionally antagonistic to one another.

A second principle of the plan was concerned with the channeling process. Rather than offering small built-up lots as was attempted in Tel-Sheva, the Ministry of Housing offered one or more relatively large vacant lots for each family within a tribal territorial framework. The inhabitants were free to build their houses according to their own budgetary limits, needs and pace. The physical and social infrastructure was prepared according to planned framework schemes. This "build-it-yourself" type of program enabled the Bedouin to adjust to the planned urban framework scheme by fulfilling their own socio-cultural needs within it. This incentive program became a success in terms of voluntary urban sedentarization.

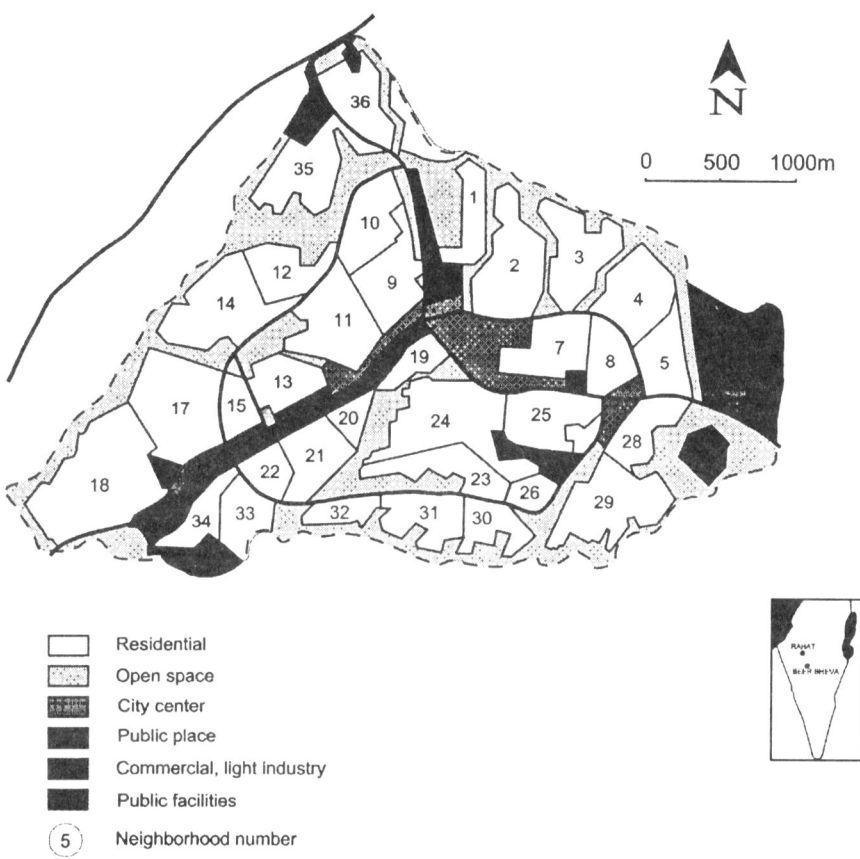

Fig. 12.4. Rahat: masterplan

The masterplan of Rahat also provided for future growth and development. The plan consisted of 36 residential neighborhoods separated by strips of public open space. The main commercial district extended linearly along the main arterial roads under the assumption that during the second phase of development, commercial land uses would be developed gradually on both sides. Along with the adaptation to the new built-up environment and further natural population increase, the vacant strips between the neighborhoods were expected to be occupied by both residential dwellings and public facilities. The changing occupation profile of the Bedouin was also integrated into the plan. For the first time, three light industrial zones were allocated within the Bedouin city limit. The areas were designed for industries such as brickyards, repair shops for trucks and agricultural machinery, and textile plants where young women would provide the bulk of the labor.

Rahat became a growth pole, stimulating new attitudes towards modern sedentarization in the surrounding Bedouin hinterland. As a result of this concept, the master plan of Tel-Sheva was revised and extended. The stages of adaptation and self-organization within the planned framework that had been observed in Rahat were also observed in Tel-Sheva. Three more Bedouin towns were planned for development in the Negev, south and east of Be'er-Sheva, all based on the same concepts and standards as implemented in Rahat. Responsive planning, which allowed the Bedouin to implement their own socio-cultural needs and which considered the spatio-social interaction pattern of the Bedouin along with the low construction density existing in their spontaneous settlement, was found to be the key to successful planned sedentarization of nomads in the Negev.

12.6
Summary and Applications in Planning

Three man-built environment projects in the Negev desert were examined in the light of preconceived regional and urban concepts. The examination demonstrated real and potential problems arising from the application of transplanted models coming from different cultural and environmental conditions to a fragile arid human ecosystem. While a simple transplantation of Western planning concepts had some success in the non-arid areas of Israel, it failed in the Negev mainly due to a basic environmental misconception. The transplanted models reflected a Zionist ideology of "back to the land" without considering the practical meaning of "land" in the dry areas. Western spatial, physical and social standards were proven unrealistic in the sparsely populated environment of the Negev. The Central Place and garden city models, and the rigid inappropriate cultural approach practiced in the early development of the Negev, in its principal town and in its first Bedouin town respectively, clearly demonstrate the failure of the application of preconceived design measures to the arid area. The shift toward a more responsive and less ideological approach reflects a flexible attitude towards planning, and has yielded better results from which several lessons can be learned.

All man-made environments are designed in the sense that they embody human decisions and choices. They thus embody specific ways of resolving the many conflicts implicit in all decision making. Desert planning which confronts a man/environment conflict (both physical and social) should therefore facilitate a framework for continuous response and interaction between the various elements of the arid system. To this end, planners must cease to be remote professionals "from above" and become responsive and sensitive to the needs and local conditions "from below." In other words, planners should move from preconceived to responsive planning. At the same time, public consciousness toward the desert built environment should be developed.

Regional development and urbanization policies should also be examined against the lessons of past experience. Comparative analysis of the accumulated experience gained through the formulation and implementation of development policies in desert environment such as the Negev, would be useful in future development projects in arid zones.

References

Alexander ER (1978) The Lakhish regional plan as applied to location theory. Growth and Change 9: 45-52

Amiran DHK, Shinar A, Ben-David Y (1979) Spontaneous settlement of Bedouin in the northern Negev. In: Shmueli A, Gradus Y (eds) The land of the Negev (in Hebrew). Ministry of Defense Publishing House, Tel Aviv, pp 562-665

Ben-Gurion D (1956) Southward (in Hebrew). The 1956 Annual Report of the Government of Israel. Government of Israel Printing, Jerusalem

Christaller W (1933) Die zentralen Orte in Sueddeutschland, translated by Baskin CW (1957) as: Central places in southern Germany. Prentice Hall, Englewood Cliffs, N.J.

Cohen E (1977) The city in the Zionist ideology. The Jerusalem Quarterly 4: 126-144

Cohen Y (1967) Urban zones of influence in the southern plain of Israel. Settlement Study Center, Rehovot

Gradus Y (1978) Be'er-Sheva. Capital of the Negev: desert function and internal structure, GeoJournal 2: 521-532

Krakover S (1979) The development of three new towns in the northern Negev: Netivot, Sderot, Ofaqim (in Hebrew). In: Shmueli A, Gradus Y (eds) The land of the Negev: man and desert. Ministry of Defense Publishing House, Tel Aviv

Lewando-Hundt G (1979) Tel Sheva - a planned Bedouin village (in Hebrew). In: Shmueli A, Gradus Y (eds) The land of the Negev: man and desert. Ministry of Defense Publishing House, Tel Aviv, pp 666-672

Musham VH (1970) Sedentarization of the Bedouins in Israel. In: Eisenstadt SN, Bar-Yosef R, Adler C (eds) Integration and development in Israel. Praeger, New York, pp 618-633

Rahamimoff A (1981) Extreme conditions and responsive architecture. In: Berkofsky L, Faiman D, Gale J (eds) Settling the Desert. Gordon and Breach

Rapoport A (1978) Nomadism as a man-environment system. Environment and Behavior 10: 215-246

Shachar A (1971) Israel's development towns: evaluation of national urbanization policy. Journal of the American Institute of Planners 37: 362-392

Shmueli A (1976) Bedouin rural settlement in Eretz Israel. In Amiran DHK, Ben-Arieh Y (eds) Geography in Israel. The Israel National Committee, International Geographical Union, Jerusalem, pp 308-326

Stern E (1977) Public transport level of-service in Israel's developing regions. Ministry of Transportation, Chief Scientist's Office, Jerusalem

Stern E Gradus Y (1979) Socio-cultural considerations in planning towns for nomads. Ekistics 277: 224-230

13 The Past as a Key for the Future in Resettling the Desert

Arie S. Issar
J. Blaustein Institute for Desert Research, Ben Gurion University of the Negev, Sede Boker Campus, 84990, Israel

13.1
Introduction

In many of the arid regions of the world, one can discover vast areas of deserted agricultural terraces, ruins of farms and ghost cities, all flourished during historical periods. In some places, one can see traces of later attempts to resettle these areas, but except for some impoverished nomad communities nothing survived. The question, which comes to the mind of the tourist walking through the streets of the deserted cities, climbing the ruined dams and terraces is whether these areas can be resettled? It is the opinion of the author of the present chapter that the answer depends on the understanding of the reasons for the desertion of these areas. As long as the answer continues to be incorrect, conclusions with regard to the way in which to resettle these regions will be erroneous too, and thus failure will be also the result of future attempts.

Until two decades ago, the blame for this desertion was put mainly on the people of the desert for their over exploitation of the natural environmental resources. Numerous reasons however, have brought many investigators to question this paradigm, and during the last two decades a new paradigm has been proposed, which says that climate changes, which caused the reduction of precipitation, brought the ancient agricultural civilizations of the deserts to a point where they had to yield either to natural or human enemies.

This paradigm, however, is not new. It was proposed in the early part of the 20th century, was accepted and later rejected and forgotten. The following chapter tells the story of the rise, fall and rise again of this paradigm, and draws conclusions with regard to the future.

13.2
A Lost Paradigm

In 1909, Yale University had granted a leave of absence from February until October, as well as a financial appropriation, to the geographer Prof. Huntington. It was given in order to enable him to tour Palestine and its neighboring countries. In the preface of his book "Palestine and its Transformation" Huntington writes:

> Extensive travels in Asia Minor, Persia, India and Central Asia led the author to adopt certain theories as to changes of climate and their relation to history. De-

scriptions of Palestine suggested that the same changes of climate have taken place there. Hence it seemed that in no other country could the theories be so well tested; for not only is Palestine so situated that climatic variations would there produce notable variations in habitability, but also its known history extends back to remote antiquity. (Huntington 1911, p.1)

During his travel in the countries of the Middle East, Prof. Huntington took upon himself the risk of a voyage into the Negev, which is the southern desert part of Palestine. He also visited the eastern part of the desert of Transjordan and southern Syria. At that time a traveler to these regions was at the mercy of the nomad Bedouins, who under the loose control of the Turkish government continued their traditional mode of life of raids on their neighbors and caravans crossing the desert.

He was impressed by the ruins of the once flourishing cities, like Petra, Ruhaiba, Auja, Palmira and Jeresh. His conclusion was that such a desertion, on such a big scale could only be explained by a climate change. In his book "Palestine and its Transformation" (Huntington 1911) and in his other many books and articles, he reported a correlation between the observations from Central Asia, such as the changes in Caravan routes, the levels of the Caspian Sea, the levels of the river Nile, the rise and decline of the ancient civilizations of the Middle East and his and others' observations and studies in many other countries. His main conclusion was that all these changes were due to climate changes on a global scale. He looked for an explanation for these in the cycles of pulses of solar activity and true to the deterministic paradigm claimed that the changes in the physical environment affected the quality of life and nature of man, and thus his history.

In his last book, "Mainsprings of Civilization," Huntington (1959) summarized his observations, a correlation and theory. In the concluding chapter he writes that the first conclusion is that we live in the midst of an intricate series of cycles, some of which are closely associated with atmospheric differences. He states that we do not know how far atmospheric electricity and ozone are causes or merely concomitants of the cycles in business and in the reproduction of animals, but clearly the field for further study is wide and alluring. He maintains that the number and intensity of ordinary cyclonic storms changes within cycles of hundreds of years long. These cycles had an impact on the history and culture of various nations, because climate appears to have influenced not only the capacity of a region to support people but also the activity of the human mind as well as the vigor of the body, for the production of food. Thus, the deterministic paradigm advocated by Huntington, claimed that the physical geographic conditions in each of the regions of our earth, mold the spiritual and physical characters of the races living in these regions. This meant that heredity, food and the geographic physical conditions expressed in temperature and climate changes determine the floods and ebbs of the tides of civilizations, which flourished during past years in this region. He backed up his conclusions with many figures, statistics and diagrams.

13.3
The Rise and Decline of the Deterministic Paradigm

The school of thought, which Huntington represented, namely the deterministic school, was very popular at the beginning of the present century, up to the beginning of the Second World War. The founder of this school was the German geographer Ratzel, who from the years 1885 to 1904 published a series of books and articles in which he expounded his views about the influence of the natural environment on man and his society. One of the terms, introduced by him was the title for his book "Der Lebensraum" published in 1904.

Since the eve of the Second World War the school of thought initiated by Ratzel and aggrandized by Huntington lost its supporters among geographers and became a story of the past. The reason for the loss of credibility of the Deterministic Paradigm was due to many reasons.

It seems that the misuse by Nazi ideologists, in the years prior to the Second World War, of the scientific conclusions and nomenclature used by the scientists of the deterministic school, like racial characteristics, "Lebensraum" etc. were instrumental in the decline of this paradigm.

Another reason was that historians and archaeologists were placing more emphasis on the human factor in opposition to the role of natural factors in changing the fate of the countries of the Near East. Among these archaeologists were people of world wide reputation like Sir Petrie, a professor of Egyptology of the University of London, one of the fathers of modern archaeology, who during the first half of the present century carried out many excavations in Egypt and southern Palestine. The archaeologist, Sir Woolley, the famous explorer and discoverer of the city of Ur in Sumer, carried out in collaboration with Lawrence (Lawrence of Arabia) a survey of Sinai and Negev, prior to the first world war. They summarized the findings of this survey in their book "The Wilderness of Zin" (Woolley and Lawrence 1936) in which they oppose the conclusions of Huntington regarding the desertion of the cities of this region, when the Moslems conquered this region.

The famous American archaeologist, Albright (1984), also denounced Huntington's conclusions. Albright conducted many excavations in numerous sites and in his book "The Archaeology of Palestine" he claimed that Huntington, blaming climate change for the desertion of the cities of the Levant, has built up "an elaborate superstructure of historical interpretation on many erroneous inferences."

The same conclusion was adopted by another American archaeologist Gluek, who conducted an extensive survey of Edom, Negev and Sinai. In his book "Rivers in the Desert." He states that the major factors which affected the course of human history, during the last ten thousand years, were anthropogenic.

Another factor causing the decrease in the popularity of the deterministic paradigm was the evolution of the science of soil conservation in America by the agricultural engineer Lowdermilk (1946). He suggested soil conservation as a counter measure to the severe processes of soil erosion, due to improper methods of ploughing and irrigation, which affected many areas in the USA. He was strengthened in his conclusion that man is to blame, while working in China, where he was sent, after the First World War, by the USA government, to help the people fight drought and famine. Lowdermilk's teaching included his "eleventh commandment" against the sin of causing wastage of land by erosion due to the improper ways of tilling. In his book "Palestine - Land of Promise" (Lowdermilk 1946), he emphasized the misconduct of man in all that relates to degradation of the land and its productivity, he concluded that the decline of the agriculture societies of the Middle East was due to the Arab invasions and the conquest of this region by the people of the desert, who lacked the knowledge of soil and water management.

The botanist Prof. Evenari and his collaborators, Shannan and Tadmor (1971), who spent many years in investigating the ecology of the Negev and the methods of irrigation used by its ancients, in their book "The Negev" supported these conclusions.

Thus the "Encyclopedia Hebraica" in 1969 in the volume devoted to Huntington concludes that "contrary to Huntington's suggestions, the present accepted opinion is that there is no proof to the occurrence of remarkable climate changes during the period of history. The level of the lakes in Palestine and Syria did not go down, while years of severe droughts, are known from ancient periods, prior to the drying up of the Levant" (p 816).

It is beyond the scope of the present article to deal with all the aspects of the geographical deterministic theory. The present discussion will concentrate on the questions, relevant to the problem of climate change, namely whether the Middle East underwent strong climate changes, during the last millennia, and whether these physical changes did influence the fate of the people in this region?

13.4
The Breaking Down of the Consensus

The unanimity of opinion about the stability of climate during the past historical periods started to be questioned during the last two decades, when more detailed investigations of the historical records in Europe, have shown that indeed climate changes did affect the environment and history of this continent. The British meteorologist Prof. Lamb in his books, especially in "Climate: Present Past and Future" (1977), "Climate History and the Modern World" (Lamb 1982) as well as in many papers recorded these changes. The most popular, well-known period is that of "The Little Ice Age" which took place between the fourteen and seventeen centuries AD (Grove 1988).

The author of the present article began to question the consensus against Huntington's determinism, when he came to the conclusion that the invasion of the sand dunes into the coastal plain of Israel, which started at ca. 800 CE was not due to an anthropogenic reason, as he had initially assumed in his Ph.D. dissertation presented in 1959. In this dissertation, the present author reported his discovery that the invasion of the sand dunes, covering a big part of the coastal plain, actually started at ca. 700 CE, namely with the Arab invasion.

The explanation, which he gave to this congruence, was in concert with the prevailing paradigm. Accordingly he blamed the invading Arabs for the destruction of the agriculture, of the coastal plain, which by its vegetation cover controlled the invasion of the dunes. He started to suspect that this anthropogenic explanation is not correct when he found out that an earlier massive invasion of sand dunes was at the end of the last glacial period, namely as a result of a climate change. The new observations, which made him change his mind, were made in connection with his investigation about the relation between the paleo climate of the Upper Pleistocene, the fossil water under Sinai and the Negev and the sediments of southern Israel (Issar 1985). These showed that during the last glacial period heavy rainstorms, which recharged the Nubian Sandstone aquifer, were also dust laden. This dust was deposited as the loess layers of the Negev. At the end of the glacial period, once the climate became warmer, deposition of the loess practically stopped and sand dunes began their invasion into the coastal plain. This invasion could be explained by an increase in the supply of sands from the Nile, a function of the strengthening of the monsoon rainstorms over subtropical Africa, from where the Nile receives a large part of its water and sediments supply. Another reason was the erosion of the Nile delta due to the rise of the sea level, a function of the melting of the polar ice, due to global warming. This new explanation brought the author to the question: whether the invasion of the young sand dunes, synchronous with the Arab conquest were not also due to climate change, rather than to human intervention?

The conventional anthropogenic explanation to the desertion of the Negev became more dubious when the author became more acquainted with the stupendous extension of the ancient irrigation and agricultural systems of the Nabatean-Byzantine cities of the Negev, as well as to their grandeur. These cities and the agricultural terraces constructed in all the river beds of the northern Negev, were deserted, more or less, at the same time when the young sand dunes advanced into the coastal plain, namely some time after the Arab invasion.

The more the author toured the area, the more he became dissatisfied with the conventional explanation offered by the archaeologists and botanists who investigated these cities and terraces and who claimed that their richness was mainly due to commerce, and the desertion was because of the collapse of the administrative and commerce systems, which supported them. He argued that if this was the only reason, then why in the first place the people of this region were so anxious to develop every small valley all over the region for agriculture? Moreover, this development did not transgress the contemporary 80-mm rain line, while the desertification did not transgress the 250-mm line. In other words, more and more data pointed in the direction that Huntington was right, and that indeed the primary reason for the desertification of the region was climate change. This desertification became, later, more extreme by the overgrazing of the animals of the nomads, who entered the region after it was practically deserted by its farmers.

Once the consensus about repudiating Huntington's explanation for the transformations of the desert of Palestine has been questioned, more data were sought, which might support or oppose the climate change hypothesis,. This brought the author to reexamine proxy data, namely indicators for climate conditions in the past. The most useful indicators are the oxygen 18 (^{18}O) composition of carbonates in deposits in lakes and cave stalagmites (Low ^{18}O content in these deposits indicates low temperatures), ancient levels of lakes (like high shores of the Dead Sea, observed by the geographer Klein), and assemblages of pollen in the deposits of the lakes and ancient levels of the sea (granted there are no local tectonic movements, which serve also as an indicator for climate change). Any rise in temperature will cause the melting of the polar glaciers and a rise in the sea level, and vice versa.

The data assembled contained many indications that indeed severe climate changes occurred during the historical period which influenced the hydrological cycle and thus the socio-economic systems and history of the countries around the Mediterranean Sea. An event of warming up and desertification, similar to that which occurred during the Arab conquest, at ca. 700 CE, was found to have happened during the Bronze Age, towards the end of the third millennium BCE. At that time the Canaanite city of Arad in the northern Negev, was deserted and sand invaded the coast and clogged up the Canaanites harbors, as observed by the marine archaeologist Raban (1987).

These observations brought the present author to question the accusation and verdict that the ancient people of Mesopotamia were responsible for the salinization of their soils, which also happened at ca. 2000 BC. The accusation of the Sumerians was brought up by the archaeologists Jacobsen and Adams (1958), who based their verdict on data from deciphered clay tablets from Sumerians archives. The data showed that although many long irrigation channels and probably also drainage channels were dug by the inhabitants of that area, between 2300 and 1800 BCE. At the same time the ratio of barley to wheat was constantly on the rise in offerings and taxes, delivered to the temples. As barley is more tolerant than wheat to soil salinity, the archaeologists concluded that excessive irrigation caused salinization. The explanation offered by the present author was that the salinization was caused because the flow of the Euphrates and Tigris diminished, due to the warmer and drier climate, and thus the inhabitants did not have enough water to irrigate and flush out the salts from their soils.

This reinterpretation of data, and history which included also the explanation of some of the stories of the Bible as connected with climate changes, were published by the author in his book "Water shall Flow from the Rock" (Issar 1990).

While working on this book the results of the investigation of the climate changes and their influence on the hydrology and history of China, which included data accumulated and processed by Chinese scientists were of special interest. During winter the climate of this country is affected by the polar continental air mass causing a northerly flow of cold dry air masses. In the summer the region is dominated by the tropical-subtropical oceanic air mass and tropical continental air mass which brings the oceanic warm and moist air from low latitude causing a southerly monsoon rains. Thus cold periods cause droughts and famine, while periods of warm climates bring more rains over China.

It was interesting to find out that the cold spell starting in the third century BCE in the Mediterranean region, which made its deserts more green and convinced the Nabatean to come down from the mountains of Transjordan to the Negev desert and start building Avdat and other cities, drove the Mongols from the cold plains north to China into the more southerly regions and forced the Chinese to build the Great Wall to protect their lands. The ensuing Moslem period, which was warm and dry in the Levant, was warm and humid in China.

During the "Little Ice Age," China became dry again and suffered from droughts and famines.

13.5
Conclusions with Regard to the Future

The general conclusion which may be drawn with regard to the future, once the old-new paradigm is adopted, is that in the case of a global rise in temperature, the Middle East, which gets most of its rain from the westerlies cyclonic climate system, will become more dry and arid. On the other hand in China and probably in all South East Asia, a warm period will mean plenty of rain, and most probably heavy floods. This is in agreement with the results obtained by the simulation runs of the General Climate Models, constructed by the meteorologists in order to simulate records of the past and forecast the future.

Taking these conclusions into consideration, the question to be asked is whether they do not annul the prospect of resettling the desert. The answer to this question depends whether one is ready to adopt a new urban socio-economic system instead of the ancient agricultural one. It is claimed that this is feasible due to the fact that agriculture, in Israel, as in many developed countries, is becoming less and less profitable, on the level of the individual farmer as well as on the national level. On the other hand the share of High-Tech and other industries as well as tourism becomes an important part of the gross national product. This causes the trend of urbanization to increase, and at the same time people's quest for higher standards of living.

Moreover, the profitableness of agriculture is going to continue to decrease as the competition concerning the allotment of water of good quality will cause its price to rise and only high profitable sources of income will be able to obtain this costly commodity. In such a situation, means of production, which can use water of lower quality, will have an advantage in regions where such water resources are available.

Although in general water of good quality is a scarce commodity in the deserts, in many places water of poor quality, is rather abundant. Today modern technology of desalination by reverse osmosis, makes this water an achievable resource, from the point of view of cost for urban supply. As mentioned, tremendous quantities of fossil water were located by the author of the present article under the Negev and Sinai.

Thus, although deserts, because their harsh climate and bare landscape, do not attract people, once water is available and urban centers in the form of city-oases are formed, the desert becomes a very attractive place for inhabitants as well as tourists. Especially so when the water of poor quality is used to support salt tolerant vegetation and game parks. Such city-oases, existed in the past, but their economy depended on agriculture and trade. In the future city-oases will flourish again, while their economy will be based on modern and advanced technology industries as well as on tourism.

Bibliography

Albright WF (1984) Archeology and the religion of Doubleday. Anchor Books
Encyclopaedia Hebraica (1969) Encyclopedia Pub. Co. Ltd, Jerusalem
Evenari M, Shannan L, Tadmor N (1971) The Negev: the challenge of a desert. Harvard University Press, Cambridge
Glueck N (1968) Rivers in the Desert. Norton & Co., New York
Grove JM (1988) The little ice age. Methuen, London
Huntington E (1911) Palestine and its transformation. Houghton Mifflin, Boston
Huntington E (1959) Mainsprings of civilization. New American Library, New York
Issar A (1960) The Geology of the groundwater of the Central Coastal Plain of Israel, PhD dissertation presented to the senate of the Hebrew University
Issar A (1985) Fossil water under the Sinai-Negev peninsula. Scientific American 253(1): 104-112
Issar A (1990) Water shall flow from the rock. Springer Verlag, Heidelberg
Jacobsen T, Adams RM (1958) Salt and silt in ancient Mesopotamian agriculture. Science 128: 1251-1258
Klein C (1982) Morphological evidence of lake level changes, western shores of the Dead Sea. Israel J. of Earth Sciences 31: 67-94
Lamb HH (1977) Climate: present past and future. Methuen, London
Lamb HH (1982) Climate history and the modern world. Methuen, Oxford
Lowdermilk WC (1946) Palestine: land of promise. Golancz, London
Raban A (1987) Alternated river courses during the bronze age along the Israeli coastline. In:Colloque Internationeaux C.N.R.S. Deplacements de Lignes de Rivages en Mediterranee (eds du CNRS Paris), pp 173-189
Woolley CL, Lawrence TE (1936) The wilderness of Zin. Jonathan Cape, London

Part Three

BUILDING AND DESIGN

14 A Desert Solar Neighborhood in Sede Boker, Israel[1]

Yair Etzion
J. Blaustein Institute for Desert Research, Ben Gurion University of the Negev, Sede Boker Campus, 84990, Israel

14.1 Introduction

The fact that Israel is a very sunny country would make one expect to find many solar buildings and solar neighborhoods in it. Surprisingly though, these practically do not exist: the number of solar buildings in Israel is small, and the first solar neighborhood has been built in this country. In general, the move towards "solar" is slow at the most. An attempt was made by the Israeli Ministry of Housing and the Ministry of Energy to promote solar neighborhood development, and indeed one such neighborhood was planned and designed through an open competition in Mitzpe-Ramon (Rahamimoff 1984), but the winning entry has not been built up to date, mostly due to lack of resources. The reasons for this lack of solar and energy efficient designs are mostly the insufficient awareness (if it at all exists) of designers and consumers to the solar design potential on one hand, and the practical non-existence of neighborhoods whose zoning takes advantage of solar rights, solar access and orientation. This, coupled with the generally small size of building lots, makes it very difficult to build solar houses in Israel.

About 60% of the area of the State of Israel is considered a hot-arid desert. In spite of a few attempts to modify the standard designs of the Ministry of Housing, building and neighborhood designs in the hot-arid part of the country are not by far different from those in the non-arid zones. The reason is that the whole area of the State of Israel is small, and the proportion of the population that lives in the arid zone, about 10%, does not "justify," technically and economically, either special urban or rural design or special building types. This is particularly important since most of the public construction in Israel is prefabricated; i.e. tends to focus on large series of identical buildings. Attempts to adjust building types to the special hot-arid zones usually resulted in additional insulation on building envelopes.

The Center for Desert Architecture and Urban Planning had an unique opportunity to design a desert-solar neighborhood when commissioned by the Israeli Ministry of Housing to design the "Bne-Beitcha" project ("Build Your Own House") in the Sede-Boker campus of the Ben-Gurion University of the Negev, where the J. Blaustein Institute for Desert Research is located. A unique set of

---

[1] Reprinted from: Etzion Y (1989) A desert solar neighborhood in Sede Boker, Israel. Architectural Science Review, 1: 103-109, with permission from the journal editorial board.

circumstances made it possible for the Unit to introduce some new design ideas and features into the project.

1. The Desert Architecture Unit is an academic, non-profit body interested in research and development in the field of desert architecture. It is also part of the national center for desert research in Israel - the J. Blaustein Institute for Desert Research. As a result, the Desert Architecture Unit enjoys a degree of professional prestige and status with the parties that commission its services. This professional prestige enables it to deviate from the prevailing "common" approach to design. In this case, the Desert Architecture Unit was commissioned by the Ministry of Housing, with the blessing of the future dwellers of the neighborhood, to design the project.
2. The majority of the members of the association that was formed in order to establish the new neighborhood consist of faculty and workers of the J. Blaustein Institute for Desert Research. Before the construction of the neighborhood started, most of them lived in alternative housing in Sede-Boker, waiting for the completion of their new houses. These people are exposed to the desert environment and to desert research, and could appreciate the need for a neighborhood specially designed for the desert conditions.
3. The fact that the majority of the future inhabitants of the neighborhood lived in Sede-Boker enabled their full cooperation and participation in the planing and the design of the project.

The target of the design was to create a modem desert neighborhood, which would be responsive to the special harsh conditions of the environment and at the same time would provide the dwellers with all modern facilities. This could partly be achieved by turning the environmental disadvantages into advantages, preferably by using natural energies.

Sede-Boker (Fig. 14.1) is located in the Israeli Negev highland, latitude 30.8°N, at an altitude of approximately 500 meters above sea level. The area is considered hot and arid in the summer: the daily average temperature is 24°C the average maximum daily temperature is 32°C, and the average minimum daily temperature is 15°C. Solar radiation is very intense, up to 7.7 kWh/m^2 (June-July) on a horizontal surface. The relative humidity during the day is low, between 20-40% during most hours, but increases significantly during the night to over 90% as the ambient temperature drops. The daily temperature swing is about 18°C, but the average temperature is within the comfort zone. Thermal comfort in buildings during the summer can be achieved by reducing heat gain during the day (insulation, small windows etc.) and increasing the thermal capacity of the structure. During the night buildings are fully opened to the prevailing winds, thus losing excess heat that accumulated during the previous day. The night wind in Sede-Boker is a breeze coming mostly from northwest, i.e. the Mediterranean Sea. Ventilation thus becomes an extremely important factor in maintaining thermal comfort in the buildings.

The winter is cold, occasionally rainy and frequently uncomfortable. The average temperature in January is 10°C with an average minimum temperature of 30°C, but night time temperatures will often drop to 0°C. The number of degree days (basis 18.3°C) is about 950. The average yearly rainfall is 80 mm, but fluctuations in rainfall are large.

Fig. 14.1. Israel; location of Sede-Boker

Solar radiation is significant also during the winter, and stands at 3.0 kWh/m^2/day (January) on a horizontal surface, almost 3.6 kWh/m^2/day on a south facing surface. These conditions make Sede-Boker almost an ideal location for solar houses: keeping a house thermally comfortable in the winter by using solar energy is certainly possible and economic.

A "Bne-Beitcha" project is a development pattern now popular in Israel. Building lots are leased by the Israeli Land Authority to people who commit themselves to building their house on the leased lots within a period of three years. Some of these projects are organized by private entrepreneurs, some by the Israeli Ministry of Housing and some by private associations created especially for this purpose. The dwellers get their lots ready for construction, and the central body (private, ministry of housing etc.) takes care of the infrastructure of the new neighborhood. To lower the high cost of building the infrastructure of the neighborhood, the lots that are leased are rather small, and in most cases do not exceed 500 m^2 per house. In a few cases, as was also the case in Sede-Boker, the lots are slightly larger, between 600-700 m^2, but even this size of lot is small and creates some unique constraints on the design of solar buildings.

14.2
The Neighborhood

The new neighborhood named Neve Zin (Fig. 14.2) consists of 78 building lots built in two stages: The first and second stages included 33 and 45 houses respectively. The total land area allocated for the neighborhood was 80 dunams (1 dunam = 1,000 m^2) to achieve all the goals, which the users and the designers have set for the new neighborhood, a few new concepts were developed to be used in the design of the new neighborhood.

14.3
Orientation

Because of local constraints, the neighborhood orientation could not be established at due south. The constraints were the master plan of the whole settlement of Sede-Boker, which assigned specific uses to land adjacent to the new neighborhood, and the road system of the settlement that had already been built before the design of the neighborhood. The final orientation was fixed at 17° west of south. Calculations indicated that this orientation would cause a reduction of approximately 5% in the availability of solar radiation for heating the buildings in wintertime. Due to the large quantities of solar radiation available in Sede-Boker even during the winter (app. 3.6 kWh/m^2/day on a south facing wall) it was decided that the reduction in the radiation due to the deviation from the south would be compensated for by increasing the size of windows and collection systems of the buildings.

14.4
Circulation

Two separate circulation networks were proposed. The first one was a combined system for both pedestrian and vehicular traffic (Woonerf, Fig. 14.2a, Fig 14.3). The second network was for pedestrians only (Fig.14.2b). Different design approaches were utilized for the two systems, in order to meet their particular use and needs.

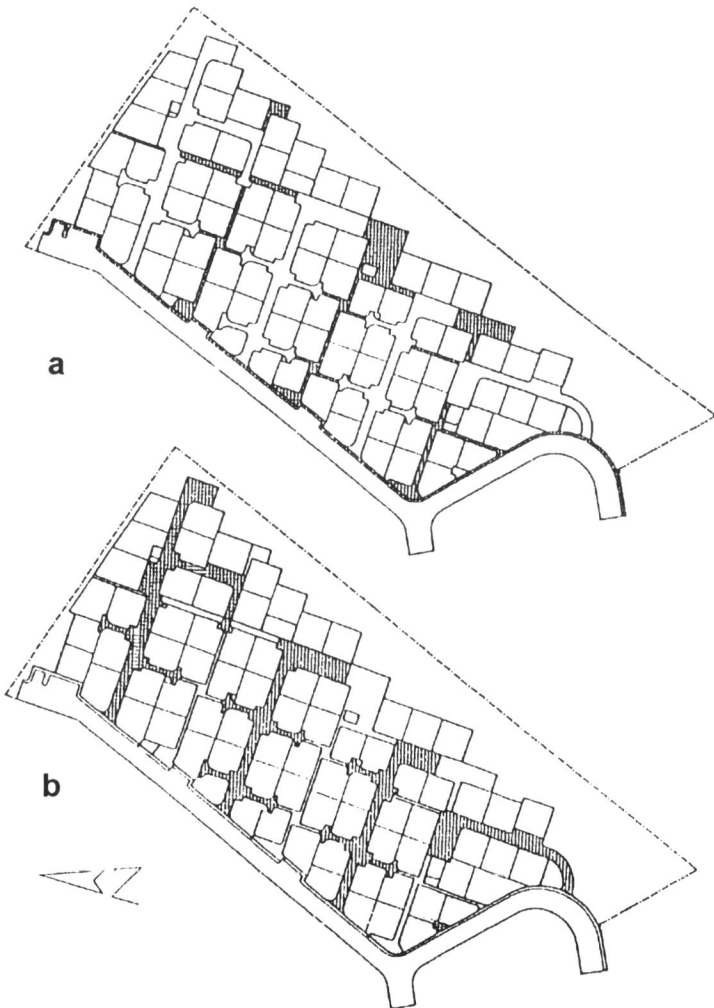

Fig.14.2. Two circulation systems in the Neve Zin neighborhood, Sede-Boker (a – the "woonerf," b – pedestrians only)

The "woonerf" system was designed on parallel east-west axes, to allow traffic to penetrate at three different points on the main road at the northwestern edge of the neighborhood into the built area. The pedestrian-vehicular roads are relatively wide (8 meters between lot lines), and allow vehicles access to each of the houses via special access extensions which are perpendicular to their main axis.

The pedestrian alleys were designed to be narrow and their orientation perpendicular to that of the "woonerf" system. They are supposed to protect the user from the sun during the summer and from the wind during the winter. It was decided to lay out the wider roads of the neighborhood in the east-west direction because of the solar-neighborhood layout geometric relations.

Fig. 14.3. A view of the "woonerf"

In general, a wide road is almost fully exposed to solar radiation during the summer, and only partly shaded during the winter. The orientation of the road axis, though, determines the duration of this exposure and the exposure of alleys perpendicular to it. This orientation also affects the intensity of solar exposure to which the houses located along these axes are exposed.

In the winter, a wide road stretched from east to west, exposes to solar radiation the south facing walls of the houses which are located on its north side, which is very desirable. In the summer, a narrow strip of shade will be cast along the southern side of this road by the houses and the fences located along its south side. This narrow strip of shade can protect pedestrians walking in the street.

The narrow pedestrian alley perpendicular to this road is shaded in the summer during the early and late hours of the day, and the west and east facing facades of the buildings cast shadow on each other to reduce along it will the enormous potential heat gain due to solar exposure coming from the east and the west.

If this road system were rotated by 90°, the wide road would be stretched from south to north and the alleys from east to west. The result would be that the winter solar gain in the south would be reduced or eliminated altogether. During the summer the east and west facades of the houses would be exposed to the heat gain of the east and the west. The alleys would be exposed to the sun during several hours in the morning and several hours in the afternoon. This orientation would be less desirable.

The "woonerf" were designed to be paved with interlocking concrete pavement tiles of light color, and street furniture was positioned at various points along them to reduce driving speed and create sitting corners, greenery islands and small activity centers for the users.

The alleys created by the pedestrian system are each about 50 meters long and are bounded by the walls surrounding the building lots, the height of which may reach up to two meters. The designer of the neighborhood was allowed by the zoning regulations to require that up to two thirds of the length of the alley will be shaded by pergolas and vines built by the owners of the two lots between which the alley is stretched.

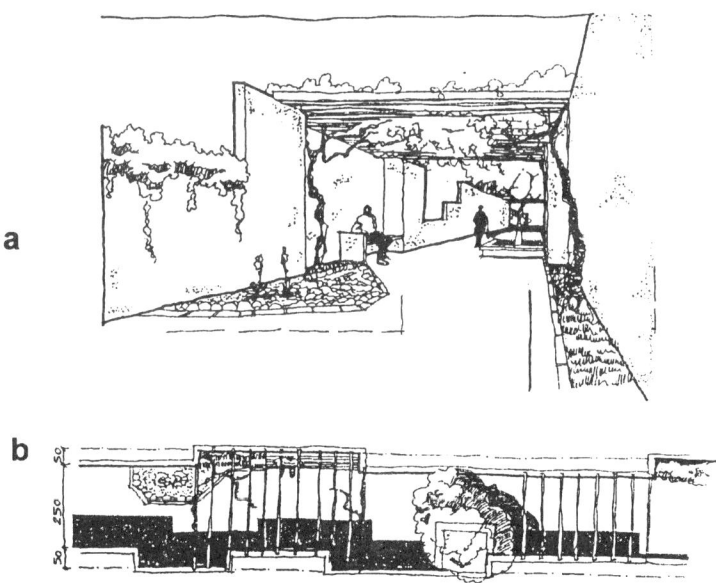

Fig. 14.4. A typical design of a pedestrian alley. a– plan; b – perspective view

Also, in agreement with the owner and the designer of the houses, the designer of the neighborhood was allowed to use a strip of private land, 0.5 meters wide by up to one third of the length of the alley, to create in it a few focal points that will reduce the feeling of walking between two narrow, straight and dull walls (Fig. 14.4 and 14.5). At each intersection of alleys a small public space was created to allow sitting, playing, etc.

Fig. 14.5. A view of an alley; pergola not yet installed

14.5
Building Clusters

The design of the building clusters (and the road system) was aimed at resolving two seemingly conflicting requirements, posed on the one hand by the future dwellers and on the other hand by the harsh climatic conditions. It is generally accepted that in order to minimize exposure to the elements, both during the summer and during the winter, desert design should be compact and buildings should be arranged densely. It was noted by the designers of the neighborhood that even though most of the future dwellers were aware of the need to lay out desert neighborhoods differently from neighborhoods in the north of the country, they insisted on having spacious lots, which would allow them extensive use of their yards, gardening opportunities and most of all, high degree of privacy. In this respect, the requirements of the Sede-Boker people were identical to those of residents in the north of the country.

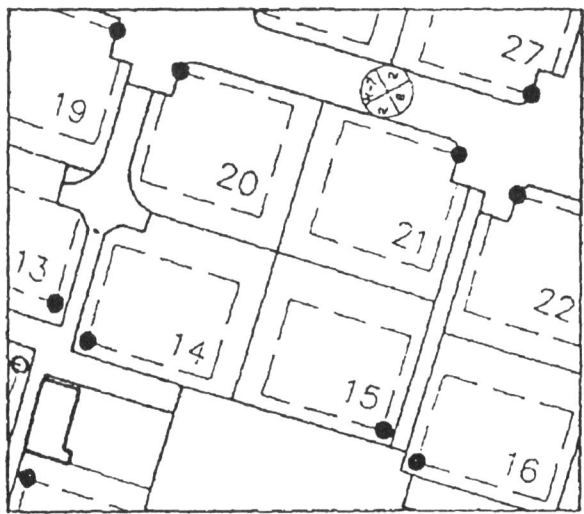

Fig.14.6. The concept of the "P" point

Fig. 14.7. The effect of the "P" point on the layout of the building cluster

It was impossible to convince people to accept smaller lots, zero setbacks, and other steps that would make the neighborhood compact, in spite of the fact that the designers advocated these measures. To respond to these two conflicting demands, building lots were arranged in sets of four, and the concept of the "P-point" was developed. The P-point was defined as a point on the setback line of the lot, on which a point of the perimeter of the volumetric part of the building (not a fence wall, for example) should rest. The P-points were located by the designers in each lot at the point on the setback lines farthest from the center of the cluster of the four adjacent houses (Fig. 14.6 and 14.7).

In most cases the horizontal projection of the house on the lot was smaller than the zoning regulations for the Neve Zin neighborhood in Sede-Boker area enclosed by the setback lines of the lot, thus leaving part of the lot area that could be built as an open space. The combined effect of the clustering and the P-point forced buildings to be placed very close to the public roads and alleys, thus creating a dense pattern of public spaces. At the same time, the private open spaces between the four buildings of the cluster were stretched to the maximum possible allowing the dwellers to enjoy the largest open space possible (Fig. 14.7). To increase the sense of openness, solid fences between the lots in the cluster were not allowed. It was recommended to the dwellers not to fence their lots at all, but in cases where fences would be an absolute necessity, they must be made of wire meshes or of vegetation. To increase the density of the public spaces, deviations from the setback lines near the alleys at the east and the west of the cluster were allowed for garages and auxiliary parts of the houses -those could be placed right at the lot line. (The dwellers rejected altogether the notion of a compulsory zero setback line for the houses themselves).

14.6
Setback Lines

Setback lines were established to permit solar access to all buildings on one hand, but also to maintain the small distances between the buildings around the public spaces on the other hand. Setbacks near the pedestrian alleys were set at 2 meters, but it was allowed to cross them with garages and auxiliary parts of the buildings. South and north setbacks were set solely on the basis of solar access. A special geometric model (Etzion 1988) was developed in order to allow the computation of the volumetric constraints over the buildings (Fig.14.8). In order not to increase the necessary distances between the houses, the volumetric restrictions allow different heights of the south facing parts of the buildings and their north facing parts: the south facing parts were allowed to be 8 meters high, the north facing were restricted to lower heights as calculated by the model.

Planting of trees was confined to locations on the lots that do not interfere with the solar access rights of the neighbor in the north. In cases where such a geometric interference could occur, the trees must be deciduous.

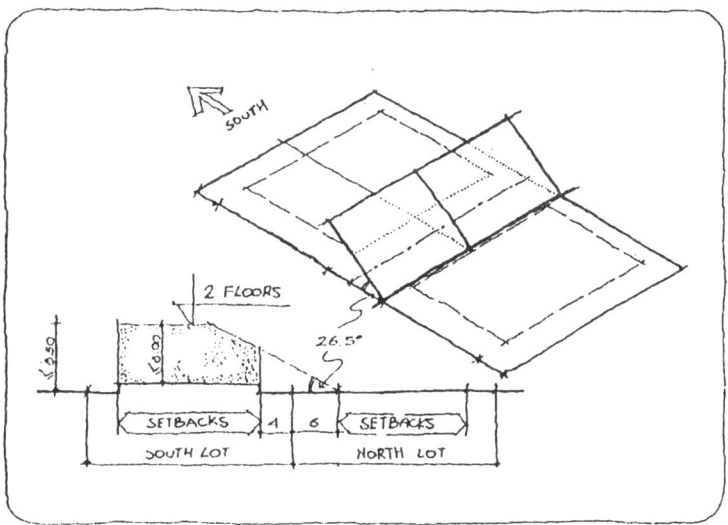

Fig.14.8. The volumetric constraints used for the design of the buildings. A section from the zoning regulations for the Neve Zin neighborhood in Sede-Boker

14.7
Water Heating

Israel is probably the world leader in domestic water heating by solar energy. Regrettably this leadership did not come without a cost - the urban skyline of the country is marred extensively by the solar water heating devices installed on the roofs. The solar water heaters used in Israel for domestic use are all of the thermosyphonic type, in which the bottom of the tank must be higher than the top of the collectors to create the thermosyphonic effect. The result is that each water heater is about 2.5 meters high, and is very hard to conceal. Until very recently, designers in Israel had not addressed this design problem seriously, and now it is already too late to reverse this problem in most parts of the country.

To avoid this dominance of the water heaters on the built landscape, it was forbidden in Sede-Boker to place the water heater in the customary manner prevailing in the country. The architects were required to provide a solution that will hide the heater from an observer standing up to 50 meters from the house. Some of the solutions suggested by the designers were to move the heater to the center of the roof, build a higher parapet around the whole roof, build a wall which is integrated with the building and even use non thermosyphonic, pump driven water heaters.

14.8
Conclusion

The neighborhood is now built. Being the first solar neighborhood in the country, it is expected that it will be examined and researched in the coming years both by its designers and by others. Some of the research planned by its designers includes monitoring the thermal behavior of a few of the houses in it, as well as the behavior of its public open spaces. Other research projects will focus on the degree of satisfaction the residents have from it. All the date collected through this research will be used to compare this new project with other projects in Sede-Boker. If successful, there is hope that this neighborhood will contribute to the evolution of housing in Israel.

References

Rahamimof A (1984) Residential cluster based on climate and energy considerations. Energy and Buildings 7: 89-107

Etzion Y (1988) A general expression for solar rights determination, Energy and Buildings 12: 149-154

15 A Bio-Climatic Approach to Desert Architecture[1]

Yair Etzion
J. Blaustein Institute for Desert Research, Ben-Gurion University of the Negev, Sede Boker Campus, 84990, Israel

Desert architecture may be characterized as "Architecture of the Extremes," being basically similar to "regular" architecture but differentiated from it by its obligation to address needs and problems of an extreme character. The problem of thermal comfort in buildings is perceived as one of the more characteristic and difficult problems that desert architecture must address, even though this is not the only problem nor necessarily the most difficult one. A typical way of addressing the thermal comfort issue in buildings is by intensive use of expendable energies, but this, of course, is not an ideal approach: it leads to waste of energy, it is expensive, and not everyone is comfortable with the thermal conditions it creates (witness the number of people who do not like air-conditioning). Various characteristics of design and construction enable the improvement of thermal comfort to be integrated into a building without the use of artificial means and expendable energy. Now, when it seems that even the drowsy Negev (the southern half of the Israeli land area, which houses only about 7 percent of the country's population) is awakening to a building surge, it is desirable to clarify these methods, and even to try to apply them in new building projects. What's more, as an ever increasing worldwide need for housing construction is evident, much of it in hot arid lands, the "right" type of building technology should be used to improve standards of living and decrease the use of purchased energies.

This article demonstrates a number of climatic and energy characteristics of building in the desert. It takes as its subject an examination of a recently completed house in the new Neve-Zin neighborhood of the Sede-Boker Campus of Ben-Gurion University of the Negev, the first real "solar" or "bio-climatic" neighborhood in Israel. The house chosen is the Etzion House (Fig. 15.1), which was designed by and built for the author of this article. It was chosen for examination here because of the author's first-hand familiarity with its design considerations and characteristics.

15.1
The Climate of the Negev

It is necessary, first of all, to introduce the Negev and its environmental conditions, and also to correct several misconceptions and false "truths" concerning the

---

[1] Reprinted from: Etzion Y (1994) A bio-climatic approach to desert architecture, *Arid Lands*, pp. 12-19, with permission from the Arizona University Press

climate in most of its regions. The relatively high Negev regions (300 m and above) are not areas as hot, for example, as the Arava or the Beit-Shean Valley (both along the Jordan Rift). The higher regions of the Negev may be characterized as having cold, uncomfortable winters and summers that are hot during the day but usually pleasant at night. The Sede-Boker Campus is located at 30.8' latitude north, 500 m above sea level. Average annual rainfall is 80 mm, but there is a considerable deviation from year to year.

The climate is considered hot and dry during the summer, with an average maximum temperature of 32°C and an average daytime temperature of 24°C. Solar radiation is very strong, and may reach 7.7 kWh/m^2 x day on a horizontal surface (during June and July). In the summer, ambient relative humidity is very low, between 20 percent and 40 percent during most of the day, but it rises considerably during the night, when the ambient temperature drops sharply, to reach 90 percent. Summer daily temperature fluctuation is about 18°C, and the average temperature is within the range of thermal comfort. Winter is cold, sometimes rainy, and uncomfortable. The average temperature in January is 10°C and the average minimum daily temperature is 3°C. The temperature at night often drops below freezing (0°C). The intensity of solar radiation during the winter is relatively high and reaches 3.3 kWh/m^2 x day on a horizontal surface, and about 4.6 kWh/m^2 x day on a south-facing vertical surface (Fig. 15.2). These conditions make Sede-Boker an almost ideal location for buildings that achieve thermal comfort in the winter by employing solar energy. Thermal comfort generated by the sun adds to the quality of life and is definitely economical.

15.2
Building Design: Sealing the Envelope

The concept guiding building design in this climate is the creation of an envelope, sealed as far as possible against the passage of energy. In this envelope should be openings, allowing desirable - but controlled, both in time and in quantity - passage of natural energy from the house outwards and vice versa. The house should be massive, with a relatively high thermal capacity. The first "truth" that should be refuted regarding building design is that the directional orientation of the building is the key to achieving thermal comfort. A house built according to the concept guiding the design of the Etzion House is largely insensitive to its orientation. If the envelope is really well insulated and has a significant thermal capacity, there is not much difference in the thermal performance of a house facing south compared to houses facing other directions, because the envelope is, practically speaking, almost sealed to the passage of energy. Orienting the openings, though, is very important, as will be explained further on. Thus the popular opinion that the bio-climatic house must have a long southern exposure is not necessarily correct: a long south-facing elevation is needed only for positioning the south-facing windows, and it should be large enough to enable just that.

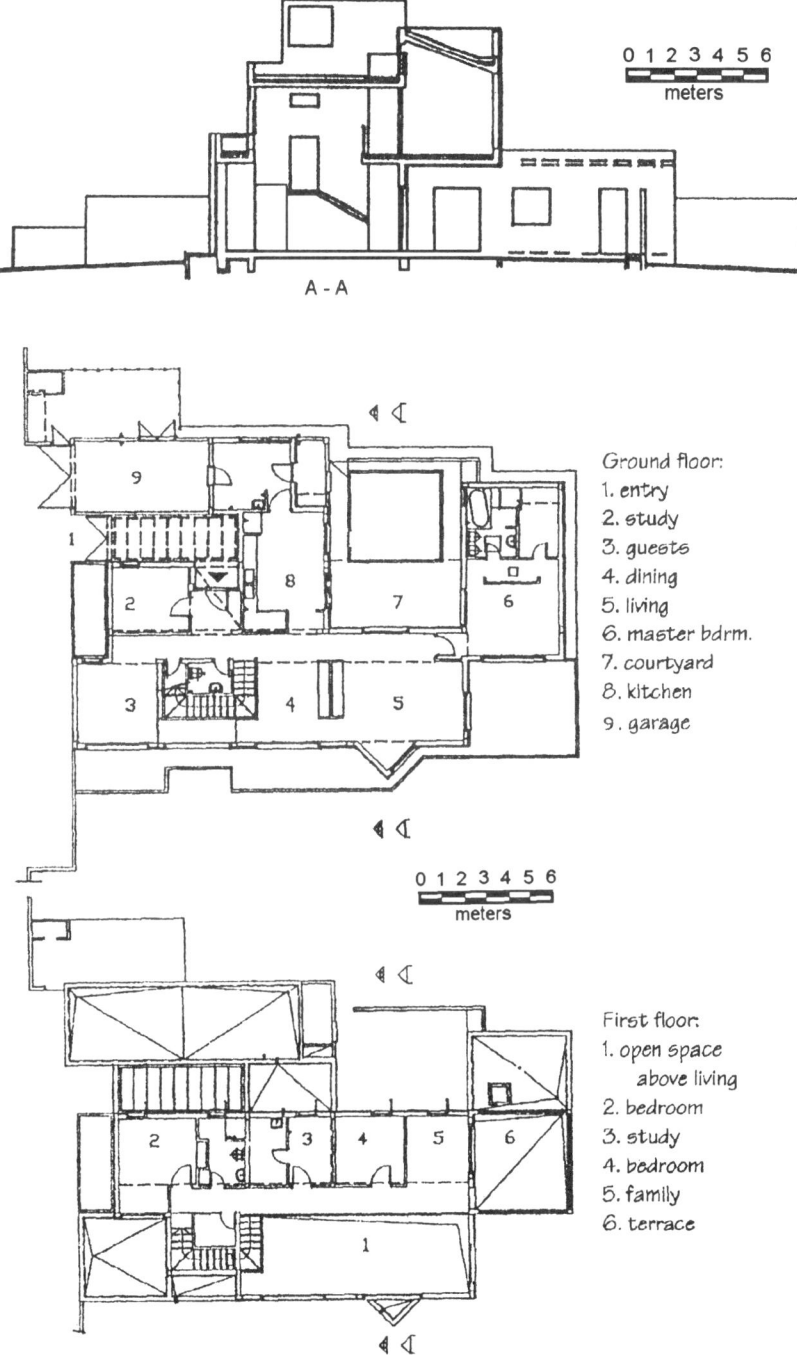

Fig. 15.1. The Etzion House: section and plan

It is possible, therefore, to say that solar or bio-climatic architecture does not have to be monotonous and boring, as many of its opponents claim.

Sealing the envelope against uncontrolled passage of energy should reduce to a minimum the possible overheating of the house in the summer and its cooling in the winter. In order to "manage" the energy economy of the house, no energy should be allowed to pass through the walls, as far as is technically and economically possible.

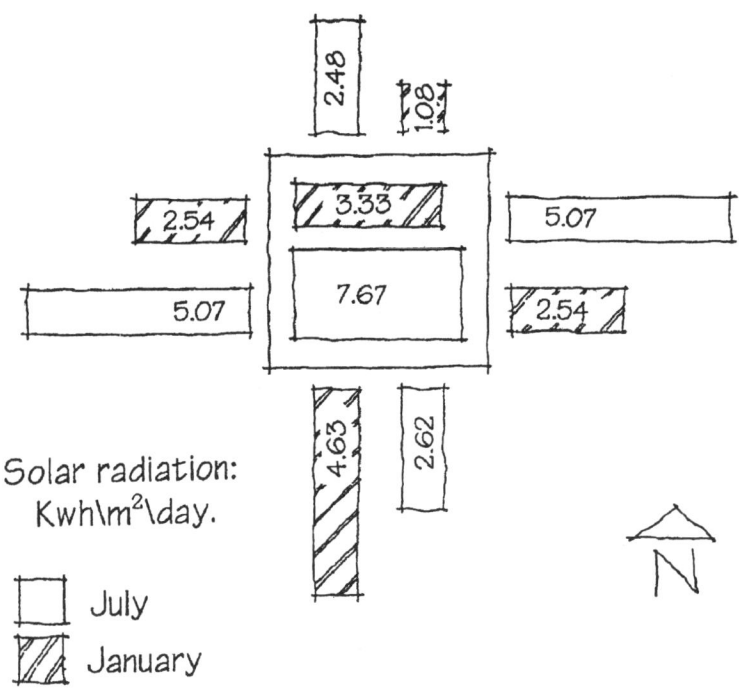

Fig.15.2. Winter and summer radiation intensities in Sede Boker on various building elevations

Desirable transfer of energy, allowing for heating the house by solar energy in the winter and cooling it in the summer by nighttime ventilation, should be done only through and by the openings. The building's significant thermal capacity should contribute to stabilizing the large daily temperature fluctuations typical of the desert and also should increase the building's thermal lag time, which is the time that passes, for instance, between the peak external temperature and the peak internal temperature.

Fig. 15.3 shows, schematically, the differences between the thermal behavior of a "light" building (such as a mobile home) and a "heavy" building.

The external walls of the Etzion House were built according to these guidelines (Fig. 15.4). The internal layer of the wall is built of massive silicate blocks, which constitute an integral part of the storage mass of the building. On the external side of the wall an insulating layer is attached, consisting of 5-cm-thick polystyrene board. The insulating layer was glued to the blocks using a special acrylic mortar. The wall was plastered over with one layer of acrylic plaster, into which a reinforcing polyester mesh, capable of resisting an alkaline environment, was embedded before hardening. A second layer of the same acrylic plaster was then applied over the first one. In order to prevent heating of the wall as a result of exposure to solar radiation, and in order to prevent the desert sand and dust from adhering to the surface (which turns almost every building in the Negev to desert colors), the external wall surfaces were kept white and smooth.

The expected energy performance of the wall used in the Etzion House exceeds the requirements of the standard established by the Israel Standards Institute (Standard 1045). The thermal resistance (R) of the wall is 2.0 m^2 (°C)/Watt. Its thermal storage capability is 60 Watt (hrs.)/m^2 (°C) and its damping coefficient is 0.21. The thermal time constant of this wall is about 160 hours. It is very important to note that the order of the layers of the wall is of supreme importance. If the order of the layers were different - in other words, if the thermal insulation were placed on the inside of the wall - only the thermal resistance of the wall (R) would remain constant (because it consists of an algebraic sum of the thermal resistance of all layers). The wall's storage capacity would be reduced almost to zero because the insulating layer would cut off the storage mass from the internal space of the building and the thermal time constant of the wall would be reduced to approximately one hour.

In the Etzion House, in order to increase the thermal storage capacity of the building, even the internal walls were built of massive silicate blocks. The roof section chosen for the Etzion House is an "inverted roof" with a large amount of thermal insulation. Owing to considerations similar to those described for the walls, insulation was placed on the external side of the roof. The inverted roof section was chosen, in spite of the problems sometimes associated with installation of various devices on it and its maintenance, because of its ability to protect the sealing layer of the roof. Most sealing materials suffer in the Negev from two phenomena that significantly shorten their life spans: exposure to solar radiation, which dries them out and makes them brittle, and extreme fluctuations of their surface temperature between daytime and nighttime (up to 60°C), which causes the layer to "work" and deteriorate. Placing the insulation on top of the sealing layer protects the latter from solar radiation and also reduces the temperature fluctuations to which it is exposed.

The roof structure of the Etzion House is made of structural concrete, which has a significant thermal capacity. Above it is a layer of light-weight concrete (drainage), a sealing layer, and a 10-cm-thick polystyrene board, the underside of which is perforated to permit water runoff under it. On top of the insulating layer was placed a layer of *Polya* gravel to a depth of 5 cm; the gravel's job is simply to keep the insulation in place. This roof has a very high R-value - 3.6 m² (°C)/Watt - (mainly to protect from solar radiation in the summer, but also to keep the heat inside during the winter), a damping coefficient of 0.3, and a time constant of almost 300 hours.

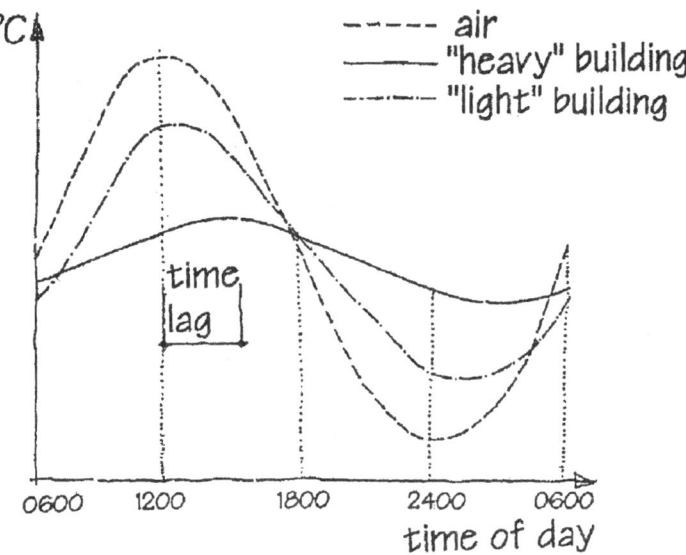

Fig. 15.3. Schematic thermal behavior of "heavy" vs. "light" structure

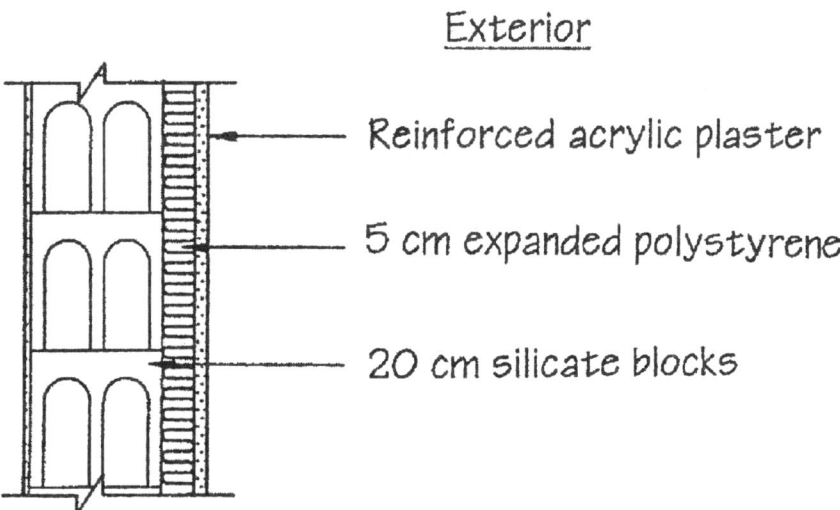

Fig. 15.4. External wall section of the Etzion House

15.3
Windows: Opening the Envelope by Design

The openings of a building can be a joy or a burden, depending on their design and location. The openings can either supply a significant proportion (sometimes almost all) of the heating required in the winter, or they may be a "hole" in the sealed envelope through which there is an uncontrolled passage of energy. In the summer, the openings can be, on one hand, well protected, allowing for very little increase in the internal heat level, or they can be a source of significant and harmful heating; it all depends on their design, placement, and quality of performance in the building.

The design of the windows should maximize solar gain in the winter and minimize such gain in the summer. Window design also should minimize the penetration of hot air into the building during the hot hours of the summer and of cold air during the winter, while allowing massive ventilation on cool summer nights. The building's significant thermal capacity should contribute to stabilizing the large daily temperature fluctuations typical of the desert, and also should increase the building's thermal lag time. From the aspect of solar radiation, the most important factor is the orientation of the windows. The large differences in levels of radiation between winter and summer in Sede-Boker (see Fig. 15.2) is mainly a result of the different altitude angles of the sun in each season, which causes different angles of incidence on the planes, and the different number of hours during which the wall and the roof are exposed to the sun.

The different solar radiation intensities from each direction show that the most preferred direction for openings is south: in winter (January), a window in a southern wall can supply about 4 kWh/m^2 x day of heating energy (which is about 90 percent of overall radiation falling on the window surface, the transmittance coefficient of traditional commercial glass being about 0.9). The northern orientation is defined as a "loser" in the winter: in most cases in Israel, more energy is lost through a northern window than can be gained through that window from the sun (in this case, only reflected and bounced radiation). Therefore, because of winter considerations, it is desirable to reduce north-facing windows to a minimum (Fig. 15.5).

From the aspect of thermal comfort in the interior space, it is recommended to double-glaze northern windows (to raise their surface temperature and to reduce radiation from human bodies to them), even though these windows may not have an unequivocal economic advantage over traditional windows.

Fig. 15.5. North and east elevations of the Etzion House; notice the small openings in the northern element

The windows facing east and west present the most difficulties. In the winter they usually will be in a kind of balance between loss and gain, and so their benefit is, at most, doubtful. In the summer (June), however, owing to the 5 kWh/m^2 x day these windows are exposed to, they are a considerable source of massive and damaging heating. The logical conclusion is that these windows should be avoided altogether, unless there is a special reason for their placement. It should be noted that in the summer there is more radiation each day per unit of area on the western and eastern walls than there is on the southern wall in the winter, and that with the radiation that falls on the southern wall during the winter it is possible to heat the house when the temperatures outside are significantly colder! The Etzion House has no windows facing west (Fig.15.6), but a few windows face east, admitting a breathtaking view of the Zin Canyon Cliffs. Blocking off this view would be a totally unpardonable "sin." Besides, one does not build a house only to save energy.

The size of the southern windows (Fig.15.7) was established by the ratio of solar radiation to load (Solar Load Ratio, or SLR). In heavy and well-insulated structures like the Etzion House, it is possible to calculate with a high degree of accuracy the necessary size of the windows based on average monthly external temperatures in the cold months. It is possible to calculate the expected heating load of the house, to decide which portion should be supplied by the sun (a primarily economic decision), and to design the size of the windows accordingly An examination of a sample of the solar houses now being built in Neve-Zin shows that the area of the southern windows usually represents 15 to 20 percent of the floor area of the house. In the Etzion House there also are a number of northern rooms that benefit from the southern winter sun, which is admitted through a few clerestory windows (see Section A-A, Fig.15.1).

Protecting the interior of the house from heating by solar radiation incident on the windows during the summer (assuming that the rest of the envelope is sealed against energy penetration) is very important for achieving thermal comfort in the house. Here, too, a number of well-known "truths" must be refuted. Figure 15.8 shows the breakdown of types of radiation as measured close to the Neve-Zin neighborhood. It appears that in the winter a large proportion of the radiation hitting the envelope is direct. In the summer, however, the direct radiation becomes a relatively small proportion of the overall radiation hitting the structure's surface (20 percent); most of the radiation is diffused from the atmosphere or reflected from the light-colored ground surface (reflection coefficient of approximately 0.3).

The conclusion that must be drawn from this is that special overhangs and permanent shading elements over glazed openings are of very limited benefit under these conditions, and that it is not desirable to count on them, especially in the Negev. The solution is exterior blinds or other movable shading mechanisms that can absolutely, or almost absolutely, prevent solar radiation from hitting the windows (Fig.15.9). It is superfluous to point out that internal blinds, such as Venetian blinds, are useless in this case, as they block only radiation that has already penetrated into the building. Since the radiation heating the structure during the summer is diffused and directionless, shutters are required in all window orientations.

The Etzion House has a number of small northern windows (area of each is about 0.65 m^2) in each of the rooms on the second floor (see Fig.15.5). In the main spaces of the lower floor there are relatively large openings, double-glazed, which also point northwards. These openings are used for cooling the house in the summer and were installed in spite of their negative performance in the winter. In the winter these windows will remain closed all day. In the summer, they will be closed during the daytime, their blinds lowered, but will be wide open in the late afternoon hours, when the outside temperature already will be below that desired inside the house; they will remain open all night. Through these windows a large volume of cool desert air should flow, which will "flood" the house during the night and remove from it both the small amount of heat that managed to penetrate the "sealed" envelope during the day and the heat that was generated in it by equipment and people.

Fig.15.6. The western evaluation of the Etzion House; notice the lack of openings

Because it is desirable to reduce the size of the windows on the one hand (winter and summer days), but on the other hand it is desirable to have large openings for ventilation (summer nights), a small window was designed whose blind opens, with the help of a special mechanism, only to be perpendicular to the wall (see Fig.15.5). This blind serves as a kind of a reflection shelf, creating some higher air pressure near the window and thus increasing the quantity and velocity of air entering through the window by about 50 to 60 percent. Assisting this mechanism is the fact that the governing breezes, mainly in the afternoon hours of the summer, always come from the northwest at about a 45° angle to the north facade.

The house was designed to achieve, in winters, internal temperatures in the range of 18 to 20°C, with the sun providing more than 90 percent of the necessary heating; summer internal temperatures average in the range of 23 to 26°C. In the summer, these expected temperatures, combined with the low daytime relative humidity (20 to 50 percent during the hot hours), promise thermal comfort within the house.

It should be pointed out that expected winter temperatures are "real," sensible temperatures, reflecting both the air temperature and surface temperatures inside the house. In contrast, in houses where heating is based on convective heating of the air within and/or radiative heating, particularly in cases where the heating is intermittent, the surface temperatures of the walls, floors, and roofs usually will be lower than that of the air, and the effective sensible temperature inside the house will also be lower.

Fig. 15.7. A view of the southern elevation on a clear winter day

15.4
The Value of the Courtyard

Another "truth" that has been proven incorrect in the Negev Highlands is that summer days are very hot and that it is difficult to protect oneself from the heat during these hours. Observations indicate that because of the large amplitude of daily temperatures in this season, good thermal conditions actually exist outdoors most of the day. A slightly simplistic, though quite accurate, calculation shows that assuming that the daily temperature amplitude is about 18°C, average hourly change of temperature is about 1.5°C. If the average daytime maximum, measured at about 3 p.m., in the hottest summer months is about 32°C, then by 6 p.m. the temperature will already be about 28°C and will remain below this value until about noon of the following day The combination of 28°C with 30 to 40 percent humidity is a very comfortable one, and residents of many parts of the country would welcome it. In fact, the beginning of the thermally comfortable period is usually even earlier, because, except for "Hamsin" days (characterized by very hot and dry easterly winds), the Negev is blessed with a daily northwesterly and relatively cool breeze, which starts at around 4 p.m. and continues until 10 or 11 p.m.

It is natural, then, that spending time outside, in the shade, during the hours after work, even during the hottest summer months, is comfortable and should be used. For this reason, the Etzion House is shaped like a three-sided, U-shaped box, built around a small courtyard whose open end faces north.

All the important spaces of the house open into this courtyard: the family room, the master bedroom, and the kitchen. The courtyard's orientation is to the north so that it will catch the comfortable summer breezes. Facing north, the courtyard is also almost totally shaded from the sun; part of it is shaded by the second floor volume, which manages, even in June, when the sun is at its highest, to keep most of the courtyard shaded during most of the day (see Fig. 15.1, Section A-A). The shading of the courtyard reduces significantly the effective temperature in it by neutralizing the element of radiation and also by maintaining the external wall surfaces surrounding the courtyard at a relatively low temperature. The result is a courtyard, very useful during most of the year, where a large portion of the inhabitants' activities can take place.

If the courtyard were to face south, most of the positive features of the northern courtyard would be lost. It is important, therefore, to differentiate between a southern exposure of glazed openings and the location of attached external spaces, whose recommended location is on the north side of the building.

15.5
Performance Monitoring

The performance of the house is evident in the data collected during the summer and winter of 1993-94. For technical reasons, the monitoring period has not been extended for weeks at a time, but a few spot measurements of a few days each were taken during both summer and winter.

Measurements were taken at various points in the house; those presented here show the temperature pattern in the living room. Measurements were taken at three different heights above floor level: 0.5 m, 1.6 m (head height), and 4.5 m. The results displayed in the accompanying charts are the daily averages of the three heights.

Fig. 15.10 shows the performance of the house during a random period at the end of the summer of 1993. It shows that during three typical days during the summer (August 8-19) the average temperature inside the house was 24.9°C, with a standard deviation of 1.61°C, while temperatures outside averaged 30.8°C, with a standard deviation of 8.6°C. The temperature inside was, on average, cooler by about 6°C than was the temperature outside, and its stability was greater: its standard deviation was less than one-fifth of the standard deviation of the temperature outside. It can also be added here that the relatively large internal amplitude stems from the intentional nighttime ventilation.

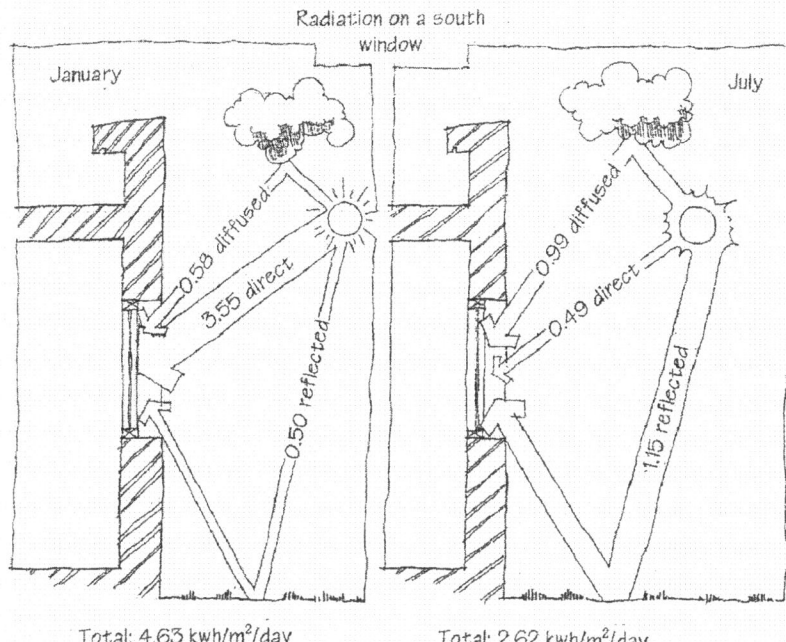

Fig 15.8. Summer and winter radiation distribution on a southern wall

Fig 15.9. A typical view of the southern elevation of the Etzion House on a summer day

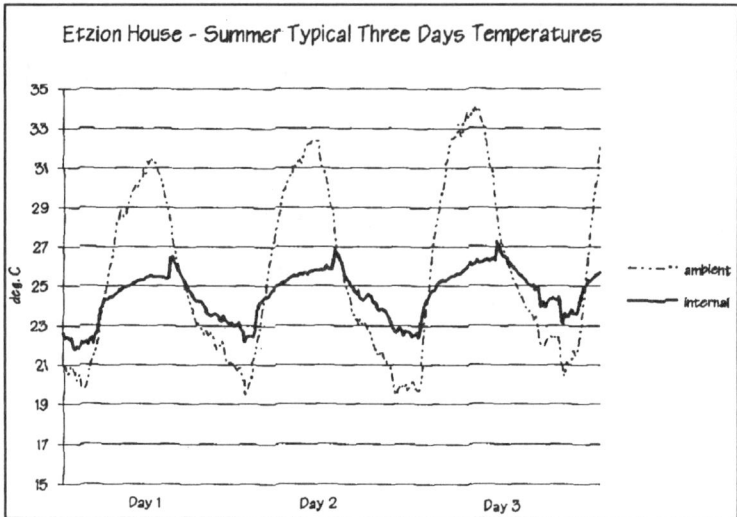

Fig. 15.10. The Etzion House: typical ambient and internal summer temperature over a three-day period

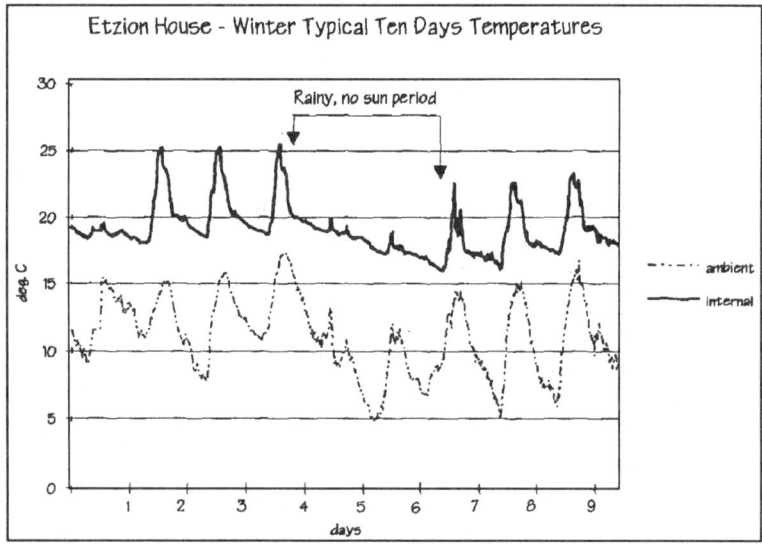

Fig.15.11. The Etzion House: typical ambient and internal winter temperature over a ten-day period

The winter measurements (Fig. 15.11) were taken during January 10-19, 1994, and show that the average temperature inside the house was 19.09°C, with a standard deviation of 1.91°C. Outside at the same time, the average was 11.1°C, with a standard deviation of 2.88°C.

It must be emphasized that these internal temperatures were obtained with no added heating at all, meaning that all the heating energy needed was obtained from the sun. The winter of 1993-94 recorded temperatures somewhat higher than the typical Sede-Boker winter (about 1°C above normal) and enabled 100-percent reliance on solar energy. It is estimated that in a typical winter the solar saving fraction will be slightly smaller. More than anything else, the Etzion House has proved to be a pleasant and comfortable house to live in. It takes some of the biggest liabilities of the desert and uses them in a manner that turns those liabilities into assets. This, from a certain point of view, might be regarded as the essence of bio-climatic architecture.

16 Urban Microclimate in the Desert: Planning for Outdoor Comfort under Arid Conditions[1]

David Pearlmutter and Pedro Berliner
J. Blaustein Institute for Desert Research, Ben-Gurion University of the Negev, Sede-Boker Campus, 84990, Israel

16.1
Urban Attractiveness and the Desert Climate

The attractiveness of a city is a matter of human perceptions, and the desert city has historically faced an image problem. The very word desert conjures up images of an inhospitable wilderness - hot, dry and dusty. By extension, the physical barrenness of the desert is often applied to urban culture, as a scarcity of life is translated to mean a scarcity of everything from grassy parks to good jobs.

It is not surprising, then, to find the typical urban settlement in arid regions built with its back to its surroundings, shutting out the harshness of the open desert and attempting to implant some urban version of a desert oasis in its midst. In traditional cities of the Middle East and other arid regions, this "enclosure" commonly evolved as a dense, tightly-knit pattern of contiguous buildings with narrow streets and small courtyards intertwined within the urban fabric. The notion that such vernacular building responses by early desert urbanists are evidence of climatic awareness is widely accepted in the literature on desert architecture:

> Urban experience in the arid zones shows that compact forms are effectively adjusted to climatic stress. The necessity of human adaptation to arid zones brought about the development of these compact urban forms, which have microclimates more moderate than those of the environs. The narrow, winding alleys and streets, which block sunlight, are relatively cool and block stormy winds (Golany 1980).

In the 20th century, the "oasis" effect has been achieved with an entirely different approach, and the image of the desert city has in some cases been reversed. Development of the "sun belt" cities in the American southwest accelerated in no small measure due to the promise of sunny weather and clean dry desert air, which attracted migrants in search of a modern urban oasis. The ability to transport water great distances and in fact convert arid land into a local non-desert, with expansive lawns and generous landscaping, has been the recipe for success in such cases – but as the limited nature of natural resources, even one as basic as clean water, is becoming more apparent, doubts concerning the long-term sustainability of this "modern" approach to settling the desert are ever-increasing.

---

[1] This chapter is based on: Pearlmutter D Street canyon geometry and microclimate: designing for urban comfort under arid conditions. In: Maldonado E, Yannas S (eds) Environmentally friendly cities: Proceedings of PLEA' 98, Lisbon, Portugal, June 1998, pp. 163-166. The paper was named "Best Paper" of the 1998 International Conference on Passive and Low-Energy Architecture.

Even in the short-term, the transplanting of temperate-zone planning models to the desert may be climatically problematic. Since the 1970's, when the movement toward environmental awareness and energy conservation gained momentum, researchers have warned against the prevailing trend by which "cities in most countries have "aped" mid-latitude styles of building and urban design, without regard for local conditions" (Chandler 1976).

16.2.1
Preconceived Planning in the Negev

The potentially hazardous consequences of such a preconceived approach are illustrated by the early planning of new towns and cities in Israel's arid southern region, the Negev. When the modern city of Be'er Sheva as well as a number of other "development towns" were being established in the 1950's, many of the new residential neighborhoods were planned according to the concept of the "Garden City." This model for urban planning was imported from Europe, where it had been developed decades earlier by Ebenezer Howard and his followers in response to the overcrowded and polluted conditions of industrial cities. In prescribing generous green space throughout the urban fabric, the Garden City approach as applied in the Negev dictated relatively large open areas and a broad dispersal of buildings throughout the landscape (Fig. 16.1).

Fig. 16.1. Attempts at creating a "Garden City" in the desert (left) resulted in a dispersal of buildings in the landscape. As an alternative, compact "Patio House" neighborhoods were designed with narrow pedestrian streets and walled courtyards (right)

As described by Gradus and Stern (1985), the result of this conceptual transplanting was a mismatch between the intentions of planners and the physical reality of the desert. Internal deserts within the city were created in the vacant areas designated for landscaping, due both to a lack of water and a lack of interest by the largely immigrant population in maintaining ornamental greenery.

Inhabitants suffered from extensive walking distances in oppressive heat, and any sense of a cohesive urban fabric was overwhelmed by the feeling of dispersion and disconnectedness. The perception of exacerbated climatic harshness, along with other social and physical factors, led many planners to conclude that the Garden City model and its consequences for the urban structure were inappropriate for the arid Negev.

In response to this perception, attempts were made to create denser, more compact residential environments with one of the goals being the improvement of microclimatic conditions (Fig. 16.1). "Patio house" neighborhoods with a compact structure of walled courtyards and narrow pedestrian paths were built as examples of the "new" approach, though in fact they echoed the traditional urban patterns typically found in the middle east (and in Be'er Sheva itself, in the "old city" built during the period of Ottoman rule). The planned city of Arad, established in the early 1960's, embraced this approach on a more comprehensive scale by integrating compact neighborhoods in the overall town plan.

16.2.2
Microclimatic Considerations

While the desirability and attractiveness of such densely built neighborhoods have varied according to social and economic circumstance, the assumption that they are responsible for enhancing the microclimate in desert cities has yet to be confirmed. To make such a claim requires a systematic analysis of the complex and often interrelated physical phenomena which determine climatic conditions and in turn thermal sensation.

Much of the research conducted recently in the field of urban climatology would suggest that compact planning could in fact prove *detrimental* in a hot climate. It has been shown that the urban "heat island" effect, and particularly its manifestation within the urban canopy (that is, the occurrence of relatively high air temperatures within the built-up area), is intensified by a *compact* urban structure, as expressed by a high ratio of building height to street width. Among the reasons for this relative overheating are the "trapping" of solar radiation (by decreases in both reflection of sunlight and reemission of heat to the sky), and the substantial impairment of ventilation, that occur within the constricted spaces of so-called "urban canyons." While the urban heat island is most pronounced on clear nights, the processes which cause overheating within such street canyons are prominent during the daytime as well (Oke 1988).

On the other hand, the internal shading of pedestrians and surrounding built surfaces that may occur in deep urban canyons provides relief from physiological heat stress, and this is not necessarily accounted for by measurements of air temperature. Shelter from undesirable winds, whether hot or cold, is also an important factor. The impact of these factors relative to the urban effects more commonly cited by urban climatologists, and particularly their implications for pedestrian comfort in an arid setting, remain to be sufficiently addressed.

The reasons for this are partially methodological, as urban climate studies more often focus on particular physical phenomena in the urban environment, rather than on the integrated comfort conditions experienced by a pedestrian. In addition, such studies rarely deal with arid conditions, because most deserts remain sparsely populated and lacking in large urban centers. Special attention to cities in such regions may be justified not only by their unique climatic characteristics, but by differences in their traditional urban morphologies (e.g. low-rise rather than high-rise street canyons) and by recent increased demand for urban development in arid zones.

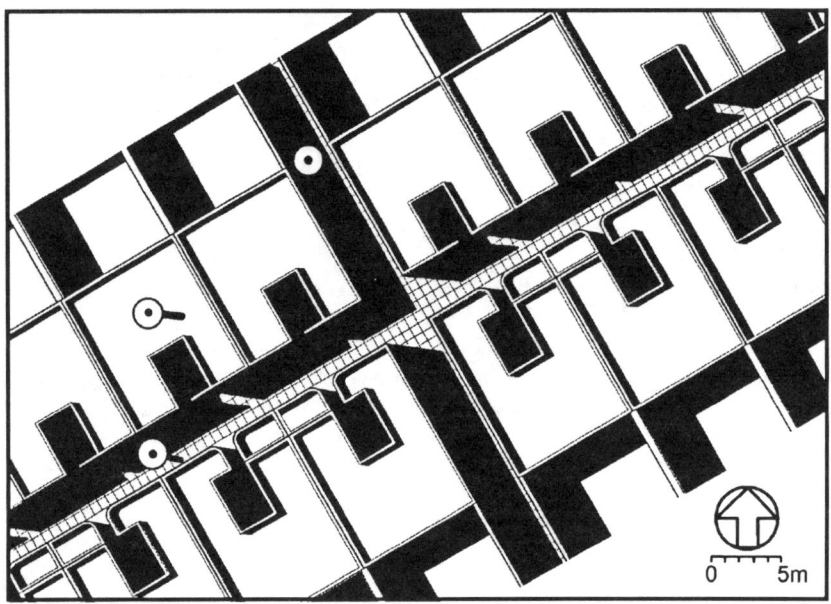

Fig. 16.2. Patio House neighborhood plan, showing points of observation and summer afternoon shading patterns

16.2
Case Study: Analyzing the Urban Microclimate

In the case study presented here, an attempt is made to understand the integrated effect which the design of urban spaces may have on the thermal comfort of pedestrians in a desert city. The study makes use of both empirical data taken from extensive microclimatic measurements in a number of low-rise urban street canyons in the Negev, and a numerical model representing the energy exchange between a pedestrian and the street canyon environment. Analysis of the monitored and calculated results reveals the relative nature of thermal stresses imposed by differences in canyon geometry on urban inhabitants of an arid region.

The site investigated was one of the previously mentioned "Patio House" neighborhoods, built in the city of Dimona (elevation 600 m) in the arid Negev Highlands. The neighborhood is composed of single-storey row houses and attached walled courtyards, with a grid of narrow (3 m width) pedestrian streets dividing the rows (Fig 16.2). The pattern of streets and housing blocks is relatively compact and of a regular, well defined three-dimensional geometry, and streets are bordered by single-storey walls.

As is typical for many arid zones, the climate of the region is characterized by wide daily and seasonal thermal fluctuations. An average daily temperature range of 20-32°C in July is accompanied by low daytime relative humidity, intense solar radiation and strong late afternoon winds, predominantly from the northwest. In winter, while minimum daily temperatures occasionally reach freezing and winds are strong, clear skies and abundant solar radiation prevail in the daytime (Bitan and Rubin 1991)

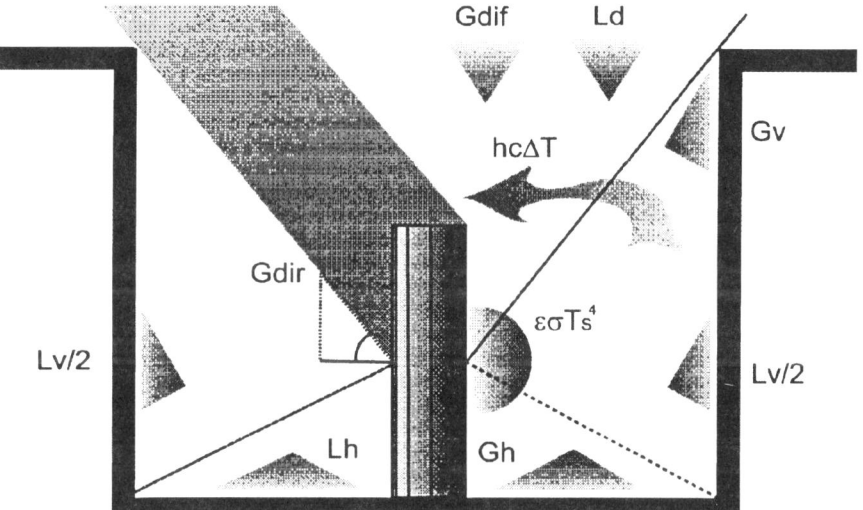

Fig. 16.3. Energy balance model expressing the thermal environment of a pedestrian in an urban street canyon. Radiative exchanges are composed of direct, diffuse and reflected solar radiation (G) and emitted long-wave radiation from the surroundings (L). Convective heat exchange is based on air temperature and wind velocity (for detail see Pearlmutter et al 1999)

Microclimatic measurements were carried out in two perpendicular street canyons, whose axes are approximately east-west and north-south, respectively, with the neighborhood grid rotated 30° counter-clockwise from the cardinal directions (Fig 16.2). Comparative measurements were taken above the roof of the single-storey row house building. Parameters monitored included air temperature, relative humidity, wind speed and direction, global radiation, net radiation, and radiant surface temperature. Continuous 24-hour monitoring was conducted during summer (June through August) and winter (December through February) seasons over a one year period.

In order to integrally characterize the thermal environment within each canyon, the *overall energy balance* of a theoretical body representing a pedestrian in the street was calculated, and compared with that of a similar body above the roof (Fig. 16.3). This energy balance model encompasses both *radiative* and *convective* thermal exchanges, while evaporative exchanges, although significant to physiological comfort and overall energy balance, were excluded from the model on the assumption that neither absolute nor relative humidity would vary significantly between points of comparison within the study area, which is largely devoid of vegetation.

The radiation balance was derived from measured radiation fluxes and surrounding surface temperatures, and the convective balance from measured air temperature and wind velocity. By calculating the overall balance of simultaneous radiative and convective fluxes, the effects of solar exposure, radiant heating, air temperature and wind were evaluated in one integrated index which expresses the total *rate of thermal energy exchange* between a pedestrian and the urban environment.

16.2.1
Summary of Case Study Results

1. *Air temperature* - From results of the monitoring, differences in air temperature based on location within the urban fabric proved to be minor. Relatively warm air is indeed found within the street canyons during the afternoon hours in summer, but with temperatures of at most 2-3°C higher than those measured above the rooftops of adjacent buildings (see Fig. 16.4).

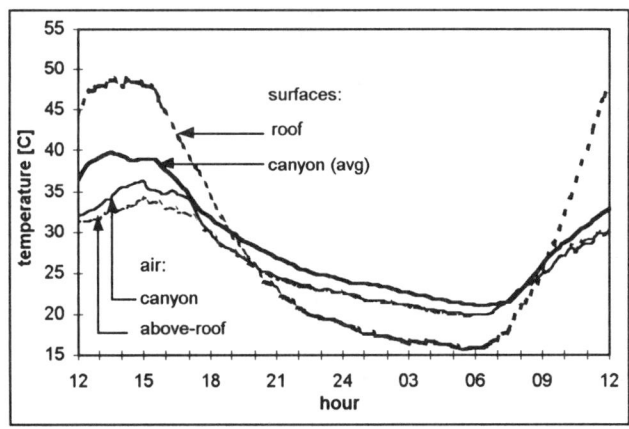

Fig. 16.4. Summer daily cycle of air and surface temperatures, in east-west street canyon and above the roofs of adjacent buildings

2. *Surface temperature* - More significant variation is observed when comparing the radiant temperatures of surfaces that define the urban space. The temperatures measured on vertical walls expectedly varied by several degrees according to orientation and height of measurement, but even these variations were small compared with the differences in temperature between vertical and *horizontal* surfaces. The flat rooftops of houses adjacent to the street, for example, were typically heated to a peak temperature of nearly 50°C on a summer afternoon, or higher by some 10°C than the average temperature of canyon walls and floor (Fig. 16.4).

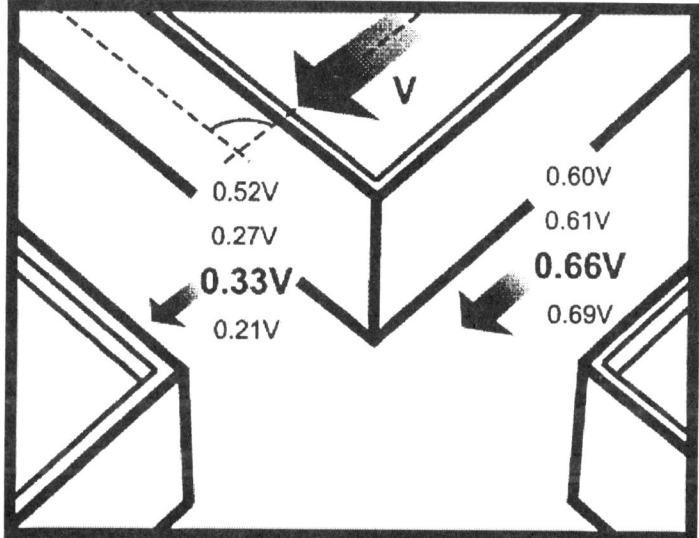

Fig. 16.5. Average wind speeds in street canyons which are parallel and perpendicular to the prevailing wind direction. Values are shown at height increments of 1m, as a proportion of the "free flow" above the urban surface

3. *Wind velocity* - A consistent pattern of wind speed "attenuation" was observed, with air flow restricted in general within the street canyons and to a greater extent in the street perpendicular to wind flow. Figure 16.5 shows the variation of wind speed by height and street orientation as a proportion of the "free flow" measured above the roof tops. It can be seen that when the wind is blowing at an "angle of attack" which is normal to the street axis, canyon wind speed at head height is attenuated to an average of 1/3 (0.33V) of the free flow, with this attenuation increasingly pronounced lower in the space. When flow is parallel to the street axis, velocity at head height is attenuated to about 2/3 of free flow and attenuation actually *decreases* with depth in the canyon. The practical interpretation of these patterns is that when the pre-

vailing wind is at a sharp angle to the street, an increasingly compact street canyon geometry (high H/W ratio) will most certainly reduce the rate of air flow, even in a low-rise street canyon. When the wind is directed along the street's length, however, air flow may be stronger in a deep canyon than it is in a shallower one.

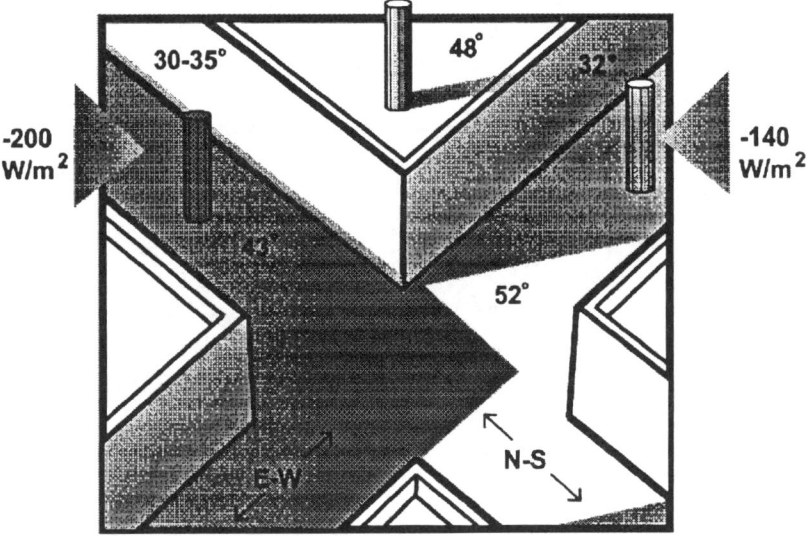

Fig. 16.6. "Snapshot" of the thermal environment of pedestrians in the street canyons on a summer afternoon: values of heat gain are shown relative to that of the above-roof body, together with shading patterns and radiant temperatures of various built surfaces

4. *Pedestrian energy exchange* – As described earlier, the measured values of microclimatic parameters were used as input for calculation of the overall energy exchanged between a pedestrian and the urban environment. In order to characterize the thermal qualities of each urban street canyon, the result of this calculation is presented as a "relative exchange," or as the difference between the energy balance of a body in the street canyon and that of a reference (above-roof) body. Figure 16.6 shows an example situation representative of most summer daytime hours, during which *a pedestrian in the street canyon absorbs less thermal energy from the environment than one in the open.* Thus while previously discussed observations of elevated air temperature and impaired wind flow within the street canyon suggest it to be a relatively overheated environment, the combination of all relevant parameters paints a somewhat different picture. Calculation of individual energy exchange components indicates that this summer daytime difference (which in late afternoon peaks at -140 W/m^2 for the E-W street and -200 W/m^2 for the

N-S street) is primarily due to shading of the body in the canyon from solar radiation (both direct and diffuse), which compensates for a reduction in its heat loss by convection (due to higher air temperatures and lower wind speeds).

In winter, a relative reduction in energy loss from the body to its street canyon surroundings (positive differential) is seen during most hours, primarily due to protection from cold winds (reduction in sensible heat loss). While overall energy loss during summer night hours was typically reduced in both canyons by about 100 W/m^2, the reduction on winter nights reached over twice that during a strong gust.

16.2.2
Discussion: Creating the Urban "Cool Island"

The above evidence suggests that like the urban "heat island" in general, the micro-scale heating effect within an urban canyon such as those analyzed here is primarily a *nocturnal* phenomenon: during daylight hours, the compact canyon is in fact a potential "cool island." This finding, however, is based on the thermal condition of a *body* in the street rather than that of the air surrounding it, and may be attributed to several factors.

The first and foremost of these factors is *shading*. As illustrated in Fig. 16.2, the street canyon's closely spaced flanking walls (particularly in the N-S street) provide both its surfaces *and* its inhabitants with considerable shade during a considerable portion of the summer day. However, the reduced exposure to solar radiation by a pedestrian in the street is not exclusively a function of shading from direct sun. A pedestrian standing at the midpoint of either street canyon is exposed not only to less direct radiation (due to shading of the body), but also to less diffuse radiation (due to a restricted Sky View Factor) and to less radiation reflected from the horizontal ground surface (due to both shading of the street and a restricted View Factor) throughout summer daylight hours. Together, these considerably offset the absorption by a body in the street of radiation reflected from vertical wall surfaces.

While a constricted urban space is known to trap heat by limiting the emission of *long-wave* radiation to the sky, this phenomenon is not prominent during daytime hours. Radiant heat gain by a pedestrian in the canyon during the day is in fact limited by the restricted exposure of a pedestrian to the *horizontal* ground surface - which of all surfaces is the greatest source of radiant heat, as it a) intercepts the highest intensity of solar radiation, and b) is often of a dark color, thus maintaining by far the highest surface temperature during the *daytime*. It is only at night that this relationship changes, and the importance of restricted radiant heat loss to the *sky* becomes dominant: since most urban surfaces have a high *thermal inertia*, heat which is absorbed during the day is stored and released primarily at night.

Ventilation is also not a dominant factor during most daytime hours, since its effectiveness in an arid region like the Negev Highlands is in any case limited by high air temperatures and low wind speeds. Ventilation is crucial, however, during the late afternoon and evening hours, when winds are consistently strong and cool, and during this period the importance of *orientation* is most evident: a space such as the north-south street in the Patio House neighborhood can provide substantial convective cooling by wind from the north-northwest, in *addition* to the previously mentioned effect of solar shading.

16.3
Conclusions

What distinguishes the desert climate, perhaps more than anything else in terms of thermal comfort, is its extreme fluctuations. Extreme heat on a summer day gives way very abruptly to the coolness of the evening, and the chill of a desert night disappears quickly on dependably sunny mornings. What most concerns the planner, then, is ameliorating the periods of most intense discomfort – and in many cases this can be achieved by taking advantage of natural forces which prevail in the interim.

The preceding discussion illustrated how a "compact" urban fabric may provide city dwellers with an environment that is more amenable than the more open surroundings, during the hottest part of a summer day. Given the desert atmosphere's rapid cooling around the time of sunset, the tendency of a compact urban street canyon to become *relatively* overheated at night presents far less of a burden than it would in a more temperate location. The combination of heat and high humidity in a Mediterranean climate, for instance, tends to lengthen the period of heat stress on a summer day and make ventilation more vital to pedestrian comfort. In this case the balance of microclimatic factors described above could significantly change – as could the implications for appropriate urban design.

In a similar way, it was found that a narrow street may provide relief from the most severe thermal stress in winter, which occurs during hours of darkness. Since in many arid regions winter nights are notoriously cold and windy, yet daylight hours are comfortable, the advantage of this planning solution is especially apparent in the desert.

It must be emphasized, however, that truly successful climatic design depends not only on such general planning decisions, but on the details as well. Beyond the question of compactness, this study emphasized the advantage of narrow streets whose axes are orientated approximately from north to south, both for shading in summer and the lack thereof in winter. In the Negev in particular, this orientation facilitates efficient wind ventilation on warm evenings as well. In addition to issues of building geometry, an amenable urban environment must incorporate proper materials, landscaping, and a host of other features which are beyond the scope of this discussion.

Going further, we should bear in mind that the research described here was intentionally focused on a very *specific aspect* of the larger question concerning climatically appropriate urban design in the desert, in the hope of drawing operative conclusions regarding this aspect. In practice, the planner is faced with a series of trade-offs between varying and sometimes contradictory requirements.

For example, considerations for improving the outdoor environment are not necessarily the same as those for optimizing comfort and energy-efficiency *within* buildings, and in the modern motorized city, the importance of pedestrian comfort has indeed been marginalized. Successful design in fact requires an understanding not only of spatial, but also of temporal, usage patterns, since the criteria for proper environmental response may vary over the hours of a day or seasons of the year.

Whether it is in the hands of the planner to determine the larger trends or the finer details, it is clear that a successful merging of planning intentions and physical realities requires a focus on the *local conditions* at hand. Even in a country as small as Israel, adjacent climatic regions differ from one another both in the scope of human settlement they can support and the urban planning approaches they demand. Lying on the periphery, it is the desert which commonly suffers from the implantation of preconceived notions – and as this study suggests, a more responsive approach is both possible to achieve and costly to avoid.

References

Bitan A, Rubin S (1991) Climatic atlas of Israel for physical planning and design. Israel Meteorological Service and Ministry of Energy and Infrastructure, Tel Aviv

Chandler TJ (1976) Urban climatology and its relevance to urban design. WMO Technical Note No. 149, Geneva

Golany G. (1980) Planning urban sites in arid zones: the basic considerations. In Golany, G. (ed) Urban planning for arid zones. John Wiley & Sons, New York

Gradus Y, Stern E (1985) From preconceived to responsive planning: cases of settlement design in arid environments. In: Gradus Y (ed) Desert development: man and technology in sparselands. D. Reidel Publishing Co, Dordrecht

Oke TR (1988) Street design and urban canopy layer climate. Energy and Buildings 11: 103-113

Pearlmutter D, Bitan A, Berliner P (1999) Microclimatic analysis of "compact" urban canyons in an arid zone. Atmospheric Environment/Urban Atmosphere (in press)

17 Adaptive Architecture: Low-Energy Technologies for Climate Control in the Desert[1]

Yair Etzion, David Pearlmutter, Evyatar Erell and Isaac A. Meir
J. Blaustein Institute for Desert Research, Ben Gurion University of the Negev, Sede Boker Campus, 84990, Israel

17.1
Introduction

Le Corbusier considered buildings "machines for living in." Modern buildings have indeed become increasingly complex, involving technologically advanced building materials, and mechanical systems for controlling interior air quality, thermal comfort, lighting and acoustics. These systems, which rely exclusively on the utilization of non-renewable energy, are often expensive to install and energy intensive in operation. This is particularly true of buildings constructed in locations with extreme climatic conditions, such as deserts, where the difference between ambient conditions and the desired interior conditions is large.

This paper illustrates a radically different approach to the provision of thermal comfort in a building. Rather than invest non-renewable energy to counteract the natural conditions, it is often possible to harness natural energies and exploit the local climate to great advantage by adapting the architectural design of the building. Climate conscious design requires a thorough understanding of the local climate, and the employment of several strategies and systems for the creation of an agreeable micro-climate with a minimal investment of energy. The success of such a design depends as much on the integration of these strategies and on the proper operation of the building by its users as it does on the individual performance of each technological system.

17.2
The Problem - Local Climatic Conditions

The building described below was built at the Sede-Boker Campus of the Ben-Gurion University of the Negev, located at 30.8° N latitude, about 480 m above sea level. The Negev Highlands - (300 m and above) - are characterized by cold and mostly sunny winters, and by summers that are hot during the day but usually pleasant at night. Average annual rainfall is 80 mm, but there is a considerable deviation from year to year. The following analysis was made of the effect of the local climate on the design during summer and winter:

---

[1] Reprinted from: Automation in Construction 6 (1997), Etzion Y et al "Adaptive architecture: integrating low-energy technologies for climate control in the desert," pp. 417-425, Copyright 1997, with permission from Elsevier Science

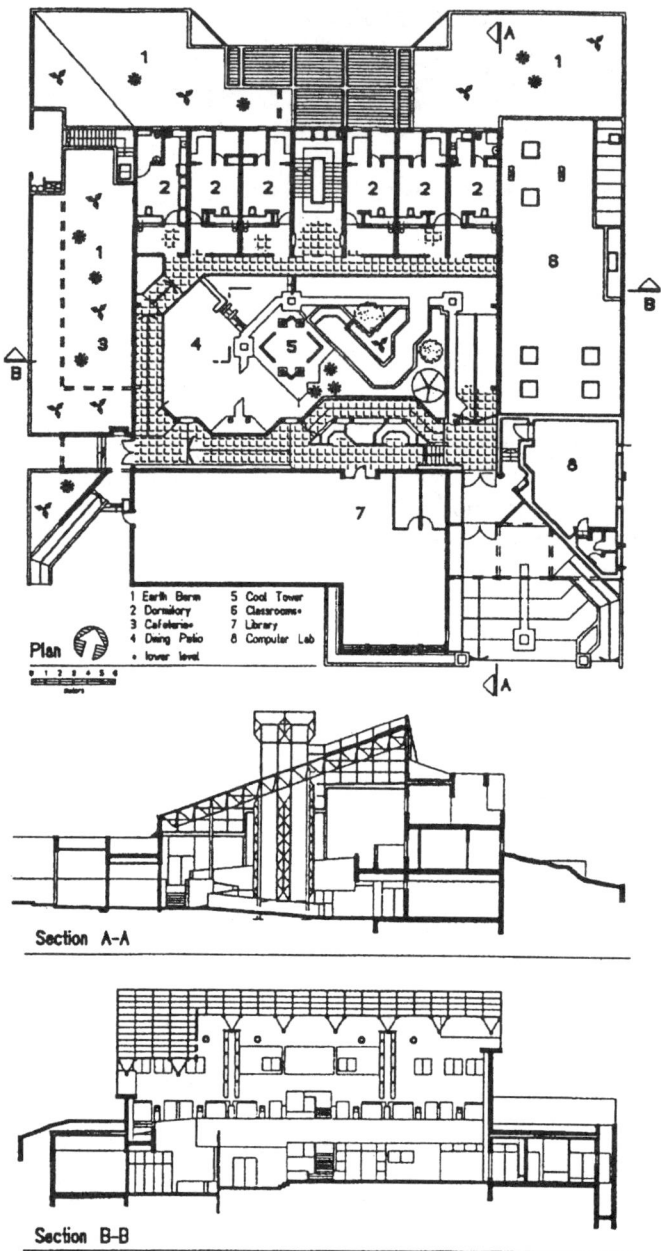

Fig. 17.1. The Building of the International Center for Desert Studies - General layout and a sections

- *Summers* are hot and dry, with an mean daily maximum temperature of 32°C (Bitan and Rubin 1991). However, nights are cool - the mean daily minimum is 17°C - so that the mean daily temperature lies within the thermal comfort range. Thus problem for the designer lies mainly in overcoming the overheated conditions prevailing around mid-day, while convective cooling is particularly effective at night due to the low ambient temperature. Solar radiation is very intense: global radiation on a horizontal surface averages 7.7 kWh/m^2 per day (during June and July). Thermal comfort in outdoor or semi-exposed areas depends not only on lowering the air temperature, but to a great extent on reducing the exposure to this intense radiation. Ambient relative humidity is very low, between 20% and 40% for most of the day, but may rise to 90% or more at night, when the air temperature drops sharply. On extremely dry days the lack of moisture in the air may cause some discomfort. However, for most of the summer, the low relative humidity extends the thermal comfort range, so that temperatures as high as 28°C may be tolerated quite comfortably.
- *Winters* in Sede-Boker are sunny but cool during the day, and cold at night. The mean daily temperature in January is only 9.3°C (Bitan and Rubin 1991). While the mean daily maximum is 14.9°C, night time minimum temperatures average 3.8°C, due to the intense radiative cooling characteristic of clear desert skies. Thus, considerable energy is required to heat buildings to comfortable levels. The same clear sky conditions result in high levels of insolation during the daytime: Global radiation averages 3.3 kWh/m^2 per day on a horizontal surface, and about 4.6 kWh/m^2 per day on a south-facing vertical surface. The abundance of solar radiation and the large number of clear days provide ideal conditions for the provision of passive heating in buildings, relying on the utilization of solar energy.

17.3
The Response: Project Overview

The Blaustein International Center for Desert Studies (BIC) building is a multi-functional complex which was designed to house the international activity of the Jacob Blaustein Institute for Desert Research (Fig. 17.1). The 1100 m^2 building is home to the Institute's library, teaching facilities, a cafeteria and lounge, and administration offices, as well as two apartments and six smaller dormitory rooms for accommodating visiting scientists and scholars. The various building elements are organized around a 500 m^2 central atrium, which straddles a main pedestrian artery linking the existing campus with its future expansion. The protection it provides from the extremes of the outdoor climate has resulted in a flourishing semi-tropical garden, which stands in contrast to the barren landscape outside. The provision of thermal comfort in this relatively large space was one of the main challenges facing the design team, but also provided an opportunity for a comprehensive approach to the climatic conditions in the building complex as a whole.

In winter, the provision of thermal comfort by passive means is achieved by two strategies – firstly, by maximizing solar heat gains (and providing the means of storing the incoming energy), and secondly by minimizing heat loss through the envelope of the building. In the case of the BIC building, the atrium was designed to function in winter as a solar greenhouse. Warm air from the atrium is drawn during the daytime to heat the adjacent spaces. Heat losses are minimized by reducing the area of exposed exterior surfaces, and by providing sufficient thermal insulation. The exterior walls have 5 cm of expanded polystyrene insulation, for a total thermal resistance of about 2 m^2 °C/W, while the roof has 10 cm of insulation, for a total resistance of about 3.6 m^2 °C/W. These values exceed considerably the minimum requirements for thermal insulation in residential buildings as set out in Standard 1045 (The Standards Institution of Israel 1989)

In summer, thermal comfort is achieved through a combination of three strategies:

1. Reduction of unwanted heat gains through careful treatment of the building exterior. Insulation of exterior walls is of primary importance, in summer as well as in winter. However, several other strategies were adopted as well:

- The exterior surface of the walls is a smooth stucco painted white, so that 70-80% of incident solar radiation is reflected, compared with about 50% for most commonly used finishes, such as limestone or textured stucco (Anderson 1977; Gubareff et al. 1960). The smooth finish was selected to reduce the adherence of airborne dust particles, thus preventing the discoloration of the external walls which occurs commonly in desert conditions.
- The orientation of window openings and glazed areas, (other than the atrium), was designed to allow ventilation, yet reduce heat gains to a minimum. There are no openings on the east or west elevations of the building, where the intensity of solar radiation in summer is higher than on any other surface except the roof (see Table 17.1). Since these elevations also enjoy less solar radiation in winter than a south facing wall, they are the least desirable orientations. Rooms in the main residential wing open onto the atrium, and have only small windows facing north, the direction of the prevailing winds in Sede-Boker during the summer, to allow cross ventilation.
- Transmission of solar radiation through the roof surface of the atrium is reduced by means of a unique selective glazing (see Fig. 17.2), and by the addition of internal shading beneath the glazed surface.

2. Use of high thermal capacity materials to maintain the interior conditions close to the daily average, which, in Sede-Boker, lies within the comfort range throughout the summer. The building's significant thermal capacity contributes to stabilizing the large daily fluctuations typical of the desert, and also increases the building's thermal lag time, which is the time that passes, for instance, between the peak external temperature and the peak internal tempera-

ture (Givoni 1969). The external walls of the building combine insulation with thermal mass. The internal layer of the wall is built of concrete blocks, which constitute an integral part of the storage mass of the building. On the external side of the wall the attached insulating layer, (consisting of 5 cm thick expanded polystyrene), is protected on the outside by a special acrylic stucco. The order of the layers in the wall is of great importance: while the thermal resistance of the wall (R-value) is simply the sum of the thermal resistances of all the layers, the thermal storage capacity depends on the degree of thermal contact between the interior air and the wall surface. Should the insulation be placed on the interior surface of the wall, the effective thermal capacity of the envelope would be greatly reduced, as would its thermal time constant. The building's exterior is covered with earth berms up to the height of the second floor windows. Previous research on earth-sheltered buildings carried out by the Desert Architecture Unit (Pearlmutter et al. 1993) and others (Davis 1979; Rahamimoff et al. 1987) has shown that earth sheltering may reduce significantly the heating and cooling loads on buildings in climatic conditions similar to those of the Negev highlands. The effect of the earth cover is twofold: first, it insulates the building's external surfaces from extreme ambient conditions, thus reducing the rates of both energy gain (summer) and energy loss (winter). Second, earth berming increases the thermal inertia of the building by increasing its heat storage capacity, thus reducing its internal temperature fluctuations.

Table 17.1 Daily incidence of solar radiation on building surfaces by orientation in Sede-Boker (kWh/m^2)

	North	East	South	West	Roof
July	2.48	5.07	2.62	5.07	7.67
January	1.08	2.54	4.63	2.54	3.33

3. A passive cooling system, the evaporative cool tower, was introduced to improve thermal comfort in a selected, high use area in the atrium, where due to the size of the space and to its exposure, the effect of other measures adopted to provide thermal comfort was deemed insufficient.

17.4
Experimental Evaluation of the Building's Thermal Performance

17.4.1
The Sunken Atrium

All building spaces are arranged around the 500 m^2 sunken and enclosed atrium (Fig. 17.1). The atrium is not only the visual and functional focus of the building, but also a thermal buffer and modifier, creating a micro climatic "oasis'" within the harsh desert surroundings. The atrium is partly shielded from the exterior environment by elements of the building itself, such as the main residential wing on the north side of the building and the library on the south. It also benefits from the effects of earth berming: The lowest level of the courtyard is 2.5 meters below grade, and the north and west wings of the building have been earth bermed against the exterior walls up to a height of 4.5 meters above grade, or 7 meters above the atrium floor.

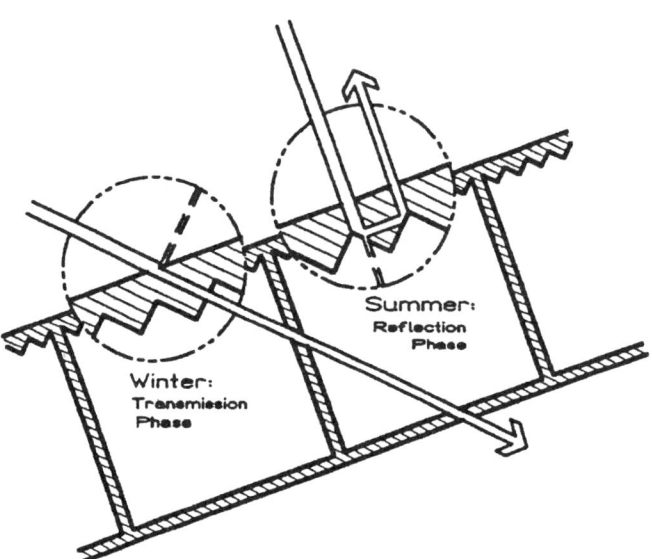

Fig. 17.2. Section through a "selective" polycarbonate sheet, showing internal reflections and selective transmissivity

The roof of the atrium is glazed with a unique double-skinned polycarbonate sheet manufactured in Israel, which is a selective transmitter of solar radiation.

The material has small, triangular prisms along the length of the interior surface ssheet, resulting in variations in the transmissivity of the material which are a function of the angle of incidence of the solar radiation. A large proportion of radiation impinging upon the surface at an angle close to the normal (±9°) is reflected, while reflection of radiation at oblique angles is lower. The roof geometry, i.e. its tilt angle (20° facing south) and the direction of the prisms (E-W), was determined so that during the hot hours of the summer most of the incident solar radiation would be reflected, while in winter most would be admitted into the atrium space.

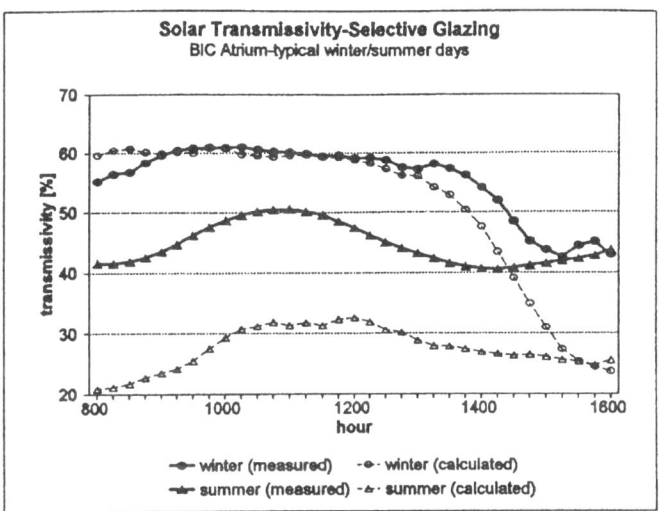

Fig. 17.3. Calculated and measured transmissivity of selective glazing in summer (Aug-Sept) and winter (Jan-Feb)

In winter, the atrium functions as a solar greenhouse. All its openings are closed, and the air within it is heated by the incoming solar radiation. The roof glazing transmits approximately 60% of incident solar radiation (Fig. 17.3), so that interior temperatures at floor level are 5-15°C higher than ambient air temperature. The temperature elevation is greatest during clear sunny days, but is evident even at night or during overcast weather. Under normal winter weather conditions at Sede-Boker, the daytime temperature in the atrium is typically 20-25°C, falling to 12-15°C at night (Fig. 17.4). The vertical temperature profile displays the effects of thermal stratification, and a temperature difference of up to 5°C between the floor and apex of the atrium was recorded on clear, sunny days (Fig. 17.5a).

In summer, thermal comfort in the atrium greatly depends on the extent to which solar heat gain can be minimized, and excess heat removed. The original design called for the reduction of solar radiation by two means:

1. Selective glazing used in the roof. The glazing panel was intended to act as a shading device in summer, but was found to transmit considerably more radiation than that calculated from the manufacturer's data (Fig. 17.3). Thus some 40-50% of incident solar radiation penetrated the glazing and increased the heat load on the building interior. The high proportion of indirect diffuse radiation, as opposed to direct beam radiation, as well as the accumulation of dust on the glazed roof, may be responsible for the reduction in the performance of the selective glazing, but this remains the subject of a separate study.
2. Light-reflecting canopy. The original design called for the installation of such a canopy parallel to the interior plane of the roof, to further reduce the penetration of solar radiation and to cut glare. Due to budget constraints, this curtain has been installed in only one section of the roof, and is now being used only as a demonstration of the original intentions.

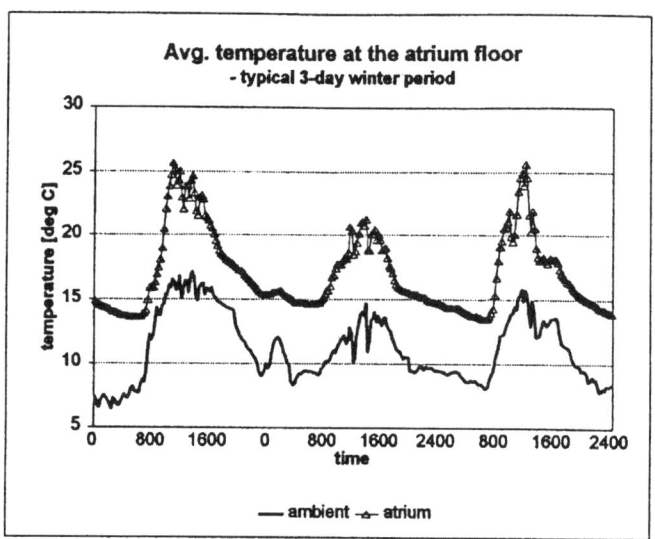

Fig. 17.4. Average temperature at the atrium floor during a typical 3-day period in winter

The removal of excess heat is achieved by two means:
1. Operable windows installed along the length of the south side of the atrium and along the north apex of the atrium are opened during the summer. The effect of these windows is to allow the removal by cross ventilation of excess heat trapped in the upper parts of the atrium.

2. A large down draft evaporative cool tower. The performance of this cool tower is described at length in a separate section of this article.

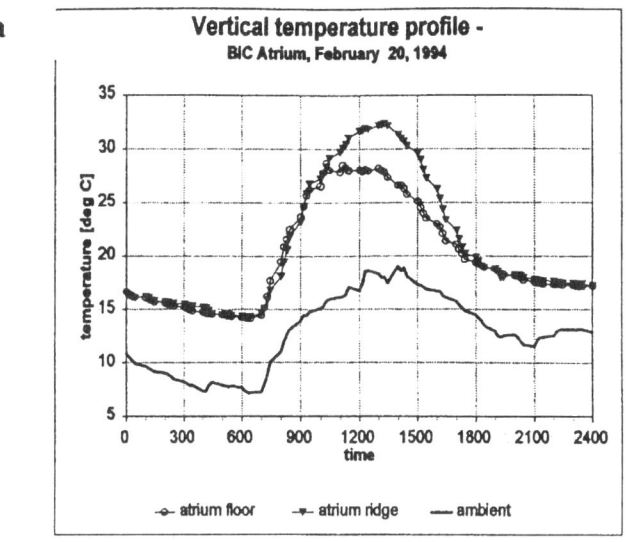

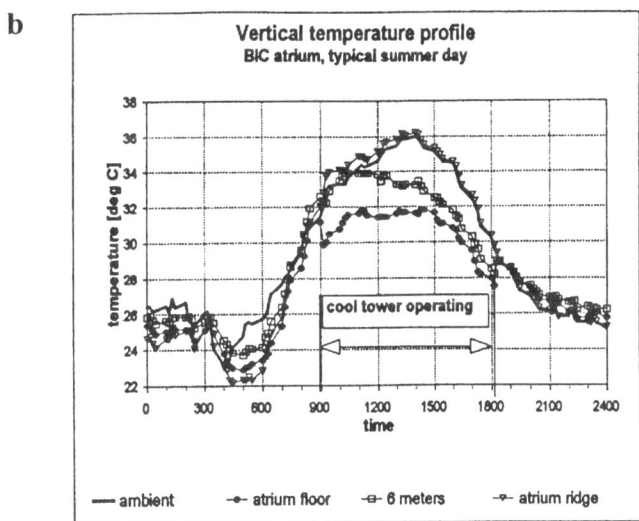

Fig. 17.5 The vertical temperature profile in the atrium on typical a) winter and b) summer days

In the configuration studied, daytime temperatures at the atrium floor level were similar to exterior air temperature, during all but afternoon hours.

Fig. 17.5b shows that for this mid-day period of highest heat stress, air temperatures at floor level were up to 4°C lower than the ambient, while those measured at the atrium's apex were slightly higher than the ambient. Thus, while the planned cooling effect is not evident in the entire atrium, its benefit is realized in the occupied area adjacent to the cafeteria, where thermal comfort is further enhanced by the airflow generated by the cool tower.

The degree of mixing of outside air with interior air in the lower parts of the atrium is a critical factor in determining the overall thermal comfort. The balance between evaporative cooling provided by the cool tower and comfort cooling provided by the natural airflow has still not been resolved, and the extent to which the atrium should be opened to natural ventilation during the summer is a matter for further study.

17.4.2
The Evaporative Down Draft Cool Tower

As previously mentioned, in summer cool air is provided to the atrium by a large evaporative cool tower. Evaporative cooling is a familiar and energy-efficient tool for space conditioning in arid regions, where daytime temperatures are high and relative humidity is low (Givoni 1994). The primary innovation of its use here is the exploitation of convective forces for the conditioning of a relatively open public space. Fig. 17.6 shows a vertical cross section of the tower: Its height is approximately 12 meters, and its horizontal section is octagonal, with a 4 meter width from side to side.

Water injectors and sprayers - which were selected empirically - saturate the air in the tower with water, causing fast and intensive evaporation and significantly lowering the air temperature. More water is injected into the air than can be completely evaporated, in order to ensure the highest possible evaporation rate and temperature reduction. The excess water falls into a collection pond at the bottom of the tower, to be recycled by a small pump to the top of the tower and to be re-sprayed into the air. The result is that no water is lost except that volume which is evaporated and which cools the air. The quantity of water which is evaporated by the air moving through the tower is about $1-1.5 m^3$ per day, depending on the ambient external conditions. At the top of the tower, a low-rpm fan was installed to supplement the natural convective down-draft that is caused by the temperature difference between air in the upper and lower parts of the tower. Calculated as a function of volumetric air flow and temperature depression, up to 120 kW of cooling power is provided by the tower in the hottest hours of the summer (Pearlmutter et al. 1996). Cool air is supplied at the lowest point of the atrium, so that it accumulates in the sitting area of the cafeteria and ascends only as it begins to warm up. As the warm air rises, it is replaced by cooler air supplied by the tower, so that the lower layer of air, which is closer to the atrium floor, remains relatively cool.

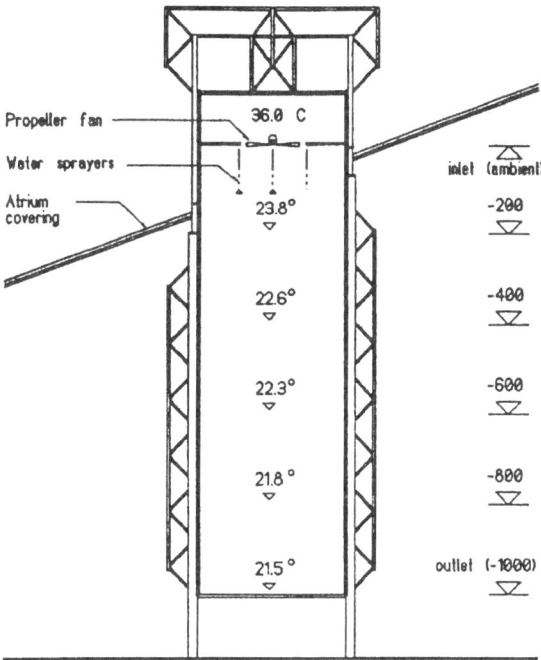

Fig. 17.6. Schematic section of cool tower, showing installation in atrium and typical temperature profile on a summer day

Fig. 17.7 shows the performance of the tower on a typical summer day. At mid-day, outside air is drawn into the tower at 35-36°C, cooled by evaporation and exhausted at 21-22°C. Although the air leaving the tower is close to saturation, upon mixing with the internal air of the atrium its humidity drops and the resulting relative humidity in the occupied sitting area is less than 65%. Calculated as a proportion of the maximum possible temperature depression obtainable by evaporation alone, the system's efficiency was found to exceed 85% during all hours of operation.

Upon observation of the cool tower's performance, a number of possible improvements were identified. The most important of these concerns the up-draft of air through the tower, due to wind generated suction at the inlet above roof level. This phenomenon, which is the result of the particular geometry of the roof and the height of the air inlet above it, counteracts the thermal down-draft and reduces the efficiency of the fan at the top of the tower. The reduction in the air flow rate through the tower results in a lower overall cooling output. Further experiments carried out at the Desert Architecture Unit (Pearlmutter et al. 1996) have resulted in the development of a wind capture unit that is to be installed at the head of the tower. The dual intent of this measure is to deflect the natural airflow above the roof into the tower, and to prevent the reverse flow observed under some conditions, thus increasing airflow and possibly reducing the system's dependence on mechanical means.

An unforeseen aspect of the cool tower's operation was the accumulation of sediment in the pool beneath it. Dust found in suspension in the ambient air is washed out by the water droplets in the tower, and deposited in the pool. Since the volume of air flowing through the tower is quite large, the amount of dust washed out is enough to require frequent cleaning of the pool. This "rinsing" effect does, however, have a decidedly positive side benefit, since it allows cleaner air to be introduced into the atrium.

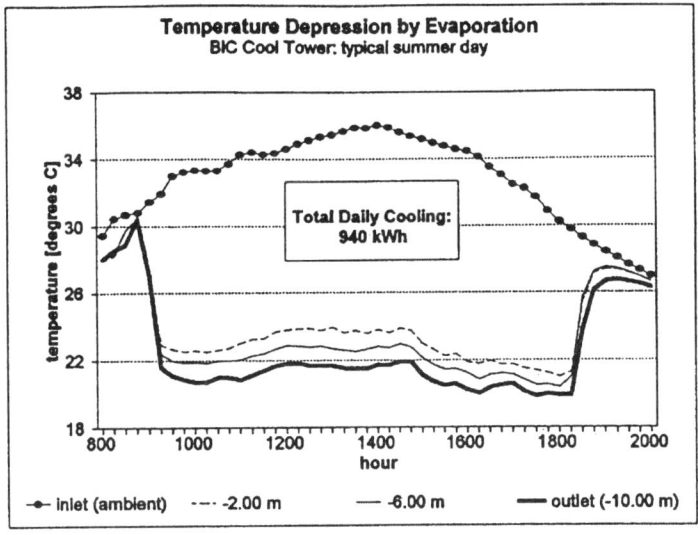

Fig. 17.7. Temperature depression by evaporative cooling in the down-draft cooling tower

17.4.3
Indirect Space Heating from Solar Heated Air

The upper stories of the north wing of the building, which serve as guest accommodations, are heated by drawing in warm air from the apex of the atrium. Air is channeled during the warm daytime hours through 15.3 cm ducts stretching from the top of the atrium into each room, drawn in by small, individually operated air turbines positioned at each of the duct outlets.

A comparison of the temperatures in two similar rooms on the same floor illustrates the effect of the heating system. A reference room, well insulated and enjoying the benefits of its exposure to the mild conditions in the atrium, but having no other source of heat, remained stable at about 16°C, while ambient air temperature on a typical day fluctuated between 8°C and 18°C (Fig. 17.8). By introducing warm air, at temperatures of up to 32°C, internal air temperatures inside the heated apartment were maintained at well over 20°C in the daytime, falling no lower than 18°C at night.

Based on the turbine's measured output of 260 m^3 per hour and observed temperature differentials of up to 8°C on a typical sunny day, the system provided a peak heating power of about 600 W and a daily heat output of 3.7 kWh, requiring only 40 W of electric power for operation. Calculation of the net heat output was based on the temperature difference between room air and the warmer air at the duct inlet, and reflects the low amount of energy required to maintain the well-insulated room at a comfortable level. Performing the same calculation for a colder room would result in a significantly higher net heat output, since the temperature of the air supplied by the duct depends only on the conditions at the apex of the atrium.

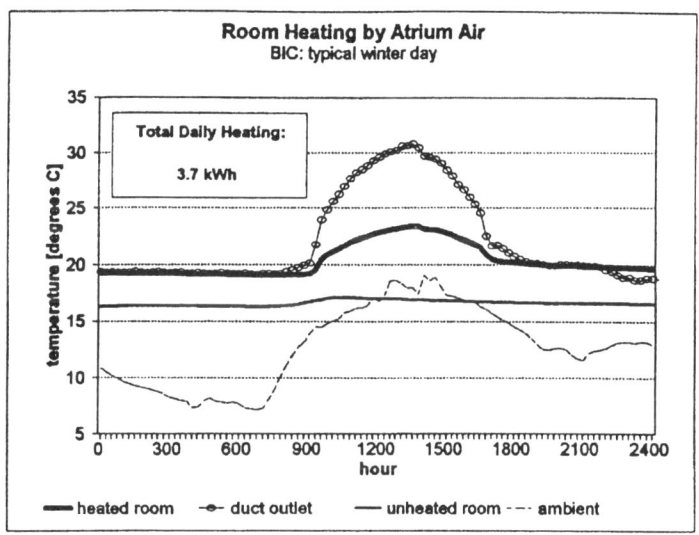

Fig. 17.8. The effect of space heating by forced convection of atrium air in winter

17.5 Conclusions

The design of the International Center for Desert Studies incorporates several innovative energy saving strategies in concert. The evaluation of these measures indicates that some, such as the evaporative cool tower and the solar heated air system appear to be very successful and cost-effective means for the provision of thermal comfort in desert climates. The performance of others, such as the selective glazing, while below the manufacturer's claims, are worthy of consideration in other projects. Most of the innovative features described in this paper, such as the down-draft cool tower, may benefit from further research and optimization. The effect of the earth cover in this building remains a subject for future investigation. It is anticipated that the completion of the building, involving the installation of the interior shading canopy underneath the atrium roof, and of a wind capture mechanism in the cool tower, should further enhance the thermal conditions inside the atrium in summer.

References

Anderson B (1977) Solar energy. McGraw Hill, New York

Bitan A, Rubin S (1991) Climatic atlas of Israel for physical and environmental planning and design. Ramot Press, Tel-Aviv University, Tel-Aviv

Davis WB (1979) Earth temperature: its effect on undergroungd residences. In: Moreland et al. (eds) Earth covered buildings: technical notes. U.S. Dept. of Energy and University of Texas at Arlington, pp 205-209

Givoni B (1969) Man, climate and architecture. Van Nostrand Reinhold Company, New York

Givoni B (1994) Passive and low energy cooling of buildings. Van Nostrand Reinhold Company, New York

Gubareff GG et al. (1960) Thermal radiation properties survey (2nd edition). Honeywell, Mineapolis

Pearlmutter D, Erell E, Etzion Y (1993) Monitoring the thermal performance of an insulated earth-sheltered structure: a hot-arid zone case study. Architectural Science Review 36(1): 3-12

Pearlmutter D, Erell E, Etzion Y, Meir IA, Di H (1996) Refining the use of evaporation in an experimental down-draft cool tower. Energy and Buildings 23:191-197

Rahamimoff A, Rahamimoff S, Silberstein A, Faiman D, Zemel A, Govaer D (1987) Design considerations for an earth-integrated education centre in the Israeli desert. Tunnelling and Underground Space Technology 2(1):69-71

The Standards Institution of Israel (1989) Thermal insulation of residential buildings - Israel Standard 1045. The Standards Institution of Israel, Tel-Aviv

Part Four

CASE STUDIES

18 Desert Settlements in Israel: Socio-economic and Physical Data

Boris A. Portnov and Wolfgang R. Motzafi-Haller
J. Blaustein Institute for Desert Research, Ben-Gurion University of the Negev, Sede-Boker Campus, 84990, Israel

This chapter features a selection of six desert towns in the Negev region of Israel. Statistical data are listed for comparing the state of development of these towns to the region in which they are located, and to the rest of the country. A short overview of the physical layout of each settlement and a short historical background is provided. Unique features and appearances typical to the towns are documented in photographs.

18.1
Be'er-Sheva

Be'er-Sheva is one of the oldest cities in the world: *"[Isaac] named the well Shibah. The city is therefore called Be'er-Sheva to this very day"* (Genesis 26:33). The location of the city on ancient crossroads and close to water sources sustained its development over hundreds of years. During the fourth millennium BCE, there were settlements in the area that are termed the "Be'er-Sheva culture." The inhabitants of these settlements were farmers and hunters, who were proficient in ivory carving and the production of bronze tools (Edri and Nir 1995).

During King David's period (1000 BCE), the city became an administrative center and the capital of the "Negev Yehuda." Remains of the city from this period indicate that it was a well-designed and fortified administrative center (*ibid.*). During the Roman-Byzantine period, until the seventh century CE, the city was a part of the defensive line (*limes*) against attacks by desert nomads. During the Arab conquest of 700 CE, the city was destroyed, and was left abandoned until the end of the nineteenth century. It was then rebuilt under the Turks. The British captured the city in 1917. During the British period, the population of Be'er-Sheva grew to 7,000 residents in 1948. Following the declaration of the State of Israel in 1948, the city was settled by demobilized soldiers, and in the beginning of 1949, they were joined by new immigrants (*ibid.*).

In the 1970s, Be'er-Sheva experienced a significant increase in population and economic growth. In the 1980s, this growth was replaced by a period of socio-economic slowdown. The lack of employment and single-family housing resulted in a major outflow of population, both to the center of the country and to the new suburban communities established to the north and north-east of the city – Omer, Metar and Lehavim (Newman et al. 1995).

In the wake of the mass immigration from the former Soviet Union in 1989-1991, Be'er-Sheva gained a development momentum. The newcomers were attracted primarily by the availability of public housing and relatively low housing prices in the area (Portnov 1998). In 1995, the population of the city stood at 150,000 residents (see Appendix). This made it the largest industrial, administrative and cultural center of the Southern district of Israel. The city hosts various governmental offices, the Ben-Gurion University, and the Soroka hospital.

The city is relatively attractive to migrants. In 1995, the migration component of its population growth exceeded greatly natural increase (see Appendix). This may have signaled the sustainability of the city's population growth in the future (see Chap. 3 of this book). Compared to other urban settlements in the area and the national average, Be'er-Sheva has a high percentage of persons with higher education and a relatively high home-ownership level (see Appendix). While the former datum shows the city's future economic potential, the latter indicator implies that the majority of the city's dwellers consider Be'er-Sheva as the place for permanent residence. The physical layout of the city is relatively compact, although wide strips of public parks and open spaces separate city neighborhoods. Two large pockets of industrial development and communal services are located in the southeastern part of the city (Fig. 18.1 – 18.4).

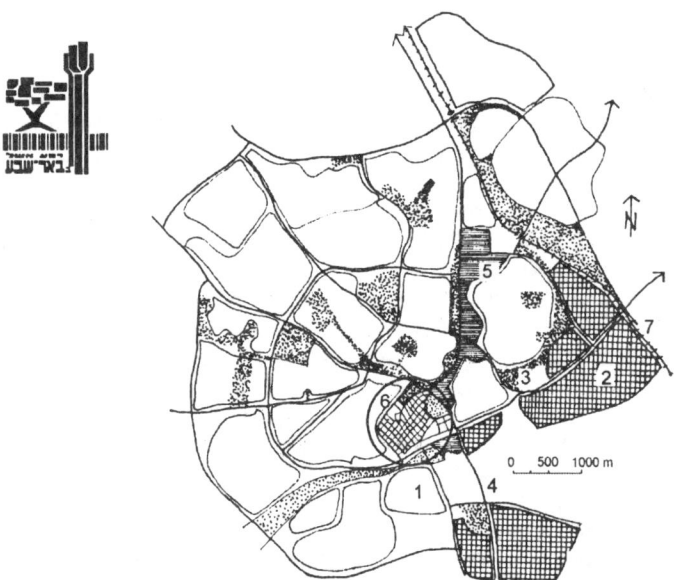

Fig. 18.1. Be'er-Sheva: general layout

1 - residential neighborhoods; 2 – industrial areas and warehouses; 3 – public parks and open spaces; 4 – main roads; 5 – public center; 6 – the Old City; 7 - railroad

Residential development in Be'er-Sheva is extremely diverse. It is formed by both low-density single-family housing in peripheral neighborhoods and high-rise buildings in the city center and adjacent areas (Figs. 18.2-18.4).

Fig.18.2. Be'er Sheva: administrative building in the downtown area

Fig.18.3. Be'er Sheva: new housing development southwest of the old city

Fig.18.4. Be'er Sheva: private houses under construction in a new suburban neighborhood in the northern part of the city

18.2
Eilat

Eilat is an international tourist resort located on the Gulf of Eilat/Aqaba. In 1995, the population of the town stood at 32,000 residents (see Appendix). Eilat's economy depends heavily on the tourist industry. In the 1960s-1970s, the port function of Eilat was supported by the government for strategic reasons. With the opening of the Suez Canal, the port sector of Eilat has steadily declined. It currently employs only several hundreds workers out of a labor force of about 15,000 people (Lithwick and Lithwick 1997). The rest of the labor force is employed in the tourist sector, in trade, food processing, services, and construction. These sectors form almost 50 per cent of the town's employment base (*ibid.*). During the late 1980s and 1990s, the construction of hotels and other resort facilities in Eilat expanded substantially, and the population of the town increased at the average of 6-7 per cent per annum (ICBS 1997). Such a rapid growth had negative consequences for the town's socio-economic development. The rapidly growing tourist sector attracted a large number of transient workers and this in turn resulted in high annual rates of out-migration (Lithwick and Lithwick 1997; Gabriel et al. 1985).

The physical layout of Eilat reflects the dependence of the town's economy on the tourist sector and service facilities: hotels, medical resort centers, and camping facilities along with the airport site occupy almost half of the built area (Fig. 18.5).

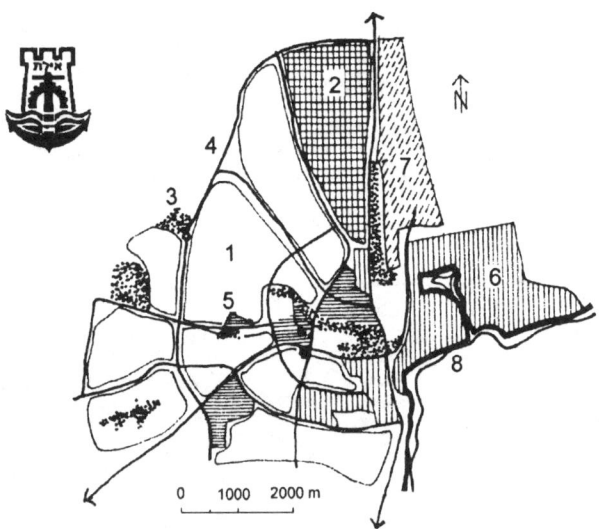

Fig. 18.5. Eilat: general layout

1 - residential neighborhoods; 2 – industrial areas and warehouses; 3 – public parks and open spaces; 4 – main roads; 5 – public center; 6 –resort facilities; 7 – airport; 8 – seashore

18.3
Dimona

Located 30 km to the southeast of Be'er-Sheva, Dimona is the largest development town in the Negev. In 1995, its population stood at 31,000 residents (see Appendix).

The town was established in 1955. Its first residents were new immigrants from North Africa and Europe who were directed to the area upon their arrival to the country. In 1958, Dimona received the status of a local council, and in 1969, it was officially recognized as an urban settlement (DCC 1996).

From its outset, the economy of the town was based on mining and manufacturing (specifically, chemical, textile and food industries) which could employ newcomers generally lacking educational background and employment skills. In the wake of recent structural changes in the national economy (i.e. a transition from low technology to high-technology industries), some of these factories were either downsized or closed which resulted in mass unemployment. The situation was partially improved after the construction of a computer equipment factory that currently employs nearly 2,000 residents of the area (*ibid.*). The predominantly industrial profile of the town is reflected in its physical layout. Large areas of industrial development and communal services are concentrated along the railroad at the northwest entrance to the town (Fig. 18.6 – 18.9).

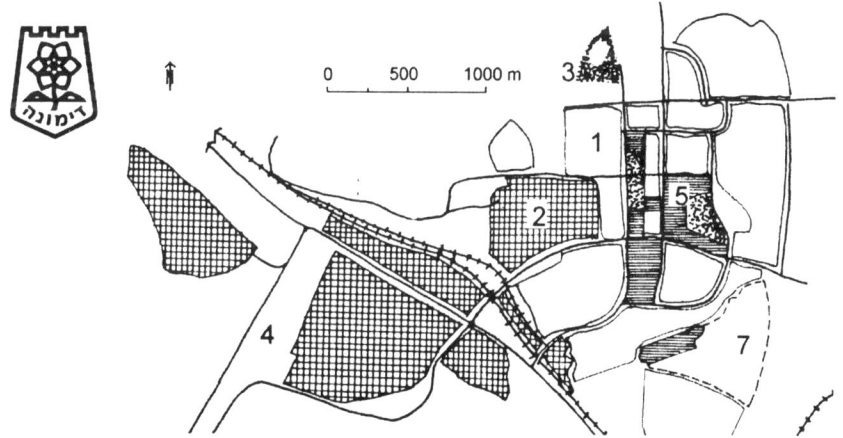

Fig.18.6. Dimona: general layout

1 - residential neighborhoods; 2 – industrial area and warehouses; 3 – public parks and open spaces; 4 – main roads; 5 – public center; 6 –railroad; 7 – development reserve

Fig. 18.7. Dimona: main street

Fig. 18.8. Dimona: pedestrian area and central market place

Fig. 18.9. Dimona: public park in the central part of the town

18.4
Arad

Arad was established as a development town in the early 1960s in a hilly area located on the border of two deserts – the Negev and Judean desert. Although remains of the ancient settlement Arad are found 10 km west of the modern town, the town is pre-designed and built from scratch by a specially formed design team. In 1995, the population of Arad amounted to some 20,000 residents (see Appendix). The town represents a relatively successful example of Israel's development town policy. In the 1970s, the town exhibited a population growth rate of over 50 per cent per annum and was among few development towns of Israel having a positive migration balance (Gabriel et al. 1985). The relative success of Arad during this period stemmed from both its proximity to Dead Sea tourist and employment sites, and attention given to the provision of amenities (*ibid.*). In the late 1980s, population and economic growth of Arad decelerated substantially. This was due to both the town's relative remoteness from the major population centers of the country (Tel Aviv and Jerusalem), and intensive housing and industrial development in adjacent urban communities, specifically in Be'er-Sheva. The hilly landscape of the area predetermined the physical layout of the town in which small hill-located neighborhoods are separated by steep slopes and greenery (see Fig. 18.10 – 18.13). The center of the town is built by medium-size apartment blocks.

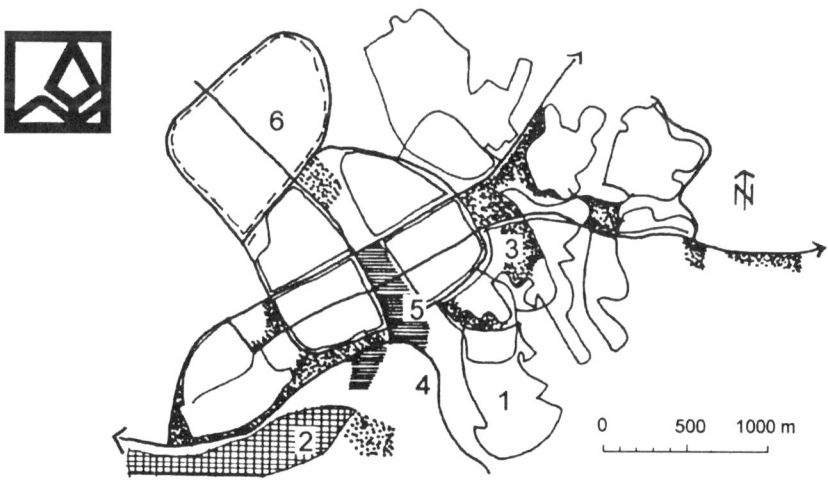

Fig.18.10. Arad: general layout

1 - residential neighborhoods; 2 – industrial areas and warehouses; 3 – public parks and open spaces; 4 – main roads; 5 – public center; 6 –development reserve

Fig. 18.11. Arad: panoramic view with the Judean desert in background

Fig. 18.12. Arad: single family terrace houses in a suburban neighborhood

Fig. 18.13. Arad: monument park at the entrance to the town

18.5
Yeroham

Yeroham is one of the small urban settlements in the Negev, 8,000 residents (see Appendix). It is located on the historic Spice Route, 32 km south-east of Be'er-Sheva. The town was founded in 1951 to accommodate both new immigrants from the Middle East and North Africa and employees of the Dead Sea Works and various public initiatives across the Negev. The percentage of Asia and Africa born immigrants in Yeroham is still much higher the national average and most of the other urban settlements in the region (see Appendix). In the 1960s, the temporary immigrant settlement was demolished and construction of the town begun. The town's name went through a number of changes: from Tel Yeroham (Hill Yeroham) in the 1950s, to Maabarot Yeroham (Transitional Camp Yeroham), to Kfar Yeroham (Village Yeroham), to today's Yeroham (Yeroham Local Council 1996). For many years, Yeroham had a negative migration balance attributed primarily to the lack of both employment opportunities and public amenities. This situation was partially reversed by the influx of new immigrants from the former Soviet Union in 1989-1991 who were attracted to the town by the availability of public housing and relatively low housing prices in the private sector. The influx of new immigrants caused some increase in housing construction and an improvement in social services and facilities. The town is stretched along the main street in the east-west direction. Two small pockets of industrial development are located on both ends of this axis. Residential development in Yeroham is represented by two types of buildings: five-storey multi-family houses, and single-family cottages (Figs. 18.14 – 18.17).

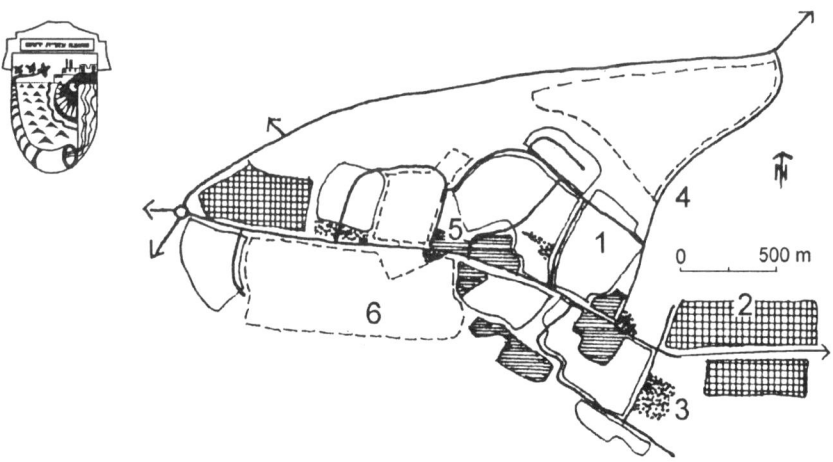

Fig. 18.14. Yeroham: general layout

1 - residential neighborhoods; 2 – industrial area and warehouses; 3 – public parks and open spaces; 4 – main roads; 5 – public center; 6 –development reserve

Fig. 18.15. Yeroham: view of new residential development from surrounding hills

Fig. 18.16. Yeroham: main square in the town center with a shaded pedestrian walkway

Fig. 18.17. Yeroham: fragment of the shaded pedestrian walkway at the town center

18.6
Mitzpe-Ramon

Mitzpe Ramon was founded in 1956 as an outpost settlement whose residents worked in the nearby stone quarries and in the construction of the road from Be'er-Sheva to Eilat. The first settlers were then joined by new immigrants. In 1964, the settlement was declared a local council (LCMR 1997).

Mitzpe Ramon is located on the northern edge of the Ramon Crater, 80 km south of Be'er-Sheva and 150 km north of Eilat. In 1995, the population of the town amounted to 4,200 residents (see Appendix). In the wake of mass immigration from the former Soviet Union in 1989-91, the population of Mitzpe-Ramon grew at a relatively high rate: up to 16 per cent per year (Gradus et al. 1993). Subsequently, the rate of population growth decreased substantially due to declining in-migration and high out-migration rates. The major reasons for low migration attractiveness of the town are the town's remoteness from the major population centers of the country and the lack of employment opportunities (Gabriel et al. 1985).

Nearly half of the town's labor force is employed in various public services and facilities (Gradus et al. 1993). The level of home-ownership in the town is less than half that in the region and that of the national average (see Appendix). This implies that the majority of the town's population do not consider it as a place for permanent residence. Physical layout of the town is relatively simple: three small residential neighborhoods built on hilltops are grouped around the compact town center (see Fig. 18.18 – 18.21).

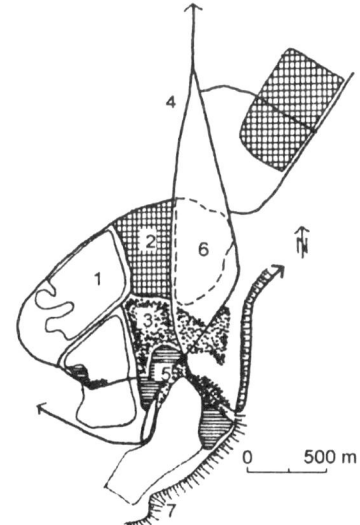

Fig.18.18. Mitzpe-Ramon: general layout

1 - residential neighborhoods; 2 – industrial area and warehouses; 3 – public parks and open spaces; 4 – main roads; 5 – public center; 6 –development reserve 7 – cliffs

Fig. 18.19. Mitzpe Ramon: panoramic view of the town center from the cliff of Ramon Crater

Fig. 18.20. Mitzpe Ramon: hotel complex next to the town center

Fig. 18.21. Mitzpe Ramon: wild Ibex roaming freely around new neighborhood

References

DCC (1996) Dimona. Dimona City Council, Dimona

Edri Y, Nir I (1995) Be'er-Sheva in the past. B. Blaustein Publications, Haifa

Gabriel S, Justman M, Levy A (1985) The development of sparsely populated arid regions: an integrative analysis with application to the Negev. In: Gradus Y (ed) Desert development: man and technology in sparselands. D.Reidel, Dortrecht, pp. 235-255

Gradus Y, Levinson E, Newman D (1993) Changes in the urban settlements of the Negev: 1989-1991 (in Hebrew). Negev Center for Regional Development, Be'er-Sheva

ICBS (1997) Statistical abstract of Israel. Israeli Central Bureau of Statistics, Jerusalem

LCMR (1997) Mitzpe-Ramon. Local Council of Mitzpe-Ramon, Mitzpe-Ramon

Lithwick H, Lithwick I (1997) Welfare economics and urban development planning: a case study of the city of Eilat. Working Paper 8. Negev Center for Regional Development, Be'er-Sheva

Newman D, Gradus Y, Levinson E (1995) The impact of mass immigration on urban settlements in the Negev: 1989-1991. The Negev Center for Regional Development, Be'er-Sheva

Portnov BA (1998) The effect of housing on migration in Israel. J of Population Economics, 11(3): 379-394

Yeroham Local Council 1996. Yeroham: history and location. B. Blaustein Publications, Haifa

Appendix. Selected socio-economic and physical data for desert towns in Israel compared to the Southern district and the national average

Development indicator	Urban settlement						Southern district	Whole country
	Be'er-Sheva	Eilat	Dimona	Arad	Yeroham	Mitzpe-Ramon		
Population size, thousands of residents	150.0	32.5	31.1	20.3	7.7	4.3	734.0	5,567.1
Land area under jurisdiction, km^2	54.6	85.1	30.6	75.9	34.1	64.5	14,107.0	21,946.0
Used area in km^2	17.6	58.7	7.2	4.6	1.8	3.6	-	-
Population density, per/km^2 of land area	2755.6	382.0	1016.7	267.6	226.4	66.1	52.0	253.7
Population density, per/km^2 of used area	8522.7	553.7	4375.0	4413.0	4277.6	1194.4	-	-
Patterns of land use, km^2/per cent:								
– Housing development	8.9/16.6	4.8/5.6	2.0/6.8	1.8/2.4	0.8/2.4	1.5/2.3	-	-
– Industry and crafts	2.8/5.2	2.6/2.7	0.7/2.2	0.5/0.7	0.4/1.1	0.6/1.0	-	-
– Public services and facilities	1.8/3.3	1.6/1.9	0.2/0.6	0.2/0.3	0.4/0.1	0.5/1.0	-	-
– Parks and public gardens	2.0/3.7	2.2/2.6	1.0/3.1	0.2/0.2	0.2/0.7	0.7/1.0	-	-
– Unused land	27.0/50.5	26.4/28.9	19.0/62.3	66.3/87.4	31.2/91.5	60.9/94.7	-	-
– Other land uses	11.1/20.7	47.5/58.3	7.6/25.0	6.9/9.0	1.1/4.2	0.3/0.0	-	-
Population make up:								
– Percentage of Asia and Africa born	20.7	14.5	26.4	11.5	27.6	18.3	18.8	11.0
– Average number of years of studying	11.5	11.5	10.5	12.2	10.6	11.6	10.8	11.6
Sources of population growth:								
– Natural growth, per cent (NG)	40.3	109.7	64.0	37.5	40.9	98.0	50.2	61.7
– Migration balance, per cent (MB)	59.7	-9.7	36.0	62.5	59.1	2.0	49.8	38.3
– MB/NG Index	1.5	-0.1	0.6	1.7	1.4	0.0	1.0	0.6
General characteristics of the labor force:								
– Participation in the labor force, per cent	53.2	75.7	49.7	58.2	46.8	47.0	53.2	59.0
– Percentage with university diploma	16.4	8.9	7.2	19.5	6.9	12.1	12.1	15.9

Development indicator	Urban settlement						Southern district	Whole country
	Be'er-Sheva	Eilat	Dimona	Arad	Yeroham	Mitzpe-Ramon		
– Percentage of unskilled workers	9.5	10.0	11.9	11.5	13.3	22.2	10.9	8.5
General characteristics of households:								
– Average household size, persons	3.0	2.6	3.3	2.8	3.6	2.8	3.2	3.1
– House-ownership level, per cent	61.5	57.6	54.7	60.2	54.6	29.3	62.3	59.9
Ownership of durable goods, per cent:								
– Personal computer	21.5	28.7	17.5	21.8	20.1	18.2	20.6	27.1
– Air conditioner	28.0	46.1	12.9	28.0	15.9	6.3	27.4	32.2
– Private car	76.9	85.0	77.4	73.7	72.3	70.1	75.3	67.9

Note: Combined from a) 1995 Census data (Israeli Central Bureau of Statistics); b) Statistical Abstract of Israel 1997 (Israeli Central Bureau of Statistics 1997); c) Local Authorities in Israel: Physical data (Israeli Central Bureau of Statistics 1998), and d) Statistical Yearbook of the Negev (Negev Center for Regional Development, Be'er-Sheva 1996). Land use data are as of 1986-1989 (patterns of land use) and 1995 (used area). Distribution of land uses for Eilat and Mitzpe-Ramon is taken from comprehensive plans of the respective towns. "-" Indicates that data are not available

Subject Index

Airborne dust
- reducing in desert cities 205-226
- transportation and deposition in urban environment 206

Arad 24, 93
- general layout 313

Architecture
- low-energy technologies for desert 291
- desert bio-climatic approach 263-277

Arid 4
- regions 7

Arid drylands
- development 155

Aridity
- categories 3
- climate causes 2

Barnaul 162, 165, 171-173, 177, 180, 186

Bedouin towns
- planning 234

Be'er Sheva 24, 26, 140, 187-204, 229, 231, 307
- comprehensive plan 233
- diagrammatic map 190
- dust composition 216
- emergency planning for immigrants 195
- general layout 308
- microclimatic variability 196
- new master plan 192
- physical layout and road pattern 234
- preconceived planning 280
- satellite rururban development 194
- urban forms 197

Climate
- aridity 2
- Be'er Sheva 188
- control in desert 291-304

Density 191-195, 201, 202

Desert
- creating urban cool island 287
- energy balance model 283
- microclimatic considerations 281
- patio house 282
- patio house case study 284
- patio house temperatures 284
- preconceived planning 280
- solar neighborhood 251-262
- thermal environment in street canyons 286
- urban attractiveness 279
- urban microclimate 282, 279-289
- wind speeds in street canyons 285

Desert architecture
- ambient / internal summer temperature 276
- ambient and internal winter temperature 276
- average temperature at atrium floor 298
- bio-climatic approach 263-277
- building design 264
- climate of Negev 263
- cool tower 301
- cooling in down-draft tower 302
- Etzion house 265
- evaporative down draft cooler 300
- external wall section 268
- indirect space heating 302
- local climatic conditions 291
- low-energy technologies 291-304
- performance monitoring 274
- solar radiation in Sede-Boker 295
- space heating by forced convection 303
- summer / winter radiation: southern wall 275
- sunken atrium 296
- value of courtyard 273
- vertical temperature profile in atrium 299
- windows 269
- winter and summer radiation 266

Desert case study
- planning theories verus reality 187-204

Desert cities
- reducing airborne dust 205-226

Desert climate
- urban attractiveness 279

Desert development 153
- role of ecology 153-158

Desert environments
- responsive planning 227-240
Desert planning
- garden city and patio house 280
- preconceived
Desert resettlement
- deterministic paradigm 243
- regarding the future 247
Desert settlements
- Arad 313
- Be'er Sheva 307
- development peculiarities, Israel 24
- development patterns 17-35
- Dimona 311
- Eilat 310
- exogenous factors 20
- Israel 307-321
- Mitzpe-Ramon 317
- policy implications 31
- socio-economic and physical data 320-321
- socio-economic and physical data 307-321
- sustainable population growth 37-59
- Yeroham 315
Desertification
- process 5
Deserts
- economic development 21
- development aspects 18
- geographic extent 4
- level of infrastructure development 21
- urban growth 6
- urban settlements 18
- urbanization 18
- urbanization in Israel 18
Development
- Israel policies 117
- paradigms 21
- peculiarities of settlements 26
- urban private construction 61-85
Development towns 24
Dimona 24, 93
- general layout 311
Dry-subhumid drylands
- development 157
Drylands
- arid development 155
- development, ecological role 158
- dry-subhumid development 157
- hyperarid, development 154
- semiarid development 155
Dust
- airborne in desert cities 205-226

- chemical and mineral composition 216
- chemical composition 216
- deposition rate 214
- design strategies reducing exposure 220
- effect of major storms 217
- experiment 209
- experimental results 213
- field methods 213
- five urban sampling cites 211
- grain size characteristics 215
- laboratory methods 213
- particle size distribution 215
- particles transport by wind 207
- rate of deposition 214
- reduction in desert cities 223
- sample locations 211
- sampling sites 210
- urban sources 218

Ecology
- role in desert development 153-158
Economic development
- capital criteria 89
- changes in employment 89
- construction criteria 90
- deserts 21
- economic output criteria 90
- labor criteria 89
- theoretical models 22
Eliat
 - general layout 310
Etzion house 263-277

Frontier settlements
- Siberia , Russia 161-186

Garden cities 191

Hyperarid 4
Hyperarid drylands
- development 154

Interregional migration
- employment and housing factors 112
- general patterns 118
- housing-employment paradigm 114, 115
- Israel 118, 125
Isolation
- effect on settlement development 87-110
Israel
- annual rates of public construction 152

Subject Index

- changes in population of core and peripheral areas 1948-95 142
- desert settlements 24
- desert urbanization 18
- desert settlements 307-321
- development policies 117
- foreign and in-country migration 119
- ideology and planning 228
- in-country migration 126
- interregional migration 118, 125
- migration curve 128
- new settlements 18
- patterns of urbanization 40
- policy of population dispersal 134
- population and residential construction 151
- population growth 1948-1995 141
- population growth in core and peripheral districts 144, 150
- population growth of periphery 146
- private construction trends 64-69
- regional development patterns 116
- Sede-Boker solar neighborhood 251-262
- settlement system 228
- socio-economic indicators 117
- urban settlements 39
- urbanization 117

Japan
- interregional migration 126
- migration balance 121
- migration curve 128
- migration in-country 120
- regional development 117
- regional development patterns 116
- socio-economic indicators 117,
- urban development 117

Krasnoyarsk 162, 165, 167, 171-173, 177, 179-181, 184, 186

Lanscape 188, 193
Lesosibirsk 162, 165, 171-173, 177, 180, 186

Microclimate
- Be'er Sheva 189
Microclimatic variability
- Be'er Sheva 196
Migration
- attractiveness model 111-131
- attractiveness of urban areas 53
- attractiveness, policy implications 129
- balance Japan 121
- behavior model 113
- curves Israel and Japan 128
- employment-housing paradigm (diagrams) 127
- generalized model 44
- in-country Israel 126
- in-country Japan 120
- influencing factors 125
- interregional Japan 126
- research method 122
Mitzpe-Ramon
- general layout 317

Negev
- climate 263
- development concept 230
- preconceived planning 280
Norilsk 162, 163, 165, 171-173, 177, 180, 186
Novosibirsk 161, 165, 171- 173, 177, 180, 186

Patio house 280, 282
Peripheral areas
- population growth, effect of public policy 133-152
- urban development 91
Peripheral settlements
- climatic harshness 103
- clusters/ distance from population center 109
- conclusions and policy implications 108
- controls 95
- development 87-110
- distribution of population and location 91
- economic development 104
- index of spatial clustering 100, 101, 102
- introduction 87
- measuring economic development 89
- population growth 88, 98
- research method 92
- research results 98
- research samples 94
- sample and population 110
- sustainability of population growth (MB/NG indicator) 101
- sustainability of population growth 99
- sustainable population growth 88
Physical environment
- settlements in Siberia 161-186
Planning
- applications 238
- Bedouin towns 234

- Be'er Sheva 233
- Be'er Sheva physical layout/ road pattern 234
- desert environments 227-240
- ideology, Israel 228
- impact on economic growth 33
- impact on population growth 33
- improved transportation 32
- investment incentives 32
- land use regulation 32
- Negev development 230
- preconceived in desert 280
- Rahat 237
- settlement system 228
- theories versus reality 187-204
- urban microclimates 279-289

Population
- and residential construction, Israel 151
- growth structure changes 27
- model of inter-urban flows 44
- policy of dispersal in Israel 134
- region 2
- regional policy evaluation 134
- settlement growth model 23
- urban based 2

Population growth
- alternative scenarios 145
- average rate of net migration 39
- components 48
- core and peripheral areas, Israel 1948-95 142
- core and peripheral districts, Israel 150
- core and peripheral districts Israel 144
- effect of public policy 133-152
- exogenous factors 56
- inequalities 39
- influencing factors 144
- Israel 1948-1995 141
- key factors 55
- MB/NG ratio 50
- measures 38, 46
- natural growth rate 39
- overall rate 38
- peripheral settlements 88
- periphery, Israel 146
- policy implications 54
- relative rate 38
- sources 1992-94 49
- strategy of redirecting priorities 57
- structural changes 27
- sustainable 37
- theoretical models 22
- urban percentage change 38

Population growth model
- net balance of migration 50
- rate of natural growth 51

Population migrations
- model 44

Private construction
- annual rate and settlement's remoteness 67
- applications in planning 80
- accessability 73
- buying power 73
- case study 74-78
- construction costs 72
- development data hierarchy 69
- effect of population size 83
- general model 72
- geographic distribution in Israel 66
- government incentives 71
- history in Israel 64-69
- infrastructure 71
- land availability 71
- location paradigm 71
- natural amenities 73
- population/migration 73
- proximity and correlation of variables 84
- research results 79
- settlement population size 68
- statistical parameters 85
- urban development 61-85

Public construction
- Israel annual rates 152

Rahat
- master plan 237

Region
- migration attractiveness model 111-131

Regional development
- Israel 116
- Japan 117

Regional migration
- attractiveness model 111-131
- behavior model 113

Regional policy evaluation
- controls 137
- cost-benefit methodologies 135
- factors affecting growth 137
- indirect methodologies 134
- model 136
- objects of intended impact 136
- partial methodologies 135
- policy targets and measures 136
- policy effects 138
- research approach 138

- research methodology 136
- results and discussion 140
Remoteness
- effect on settlement development 87-110
Responsive planning
- desert environments 227-240

Suburban development 190-191
Sede-Boker
- building clusters 258
- circulation 255
- circulation systems 255
- concept of "p" point 259
- location in Israel 253
- "P" point and layout 259
- solar neighborhood 251-262
- volumetric constraints 261
- water heating 261
Semiarid 4
Semiarid drylands
- development 155
Settlement development
- climatic harshness 30
- distance to urban center (remoteness) 30
- effect of remoteness and isolation 87-110
- influence of desert 30
- isolation (grouping) 30
- population 30
Settlements
- access to metropolitan center 23
- annual rate of construction 26
- annual population growth 1992-93 40
- availability of skilled labor 21
- climatic conditions 22
- climatic harshness 23
- development clusters with no urban core 31
- development clusters with urban core 31
- development peculiarities 26
- distance for daily commuting 22
- economics of transition 166
- exogenous factors influencing index of prestige 174
- factors influencing attractiveness 180
- factors influencing index of prestige 173
- geographic location of samples 25
- harsh climate 20
- immigration 42
- incountry migration 43
- index of prestige 174
- Israel 18
- lack of previous development 20
- land availability 22, 23

- level of attractiveness, expert's and resident's 176
- level of urbanization 23
- migration/natural growth ratio 26
- overall population growth 26
- physical environment, Siberia 161-186
- physical parameters/index of prestige 173
- planning policies 31
- planning strategies 31, 32
- population growth, migration 23
- population growth model 23
- prestige and market value 180
- rates of private construction 28
- remoteness 20
- selection for comparative analysis 25
- Siberia, components of attractiveness 173
- Siberia, geographic location 162
- Siberia, residential land value 184
- Siberia, social factors/index of prestige 182
- Siberia, the cities 165
- Siberia, the region 163
- Siberia, topological groups of territories (TGT) 172
- social attractiveness, Siberia 161-186
- socio-demographic parameters 186
- spatial patterns/district attractiveness 170
- spatial isolation 20
- structure of annual population growth 26
- urban 25
- urban patterns 9
Social attractiveness
- settlements in Siberia 161-186
Solar neighborhood 251-262
- building clusters 258
- circulation systems 255
- circulation 255
- concept of "P" point 259
- location 253
- orientation 254
- pedestrian alley 257
- view of alley 258
- view of woonerf 256
- volumetric constraints 261
- water heating 261
Subhumid 4
Sustainable development
- ecological role 158
- economic aspects 38
- environmental dimension 37
- socio-demographic aspects 38
Sustainable population growth
- definition 37

- MB/NG ratio 50
- measures 46

Urban climate 207
- rainfall 208
- temperature 208
- wind regime 208

Urban development
- environmental management 62
- housing 62
- infrastructure 62
- local government 62
- migration balance vs. private residential construction 63
- peripheral areas 91
- policy implications 31
- population distribution and residential building 64
- private construction 61-85
- socio-economic 62
- transport 62

Urban environment
- transport and deposition of dust 206

Urban localities
 - population growth 26

Urban microclimate
- arid conditions 279-289

Urban settlements
- categories for grouping 52
- clustering and private construction 106
- clustering and unemployment 107
- Israel 39
- migration attractiveness 53
- rate of private construction 105
- sustainable population growth 37-59, 88

Urbanization
- deserts 6, 18
- Israel 117
- patterns and policies 40

Yeroham
- general layout 315

Author Index

Abe H 73, 82, 113, 118, 120, 122, 130
Adams RM 246, 248
Albright WF 243, 248
Alexander ER 229, 239
Alonso W 71, 82
Amiran DHK 235, 239
Anderson B 294, 304
Andoh K 71, 82
Anson J 41, 58
Appleyard D 195, 203
Armstrong H 61, 82, 90, 109, 113, 130
Aynsley RM 209, 225
Azmon A 217, 218, 225

Bagnold RA 206, 225
Balchin PN 135, 148
Bar-Cohen A 18, 34
Barff R 131
Barkai Z 18, 19, 34
Ben-Arieh Y 189, 203
Ben-David Y 239
Ben-Gurion D 228, 239
Benzaquen J 65, 82
Berliner P 12, 279, 289
Bitan A 30, 34, 46, 58, 82, 96, 109, 122, 123, 130, 209, 225, 283, 289, 293, 304
Boeken B 219, 224, 225
Bonaiuto M 185
Bonnes M 163, 185
Borjas GJ 113, 130
Bourne LS 61, 82, 91, 96, 109
Breines S 194, 203
Brenner S 225
Brown LR 38, 58
Buchman M 223, 225
Burnley IH 113, 130

Central Bureau of Statistics 46
Champion AG 139, 148
Chandler TJ 208, 209, 225, 280, 289
Christaller W 229, 239
CIA 117, 130
Clark C 66, 71, 82, 103, 109

Clawson M 61, 82, 91, 109
Clealand CB 17, 18, 20, 34
Cloudsley-Thompson JL 1, 5, 13
Cohen Y 229, 239
Cohen E 228, 229, 239
Comay Y 19, 35, 41, 58, 95, 97, 110

Davenport AG 200, 203, 209, 225
Davis WB 295, 304
DCC 311, 319
De Jong GF 23, 34, 39, 47, 58, 95, 96, 109, 112, 114, 123, 124, 130, 143, 148
Dean WJ 194, 203
Di H 304
Diamond DR 61, 82, 89, 109, 135, 148
Donagi AE 225
Doxiadis K 96, 103, 109
Drabkin-Darin H 19, 34, 65, 82, 133, 148

Edri Y 307, 319
Ehrlich AH 47, 58
Ehrlich PR 47, 58
El-Shakhs S 220, 225
Ellis M 131
Encyclopedia Hebraica 248
Ercolani AP 185
Erell E 8, 10, 11, 12, 17, 64, 71, 83, 87, 92, 97, 104, 110, 115, 122, 131, 137, 138, 149, 194, 203, 205, 291, 304
Etzion Y 12, 13, 133, 194, 203, 204, 225, 251, 260, 262, 263, 291, 304
Evenari M 34, 244, 24
Ewers HJ 139, 148

Faiman D 204, 225, 304
Fawcett JT 23, 34, 39, 47, 58, 95, 96, 109, 112, 114, 123, 124, 130, 143, 148
Fedorovskay E 185
Fergusson JE 206, 216, 225
Fialkoff C 41, 58, 65, 72, 82
Fischer CS 23, 34, 95, 96, 109
Foner HA 225
Frey WH 115, 130

Friedmann J 57, 58, 112, 130, 139, 148
Fugitt GV 130
Fulton JA 115, 120, 130

Gabriel S 310, 313, 317, 319
Ganor E 156, 158, 210, 213, 216 217, 218, 225
George P 112, 130
Gerson 19
Gerson M 34
Gibson RM 130
Givoni B 209, 220, 222, 225, 295, 300, 304
Glass D 163, 185
Glueck N 243, 248
Golani Y 192, 193, 194, 200, 203, 204
Golany G 8, 13, 18, 19, 20, 34, 194, 203, 279, 289
Golovatskaya N 166, 185
Goosens D 206, 207, 219, 225, 226
Govaer D 304
Gradus Y 12, 18, 19, 34, 35, 54, 58, 59, 95, 110, 133, 137, 139, 148, 204, 227, 231, 236, 239, 240, 280, 289, 317, 319
Gravetter FJ 163, 185
Green HC 18, 20, 34
Greenwood MJ 112, 114, 130
Grove JM 244, 248
Gubareff GG 294, 304

HABITAT 62, 82
Hall P 61, 82, 91, 109
Hanlong L 163, 185
Hansen NM 57, 58
Hare KF 2, 3, 13
Harison J 163, 185
Haughton G 37, 58
Haussmann Architects Consultants 200, 203
Henderson RA 89, 109
Herzog Z 188, 203
Hester RT Jr 170, 185
Hillier B 163, 185
Holdren JP 47, 58
Hopf M 157, 158
Howard E 191, 203
Hunter C 37, 58
Huntington E 242, 243, 244, 245, 248

ICBS 34, 46, 58, 64, 65, 69, 74, 80, 82, 89, 91, 98, 109, 116, 117, 118, 130, 140, 148, 150, 151, 310, 319
IMF 72, 82, 133, 148
Israeli A 188, 203

Issar AS 8, 12, 241, 245, 246, 248
Isserman AM 135, 149

Jacobsen T 246, 248
Jacobson J 38, 58
Jain JK 5, 13
Jenner A 130
Johnson DL 13
Johnson-Haring K 13
Joseph JH 156, 158
JSB 116, 117, 120, 121, 123, 130
Justman M 319

Kaganova O 166, 185
Kamon E 18, 34
Kark S 158
Kates RW 1, 6, 8, 13
Keane C 163, 185
Keane 163
Kirschenbaum A 19, 35, 41, 58, 95, 97, 110
Klein C 245, 248
Kneese AV 17, 20, 26, 35, 97, 110
Konya A 220, 221, 225
Koppen W 3
Krakover S 133, 148, 204, 229, 239
Krushlinskiy V 173, 185
Kuklinski A 57, 58

LaLonde RJ 113, 129, 130, 144, 149
Lamb HH 244, 248
Landsberg HE 207, 208, 209, 225
Lasurenko S 185
Lavi N 225
Lawrence TE 243, 248
Layton AP 26, 35, 61, 82
LCMR 317, 327
LCT 3, 4, 13
LeCorbusier 187, 204
Leeuw F de 61, 82
Lerman E 41, 58, 133, 149
Lerman R 41, 58, 133, 149
Levinson E 35, 49, 95, 110, 319
Levy A 319
Levy JM 26, 35, 61, 82, 90, 91, 110
Lewando-Hundt G 235, 239
Liang Z 120, 131
Lin N 163, 185
Lipshitz G 63, 64, 82, 101, 110, 112, 113, 131, 133, 137, 138, 139, 140, 144, 149
Lithwick H 310, 319
Lithwick I 310, 319

Lowdermilk WC 243, 244, 248
Lynch K 163, 185

Maddock T 17, 35
Mamane Y 218, 225
Markusen A 61, 82, 118, 122, 131
Maslovskiy V 166, 183, 185
McDowell JM 112, 130
McGranahan D 26, 35, 61, 69, 82
MCH 66, 82
Meir IA 12, 187, 193, 196, 201, 203, 204, 220, 221, 223, 225, 291,304
Melbourne W 225
Merrifield JD 135, 149
Mertens H 194, 204
Messinas EV 196, 204
Michel F 112, 114, 131
Middleton N 3, 4, 5, 6, 13
Mills ES 26, 35, 61, 73, 82
Milne W 115, 122, 131
Moore B 135, 149
Moore EG 23, 35, 47, 59, 95, 96, 110, 113, 114, 123, 131
Motzafi-Haller WR 13, 307
Murphy PA 130
Musham VH 235, 239
Muth RF 71, 83
Neeman E 225
Newman D 24, 35, 41, 59, 95, 97, 110, 307, 319
Newman P 71, 83
Nijkamp P 139, 148
Nir I 307, 319
Novitskiy I 185
NSF 3, 4, 13

OECD 194, 204
Offer Z 207, 217, 218, 219, 225, 226
Ohta M 71, 82
Oke TR 208, 209, 226, 281, 289
Orev Y 222, 226

Paciuk M 200, 204, 209, 225, 226
Pearlmutter D 10, 11, 12, 28, 37, 61, 63, 70, 83, 88, 95, 96, 101, 103, 110, 113, 114, 115, 120, 124, 131, 133, 137, 138, 140, 146, 149, 196, 204, 279, 283, 289, 291, 295, 300, 301, 304
Penman HL3
Pennlaker J 185
Perrot A 131

Perroux F 112, 131
Poot J 113, 131
Poreh M 200, 204, 209, 226
Portnov BA 1, 8, 10, 11, 13, 17, 28, 37, 54, 59, 61, 63, 64, 70, 71, 83, 87, 88, 90, 92, 95, 96, 97, 101, 103, 104, 110, 111, 113, 114, 115, 120, 122, 124, 131, 133, 137, 138, 140, 146, 149, 161, 164, 166, 183, 185, 307, 308, 319
Pye K 206, 207, 219, 226

Raban A 246, 248
Rahamimoff A 227, 239, 251, 262, 295, 304
Rahamimoff S 304
Rahman OMA 163, 185
Rapoport A 170, 185, 235, 239
Rhodes J 135, 149
Richardson HW 71, 83, 112, 114, 131
Robertson M 163, 186
Rosenberg MW 23, 35, 47, 59, 95, 96, 110, 113, 114, 123, 131
Royal Dutch Touring Club 195, 204
Rubin S 30, 34, 46, 58, 82, 96, 109, 122, 123, 130, 209, 225, 283, 289, 293, 304

Safriel UN 11, 153, 156, 158
Sage C 38, 59
Saini BS 8, 13, 18, 20, 35, 71, 83, 205, 221, 226
Sapir S 189, 203
Schechter J 18, 35
Shachak M 219, 224, 225
Shachar A 229, 239
Shanan L 34, 244, 248
Shantz 4
Shefer D 133, 149
Sheffer G 19, 35
Shilony Z 191, 204
Shinar A 239
Shmueli A 235, 240
Show D 164, 185
Silberstein A 304
Singer J 185
Skibin D 208, 226
Smardon RC 163, 185
Smith WF 26, 35, 61, 83, 90, 91, 110
Spence NA 61, 82, 89, 109, 135, 148
Stern E 19, 34, 54, 58, 188, 204, 227, 229, 236, 240, 280, 289
Stock R 114, 130

Tadmor N 34, 244, 248

Taylor J 61, 82, 90, 109, 113, 130
Teklenburg JAF 163, 186
The Standards Institution of Israel 294, 304
Thisse JF 131
Thomas D 3, 4, 5, 6, 13
Thornwaite CW 3, 13
Timmermans HJP 163, 186
Topel RH 113, 129, 130, 144, 149
Tsoar H 12, 204, 205, 207, 219, 226
Tugnutt A 163, 186
Turner RK 37, 38, 59

UN 18, 35
UNCOD 2, 4, 5, 6, 13
UNEP 1, 5, 6, 13, 153, 158

Vickery BJ 225
Vining DR Jr 26, 35, 131, 139, 149
Volis S 158
von Schwarze DG 192, 193, 203

Wagenberg AF 186
Walker R 115, 120, 131
Wallnau LB 163, 185
White FG 6, 13
White MJ 120, 131
Wiebenson D 191, 204
Williamson JG 117, 131
Wittick A 170, 186
Wong C 61, 62, 83, 87, 110
Woods S 194, 204
Woolley CL 243, 248

Yeroham Local Council 315, 319

Zangvill A 217, 226
Zemel A 304
Zohary D 157, 158

Printing: Saladruck, Berlin
Binding: Buchbinderei Lüderitz & Bauer, Berlin